Plant Pathology in the 21st Century

More information about this series at http://www.springer.com/series/8169

Maria Lodovica Gullino • James P. Stack
Jacqueline Fletcher • John D. Mumford
Editors

Practical Tools for Plant and Food Biosecurity

Results from a European Network of Excellence

Volume 8

Springer

Editors
Maria Lodovica Gullino
Università degli Studi di Torino
AGROINNOVA
Grugliasco (TO), Italy

Jacqueline Fletcher
Department of Entomology
and Plant Pathology
Oklahoma State University
Stillwater, OK, USA

James P. Stack
Department of Plant Pathology
Kansas State University
Manhattan, KS, USA

John D. Mumford
Centre for Environmental Policy
Imperial College London
Ascot, Berks, UK

Plant Pathology in the 21st Century
ISBN 978-3-319-46896-9 ISBN 978-3-319-46897-6 (eBook)
DOI 10.1007/978-3-319-46897-6

Library of Congress Control Number: 2016963796

Printed on acid-free paper

This Springer imprint is published by Springer Nature
The registered company is Springer International Publishing AG
The registered company address is: Gewerbestrasse 11, 6330 Cham, Switzerland

Foreword

In a world facing a growing risk of man-made and natural crises and disasters, the security of citizens and critical infrastructures and the environment protection have become a high priority in the European Union.

Strengthening capacities in crisis management and improving resilience represent key policy and research challenges. To better protect citizens and national infrastructures, the race is now on improving Europe's preparedness and prevention to man-made and natural threats, as well as reinforcing operational response capacities in case of emergency situations.

This book is the outcome of the "Plant and Food Biosecurity" project, a Network of Excellence funded within the security thematic area of the European Seventh Framework Programme for Research and Technological Development (FP7), aiming to invest in knowledge and develop further technologies in order to protect citizens from man-made (accidental or intentional) and natural threats.

Within this framework, the project tackled the threat of and damage from biological incidents of accidental, natural or intentional origin, including acts of bioterrorism, defined as the intentional release of harmful biological agents such as bacteria, viruses or toxins to cause fear, illness or death of people, animals or plants and/or disrupt social, economic or political stability.

The project scope embedded the overall risk management cycle, from preparedness, prevention, detection and surveillance to response and recovery in the topic areas of plant biosecurity and food safety, taking also into account the need to ensure a proper transfer (and implementation) of research outputs – including "practical tools" – to users, namely, producers, policy-makers, scientists, agri-food industry and field practitioners.

A proper and tailor-made exchange of information about research project results is essential to enhance the transfer of research solutions to users in a timely and relevant fashion in order to enable a response to potential agroterrorism threats. Such exchanges are also needed to identify and address users' needs regarding research, technologies and policies, especially in a field where EU capabilities to detect and respond to agroterrorism, or biocriminal acts, are ruled by a number of international, EU and national policies divided among many different organisations.

The book addresses the result of tasks accomplished by 13 partners located in eight different countries, in Europe and beyond: it outlines and characterises threats and gaps in plant biosecurity and food safety areas, analyses the relevant policy framework and the lessons learned from the practice and identifies the most promising tools and methods for risk assessment, detection, diagnostic and containment.

In addition, the authors are also making reference to capacity building, research networking and knowledge transfer, as well as to opportunities for further collaboration in addressing the full spectrum of global biosecurity concerns. As a consequence, this book will be a helpful tool both in becoming more acquainted with the issue of plant and food biosecurity and also in being aware of the possible ways to implement further research and analysis on these subjects.

European Commission
Directorate-General Migration and Home Affair
Rue de la Loi 200, B-1049 Brussels

Philippe Quevauviller

Preface

Biosecurity is a strategic and integrated approach for analysing and managing relevant risks to human, animal and plant life and health and associated risks to the environment. Plant biosecurity aims at protecting all plant resources and the food supply from the natural or intentional introduction, establishment and spread of plant pests, pathogens and noxious weeds. Although most plant disease outbreaks have natural causes or are the result of inadvertent introductions of pathogens through human activities, the risk of a deliberate introduction of a high consequence plant pathogen cannot be excluded.

This book is part of a series of volumes on plant pathology in the twenty-first century, and it stems from Plant and Food Biosecurity (PLANTFOODSEC), a Network of Excellence running from 2011 to 2016 and funded under the European Seventh Framework Programme for Research and Technological Development (FP7). PLANTFOODSEC focused on biological threats having the capacity to affect and damage agriculture, infect plants and ultimately affect food and feed at any stage in the supply chain. The project aimed to develop and implement a virtual centre of competence to prevent, respond to and recover from both intentional (agroterrorism) and unintentional biosecurity threats to EU agriculture, farming and the agri-food industry.

PLANTFOODSEC encompassed plant biosecurity and food safety areas, focusing not only on enhancing capabilities for prevention, detection, response and recovery from threatening plant pathogens but also on mycotoxins and on the contamination of fresh produce and other plant-derived foods by human pathogens on plants (HPOPs) – primarily enteropathogenic strains of *Escherichia coli* and *Salmonella* spp. – that can colonise and contaminate plants at any point along the food production and distribution chains, creating possibilities of outbreaks of foodborne illness.

The considerable amount of research promoted by the European Union – which has also involved non-EU countries such as the United States, Israel and Turkey – has made possible the development of a comprehensive set of tools covering the entire risk management cycle, from prevention to preparedness, detection, response and recovery, which are presented in this book.

In particular, the different chapters cover the identification and regulatory analysis of biosecurity challenges, pest risk assessment, experimental and modelling approaches applied in plant disease epidemiology, decision tools and microbial forensics, diagnostics and detection tools. Moreover, training, dissemination and networking subjects are also covered.

We believe that, besides representing a written testimony of PLANTFOODSEC project, this book will be useful for all the stakeholders in the agri-food chain, including producers, researchers and authorities responsible for plant health and food security interested to go in depth into the world of intentional and unintentional threats to plant biosecurity and to food safety.

We would like to convey our appreciation to all the colleagues who accepted to be part of this book, Zuzana Bernhart and her group at Springer for their kind support and Laura Castellani for her skilful technical assistance.

Grugliasco, Italy Maria Lodovica Gullino
Manhattan, KS, USA James P. Stack
Stillwater, OK, USA Jacqueline Fletcher
London, UK John D. Mumford

Contents

Chapter 1
Considering Vulnerabilities, Threats and Gaps in Plant and Food Biosecurity

Paul Robb

Abstract Whilst the majority of plant derived foods produced for human or animal consumption are safe and wholesome, sometimes complex production and distribution systems are not immune to vulnerabilities, threats and gaps in biosecurity as a number of examples will show. We live in an ever changing world so vigilance is required to identify and prevent new and emerging issues that could impact on production capacity, plant biosecurity or food safety and food chain resilience. Rather than list already well known issues, a number of generic approaches to considering vulnerabilities will be described encompassing natural, accidental and malicious events. Tools such as HACCP, TACCP, PESTLE and plant risk assessments help managers suggest how vulnerabilities and threats in food and plant biosecurity can be managed to tolerable levels. Tools and datasets developed within PlantFoodSec that support a proportionate response are included in discussions to identify predictable issues by stakeholders at all levels.

Keywords Plant food chain • Vulnerability • Risk • Threat assessment • TACCP • PESTLE • Lessons natural and malicious

1.1 Introduction

The vast majority of plant derived foods which are produced for human or animal consumption are safe and wholesome. However, often complex production and distribution systems are not immune to a range of potential threats and imperfections in the "seed to salad on the plate" food chain. There are a wide range of protective systems in place to prevent the adverse consequences of natural, accidental or malicious contamination including disease outbreaks affecting both food plants and consumers. Many of these protective measures have been established following significant outbreaks either in the plants themselves or because of an adverse effect on consumers. The strong science base that exists in this field has built upon the

P. Robb (✉)
Fera Science Ltd, York, UK
e-mail: paul.robb@fera.co.uk

M.L. Gullino et al. (eds.), *Practical Tools for Plant and Food Biosecurity*, Plant Pathology in the 21st Century 8, DOI 10.1007/978-3-319-46897-6_1

need to prevent and respond to such events and has made major contributions in particular to the prevention and control of disease and contamination at all stages of this arm of the food chain.

The PlantFoodSec project (https://www.plantfoodsec.eu/) has brought together key members of the international scientific community who understand plant production methods and have experience of developing, establishing and using tools to enhance biosecurity and safeguard the plant food chain. As part of this project team, a small group of specialists (the security panel) provided internal review and guidance to the project teams on matters which may have potential to be misused for malicious purposes. In reality this function provided reassurance to the teams that their work should be published and disseminated as their outputs reinforced protective measures rather than highlight major gaps in knowledge and vulnerabilities in the food chain. The panel also encouraged collaboration with other agencies and promoted project outputs to those engaged in emergency response and in particular with protection of agricultural food production systems. Although the project had a plant focus, the vulnerabilities considered and gaps filled by the project have wider applicability which this chapter aims to demonstrate.

As we live in an ever changing world, vigilance is required to identify new and emerging issues that could impact on production capacity, plant biosecurity or food safety and food chain resilience. In this chapter we will explore a number of ways in which vulnerabilities can be identified, threats evaluated and suggest how gaps in food and plant biosecurity can be managed to acceptable levels. Many of these approaches refer to tools and datasets that have been developed within PlantFoodSec to support a proportionate response to any predictable issues. Rather than highlight particular weaknesses, this chapter seeks to explain some of the many approaches available to stakeholders at different levels to identify gaps in food chain biosecurity.

In this chapter the term "food security" is used in the context of guarantee of supply and "food defence" in the context of safeguarding the food chain from malicious intervention with "food safety" being used in the context of ensuring food is wholesome and can be consumed safely.

1.2 Vulnerabilities

One definition of the term vulnerability is:

"Exposed to the **possibility** of being attacked or harmed, either physically or emotionally" (http://www.oxforddictionaries.com/definition/english/vulnerable). In this context, emotional impacts will include public perception encompassinging, at times, the adverse consequences of intervention. Perception of food safety risks is a topic outside the scope of this work but there is a large literature describing the importance of the topic (e.g. Lobb et al. 2007; Redmond and Griffith 2007; Verbeke et al. 2007).

There is little doubt that closer links between the natural and social sciences are developing with mutual benefits but there are still challenges in developing a common lexicon and shared understanding in this area. Managing stakeholder expectation will continue to be a key aspect of consequence management of unexpected events.

It would be naïve to suggest that the food chain or plant production systems are not open to the possibility of damage or could suffer harmful impacts from natural, accidental or malicious actions but the detailed examination of vulnerabilities (and mitigation measures) is conducted at many levels. For example,

- at the operational level producers or food processors may consider production of single products or crops,
- at the tactical level, larger businesses might consider production and storage options to maximise retailer choice and shelf-life,
- strategically, international businesses or Governments may consider a wider international food chain, cross border issues and multiple supply chains to guarantee supply.

These are not rigid examples but hopefully demonstrate the complexity of assessing vulnerabilities in the food chain and the need to consider a very wide range of stakeholder requirements. Vulnerabilities can arise for a number of reasons and it is convenient to consider these as natural, accidental or malicious.

1.3 Natural Vulnerabilities

Plants for food or feed are rarely grown aseptically outside of specialised research institutes (some hydroponic systems may be near to this) and the growing environment is itself vulnerable to a range of naturally occurring events that impact upon food/feed plants.

Perhaps the most obvious natural vulnerability is susceptibility to disease outbreaks (e.g. Johnson and Cummings 2015) or pest infestation (e.g. http://www.fao.org/emergencies/emergency-types/plant-pests-and-diseases/en/) which can affect yields, impact upon availability or affect nutritional value with an impact on food security (guarantee of supply), especially in those countries where alternatives are scarce or uneconomically viable to access.

Water security is increasingly becoming recognised as being a key vulnerability in some countries with impacts on irrigation as well as biosecurity, e.g. where disinfection or processing of water is needed before use. Control of water will become more important if recent changes in weather patterns continue to develop with a shift in deposition causing a change in drought and flooding patterns across the globe.

Other natural vulnerabilities include events such as the eruption of the Eyjafjallajökull volcano in Iceland in April 2010 which received widespread press coverage (http://news.bbc.co.uk/1/hi/world/europe/8634944.stm). Significant impacts of this event included restrictions on air travel from the resulting ash cloud which impacted across many parts of Europe. Whilst direct impacts on food were limited to potential increased fluoride levels in deposited ash affecting nearby pasture and grazing, indirect impacts were felt on transportation of short shelf-life produce which is mainly conducted by air. Not only does a freeze on air travel result in financial losses as perishable goods deteriorate but also a potential biosecurity challenge and waste disposal issues.

Without needing to engage in the climate change debate (http://www3.epa.gov/climatechange/science/overview.html, http://www.worldbank.org/en/topic/climatechange/overview), the world is clearly undergoing a series of weather variances that are impacting on plant production with increased vulnerability to weather extremes, changes in growing seasons and increased prevalence of diseases that previously would have been classified as being exotic being observed.

In addition, natural evolutionary change in organisms has caused problems that have impacted across Europe. For example an outbreak of Shiga-toxin producing *Escherichia coli* (STEC), serotype O104:H4 (Karch et al. 2012) originally reported in Germany (European Food Safety Authority 2011) in May 2011, proved to be a significant event. This was initially associated with consumption of fresh vegetables although later, this was linked to consumption of seed sprouts. Assignment of the source of infection in consumers was initially flawed and attributed in error to cucumbers grown in Spain where German laboratories detected *E. coli* in imported cucumbers. Application of the precautionary principle (http://eur-lex.europa.eu/legal-content/EN/TXT/HTML/?uri = URISERV:l32042&from = EN) meant that, in the absence of data to the contrary at the time, large volumes of cucumbers in Spain were consigned to landfill (as well as other salad vegetables which consumers felt were at risk). Compensation payments from the EU to the affected producers of €210 million did not meet all losses with substantial reputational loss by a major industry. Further examination of the organisms detected in the cucumbers identified a different strain of E Coli to that causing serious health issues.

Consumption of sprouted seeds was subsequently associated with occurrence of an identical outbreak in France in June of 2011 with evidence suggesting a common source. Eventually, tracing suggested that the contaminated material most likely arose from a specific consignment of fenugreek seeds imported from Egypt.

The situation was complicated by the fact that STEC O104:H4 was a very rare serogroup in humans in the EU and indeed worldwide with only low single figure cases being reported before the outbreak. At the end of the outbreak, a total of 3911 cases had been reported to the ECDC and WHO.

This is a good example of how vulnerable the plant food supply chain can be from naturally evolving organisms. It is not uncommon for assignment of the causative agent for food poisoning to be made from clinical isolates rather than from examination of the foods consumed. It is of course a key protective measure that the food industry tests routinely for microbial contamination in produce. Nevertheless genetic mutation of *E. coli* O104 impacted on the assays used by National and EU Community Reference Laboratories (NRL and CRL) but rapid diagnostic method development by the CRL allowed NRL to begin testing with minimal method development which aided public reassurance and eventual control of a complex situation. This infrastructure was a significant resource used to manage the outbreak.

It may be worthwhile describing the precautionary principle used by regulatory authorities across Europe to safeguard consumers. This is invoked "when a phenomenon, product or process may have a dangerous effect, identified by a scientific and objective evaluation, if this evaluation does not allow the risk to be determined with sufficient certainty" (http://eur-lex.europa.eu/legal-content/EN/TXT/HTML/?uri = URISERV:l32042&from = EN).

This principle may only be invoked when the following three conditions are met after a suitable risk assessment where:

- Adverse effects or potentially adverse effects have been identified;
- Scientific data available has been evaluated and
- Scientific uncertainty has been taken into account.

In addition, any response should,

- Ensure proportionality between the measures taken and the chosen level of protection;
- Maintain non-discrimination in application of the measures;
- Require consistency of the measures to be taken with similar measures already used in similar situations or using similar approaches;
- Include an examination of the benefits and costs of action or lack of action;
- Review the measures in the light of scientific developments.

The legislation notes that "in the case of an action being taken under the precautionary principle, the producer, manufacturer or importer may be required to prove the absence of danger."

Producers/suppliers requiring additional testing to demonstrate lack of hazard will add to pressures on finite laboratory resources and in practice collaboration between authorities and producers can be mutually beneficial.

In general, food production chains are protected by well-established mechanisms operating at local, national and international levels so as to safeguard products from a range of challenges throughout their life cycle.

The majority of plant based foods are grown in environments which are controlled to a greater or lesser extent by human activity. Growers will tend crops with the aim of maximising yields which can be a driver towards increased biosecurity (prevention of infection/infestation) and biosafety (prevention of harm arising from a biological infection).

1.4 Accidents

Accidental contamination of food plants occurs from time to time from man-made or natural events but in general, accidental chemical contamination occurs much more frequently than biological. However, one of the more common sources of accidental biological contamination arises from non-ideal storage of harvested crops. There are many examples of this resulting in fungal growth with generation of toxin.

One example of popular interest concerns recent theories regarding the Salem Village (USA) Witch Trials in the late 1690s. Environmental conditions in the village of Salem Massachusetts in 1691–3 were possibly favourable for growth of the fungal contaminant, ergot, producing LSD like compounds which could induce hallucinations and symptoms thought at the time to be associated with demonic possession (Caporael 1976). Whether this was the case or not, it remains a credible example of possible accidental food poisoning with disastrous consequences for those affected.

Food poisoning from preparation of regional delicacies can be due to carelessness, poor hygiene or in some cases an unfortunate combination of events. For example, the Indonesian delicacy Tempeh Bongkrek is made by fermenting coconut presscake or coconut milk with the fungus *Rhizopus oligosporus*. When the mould grows, the mycelia physically bind the coconut together to form a cake. However, if the product is contaminated with *Burkholderia cocovenenans*, an aerobic gram-negative bacteria, then serious poisoning can occur with 34 deaths per year being reported in the ostensibly plant based product between 1951 and 1975, at which time it was banned (although illicit kitchens were suspected as still producing the delicacy). *Burkholderia cocovenenans* has some interesting biology and when particular nutrient combinations are available, the organism will produce toxins with bongkrekic acid being the main toxin produced (Garcia et al. 1999; Scotter et al. 2015).

Accidental release from experimental facilities remains another vulnerability, albeit such facilities operate under tight controls. Research into highly infectious diseases is normally conducted in specialised high containment laboratories or assessing invasive species in tightly controlled environments. Such facilities are tightly managed and will include measures to prevent accidental release, safe disposal of wastes and fumigation routines to mitigate the risk from spills or other adverse events. Some plant pathogens do not require such high containment (biosafety level 3 or 4) but if they are exotic (not endemic in the area where research is being conducted) then additional precautions offered by such containment facilities (biosafety level 3 for example) may be useful in risk mitigation.

Many generic biosecurity measures are aimed at limiting the impact of accidental importation or releases of plant disease or pests. For example, the UK plant biosecurity strategy published in 2014 (https://www.gov.uk/government/uploads/system/uploads/attachment_data/file/307355/pb14168-plant-health-strategy.pdf) describes a number of the key considerations that need to be taken:

- activity should be directed at priority pests and pathways and be informed by comprehensive risk assessment
- includes plant pathology, population dynamics, and
- epidemiology, as well as the social sciences to understand the values at stake
- meets EU and international obligations, to enable businesses to trade in clean material and grow

- ensures everyone (government and its agencies, industry, NGOs, landowners and the public) shares a common understanding of biosecurity and their role and responsibilities
- ensures that those who benefit from plant biosecurity activity should, where appropriate, be responsible for that activity and bear the cost of it
- ensures the Plant Health Services are able to respond effectively to new and emerging threats
- ensures GB as a whole is resilient, capable and prepared to respond flexibly to new and emerging threats
- ensures GB production has a good reputation to allow exports of plants and plant products to develop, with consequent economic and social benefits

As part of this strategy, work is taken:

- pre-border through collaboration with international authorities to share understanding of disease movement through Europe and understanding of novel threats,
- at the borders to assess incoming plant material (and some soils) to mitigate the risk of accidental importation of invasive species,
- inland to detect any new infections quickly and develop/exercise eradication contingency plans.

Similar approaches are taken across Europe although managing plant material movements across land borders has additional challenges.

Good biosecurity is key to management of accidental outbreaks. Biosecurity is very much scenario dependant but hinges on good hygiene, high levels of diligence in plant product inspection, effective record keeping to aid tracing and importantly shared risk assessments on specific hazards. As an example the UK plant pest risk register (https://secure.fera.defra.gov.uk/phiw/riskRegister/downloadEntireRiskRegister.cfm) contains over 800 pests affecting food and decorative plants.

Vulnerabilities are not limited to agricultural crops or imported material and even the so called "free foods" (wild fungi, fruits, berries, etc.) are not immune to natural disease outbreaks and disease reservoirs in companion plants (e.g. in hedgerows) can be an important factor in risk evaluation, mitigation and outbreak recovery.

There may be overlap between natural and accidental vulnerabilities and it may not always be easy to identify malicious events if they are covert in nature.

1.5 Malicious Actions

Fortunately, malicious attacks against plant production are relatively rare. Nevertheless they do occur. Gardening competitions such as "Britain in Bloom" can attract unwanted addition of pesticides by rivals to flower baskets/beds with disastrous consequences for competitors. However, although such events and domestic dispute equivalents are reported from time to time, there are few malicious attacks either against food plant production capacity using biological agents or using food plants as a delivery mechanism. However, there has been a widely publicised attack using salad vegetables as a delivery vector in an attempt to affect human health on a local population.

In 1981, the Rajneeshee cult bought a 64,000 acre farm in Oregon USA as part of a plan by their leader, an Indian philosopher Bhagwan Shree Rajneesh, to build a Utopian city on their new land. Having taken over control of the local town council through elections, the cult was able to gain permissions to undertake limited development but was still unable to obtain the regional planning consents they required to expand their development into a new city.

In the middle of September 1984, several locals became ill from salmonella food poisoning with all having eaten at a local restaurant. *Salmonella typhimurium* was quickly recognised as the causative agent and those affected recovered after treatment with normal therapies (Torok et al. 1997). Initial views of investigating authorities were that this was a natural event with poor food handling being suspected as the root cause.

However, a week or so later, the total number of affected persons in the outbreak reached over 750 in a biphasic epidemic. A major response was initiated with local hospitals dealing successfully with 45 hospitalised casualties but fortunately again there were no fatalities. At that time, there was no evidence of deliberate contamination. Once again poor food handling practices were considered as being the cause although the relatively high number of restaurants involved was unusual. The incident would have remained a natural/accidental contamination event but a year later, a disaffected member of the cult alerted the authorities to the possibility that the food poisoning was deliberate.

On further investigation, US authorities found covert laboratories within the cult premises with identical *Salmonella typhimurium* to the outbreak strain being found. Prosecutions resulted and it later became clear that cult members had been encouraged to avoid restaurants during the period in which salad bar items were contaminated. This was to reduce the number of non-cult voters who would attend the polls at a local election thus influencing the election results in favour of cult members. This example highlights that detection of covert biological attacks is challenging although response (health management) processes are virtually identical for covert and overt releases.

1.6 Assessing Vulnerabilities and Gaps

There are well documented approaches to assessing food chain vulnerabilities (e.g. http://www.sigmachain.eu/uploads/dateien/fp6-518451_stakeholders_guide_on_vulnerabilities_web.pdf; http://www.springer.com/978-90-481-9557-2) which may also apply in general to plant production systems. Plant and wider food production systems encompass a "farm to fork" process which can be extremely variable in scale and complexity.

At the simplest end of this spectrum, production can be at a local level with the aim of growing food for personal consumption. At the other end of the spectrum, large industrial scale facilities may be producing millions of units daily (e.g. billions of loaves of bread annually from the 731.6 million metric tons of wheat produced each year) which often feed into broad distribution and retail networks from which consumers make an informed choice. In the latter instance the food chain is not widely vulnerable to a short term disturbance of a few days in one particular location (e.g. a spoilage problem caused by transport disruption in one country). Whilst this could impact locally, an international supply chain would support larger scale users who could switch suppliers to overcome limited timespan shortages. However, should a plant disease outbreak occur in the major wheat producing countries affecting yields, then this could have a much wider impact, especially if the genetic pool of plants used is common amongst producers and is susceptible to the same diseases.

Products themselves can be complex involving large numbers of ingredients and the growing demand in industrialised nations for "ready to cook" products means that a single unexpected contaminant in a common ingredient can have major consequences. In the UK, a major product recall in the period 2003–2005 (http://tna.europarchive.org/20111030113958/http://www.food.gov.uk/safereating/chemsafe/sudani/) was initiated because widely used ingredients (chilli powder) had been coloured with non-permitted Sudan Dyes to make them more visually attractive to users (perhaps based on adding a red coloured chemical to make the chilli seem hotter.).

The recall included contaminated spices themselves, sauces made from them in products destined for retail consumption and for use in commercial production facilities or in pre-prepared foods. With around 600 retail and wholesale product types being recalled by UK and EU authorities because of a potential health impact from the genotoxic and carcinogenic contaminant, there was a significant impact on regulators, producers and significant concern for consumers.

Mislabelled foods and fraudulent descriptions are all known vulnerabilities but much work has been conducted to improve traceability of foods and in effective labelling to ensure authenticity of products (Kelly et al. 2011; Vemireddy et al. 2015; Phelan and Jonker 2015).

Whether a production system is simple or complicated, it is important to consider and document considerations of vulnerabilities and the different compartments in the food chain often include quite specific production and distribution networks. It is common practice for each link in the chain to consider relevant microbiological, chemical and physical hazards and to establish and document effective interventions using the Hazard Analysis and Critical Control Point (HACCP) approach. More recently, Threat Analysis Critical Control Point (TACCP) and Vulnerability Analysis Critical Control Point (VACCP) approaches have become parts of the method by which the food chain can be reviewed, allowing high risk activities to be mitigated and safeguards introduced to prevent rather than manage the risks.

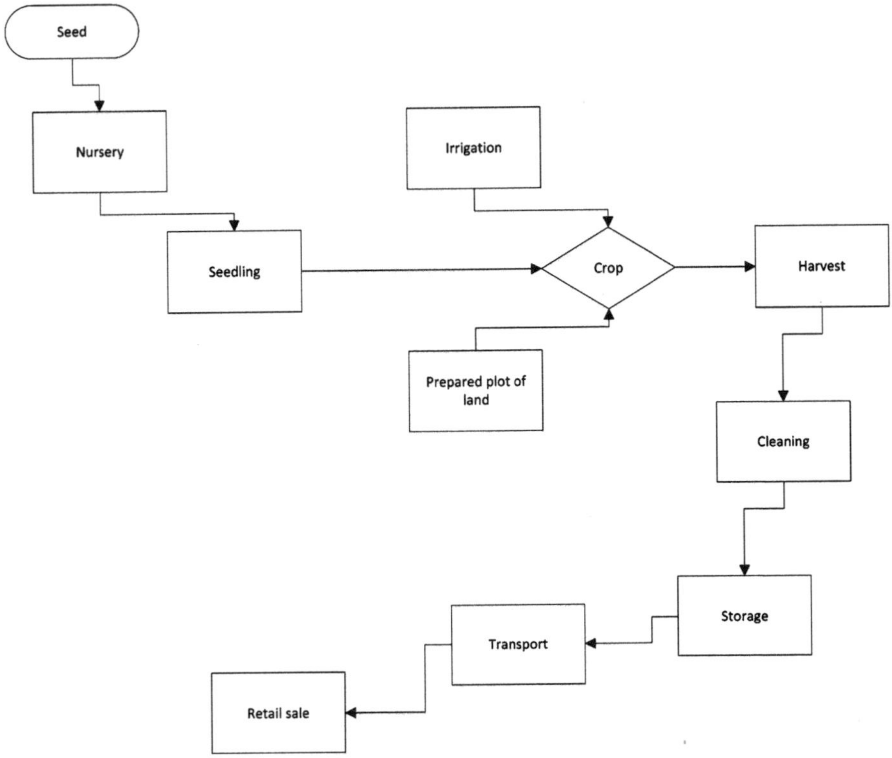

Fig. 1.1 Process map for a generic plant based food production system

A variant, Risk Analysis Critical Control Point analysis (Serra et al. 1999), has also been reported in which the consequences of product or process variation on the consumer/end user are assessed but this is not, as yet, in common use.

These approaches are not developed specifically for any one part of the food chain but can be applied generically both vertically (up and down a food chain) and horizontally (encompassing the detail of a particular element of the chain). Whilst the level of detail will vary from a farmer producing a single crop to a retailer ensuring that multiple short shelf-life product lines remain available for consumers, similar approaches are possible.

There is no intention to describe HACCP in detail in this chapter as the wide literature on the subject is easy to obtain with formal training courses being readily available from a range of providers. Suffice to say that in common with other critical control point assessments, the first stage is to map the process under consideration. A simple process map or process flow is shown in Fig. 1.1 as an example. Process maps will vary in detail but it is important to prioritise efforts in complex systems, for example, work undertaken which identified agents of concern has been extended to include naturally occurring diseases and food crops (Suffert et al. 2009).

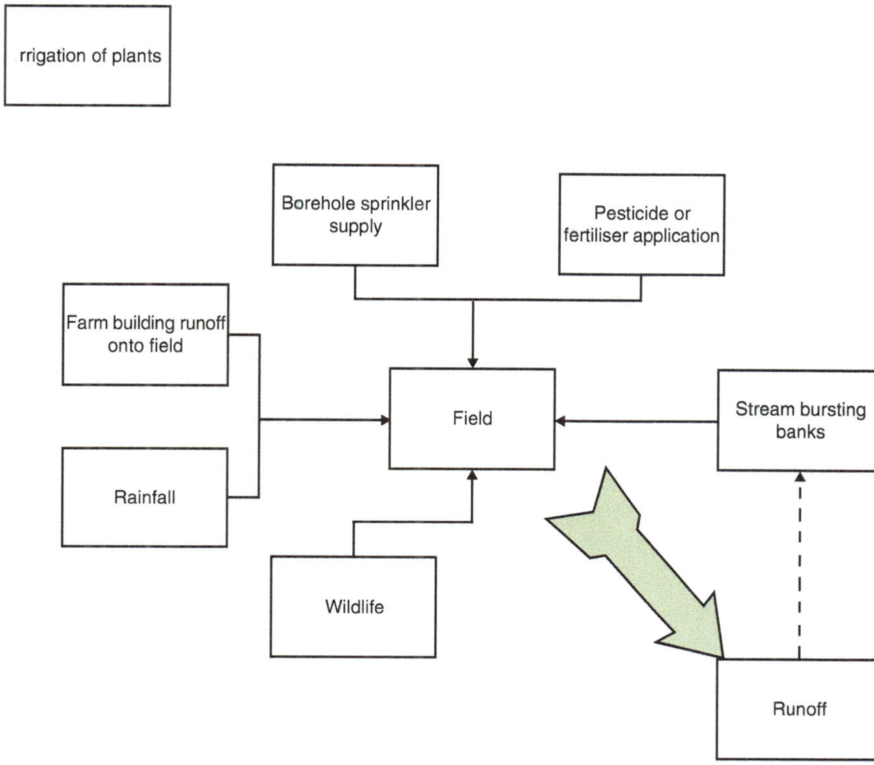

Fig. 1.2 Process map for one element of a plant based food production system

From a HACCP perspective, this may be too high a level of detail with each box in Fig. 1.1 having to be broken down into more detail to identify intervention points that would reduce the possibility of a biosecurity breach. Figure 1.2 shows a more detailed breakdown of the irrigation element of the system outlined in Fig. 1.1.

Clearly each of the boxes shown in Fig. 1.2 can be broken down further, for example, the borehole sprinkler system box may need to include microbiological loading of the water, whether there is an intermediate storage tank for the irrigation system, dead legs in the system, etc.

In the system described in Fig. 1.1, ensuring that certified disease free seed is used may be a critical control point, or perhaps ensuring that the post-harvest cleaning process does not produce a reservoir of disease in a dip tank may be another. Ensuring the soil used for planting is clean of disease and the field margins are free of other plants that can harbour crop disease could be other key steps. For each critical node in the process, a monitoring and control plan should be developed with associated record keeping and management infrastructure to develop and inform contingency plans for dealing with anomalous occurrences. Importantly this can be used to reassure customers (wholesalers, retailers and consumers) that production is under control.

1.7 Learning Lessons

Despite best endeavours, food poisoning and plant disease outbreaks will occur from time to time but understanding the reasons for outbreaks is vital in identifying vulnerabilities in the food chain and informing risk assessments.

For example, in 2006, an *E. coli* outbreak in the USA (Grant et al. 2008) was found to have been caused by contamination of spinach leaves in retail "ready to eat" salad leaf packets. In that outbreak nearly 200 persons were affected with 3 fatalities and haemolytic-uremic syndrome was observed in a number of infections.

The disease was identified in 13 samples of product from a single production run (as shown by a common batch code which highlights the importance of product traceability) but the impact was felt across the USA (http://www.cdc.gov/ecoli/2006/spinach-10-2006.html). Sourcing staple foods from a wide geographical area is not uncommon and a faulty product could be quickly spread over a wide geographical area (perhaps controlled by different regulatory authorities) which could make epidemiology based tracing based on clinical cases challenging.

However, in this instance, tracing of infected material using production batch codes suggested contaminated packages had used plant material sourced from potentially 4 ranches. Investigation teams visited these premises and relatively large numbers of feral swine were observed on at least one of them (Jay et al. 2007). The teams took swab samples from captured feral animals and on one ranch in particular, the strain of disease found was very similar to the outbreak strain. With 149 animals (estimated) on these premises, this was considered to be a likely cause of the outbreak. Major incidents are quite often the result of multiple factors and although *E. coli* was not observed in local waterways (a common vector in the environment to plant transfer chain), faecal contamination by feral animals direct onto the plants or adjacent soil was also considered possible.

The fencing used around production fields was not sufficient to prevent ingress of animals onto the fields (swine can dig under fences) and signs of rooting were observed in the soil where plants were grown. In addition, the machine used to harvest baby spinach could also pick up faecal material on soil along with the plants harvested and thus could have contributed to contamination of the produce.

Lessons from this outbreak would suggest that enhanced monitoring of water sources, improved physical separation of large wildlife from spinach fields, a different harvesting approach and improved washing/process water testing with an increased sampling rate for final product testing may be worth considering. Such lessons are invaluable in highlighting issues that might have wider applicability and are a major resource for those wishing to improve food chain resilience.

Additional monitoring of the finished product gives extra reassurance for consumers and may increase the probability of finding contamination "hot spots". However in large scale production systems finding spot contamination in time to be of use is a significant challenge given finite analytical resources, some relatively lengthy analytical turnaround times and a short shelf-life product.

Focussing the use of finite resources to key control points is a significant benefit of a HACCP approach both in terms of cost effectiveness and consumer protection. In the above example, it may well be that monitoring the wash water used to clean multiple plants could show the system was under control compared to the benefits of extending finished product examination. Each compartment in the process flow is potentially specific to that scenario and assessments need to be undertaken by staff trained in risk assessment who fully understand the processes under consideration and the limitations of microbiological examination methods.

Irrigation water is a significant potential source of contamination; particularly for those crops which undergo limited processing (crops undergoing heat treatment may be less vulnerable). Other control points worth considering in HACCP assessments would include operator hygiene, machine cleaning regimes, process/cleaning water condition and storage conditions.

Protecting plants growing in the fields from infection by plant pathogens is also critical to ensure a satisfactory yield and quality of product. This will require consideration of seed quality – is the seed stock from disease free sources?, is there a need to use coated seed and are there associated risks, if the farmer chooses to use young plants from a nursery?, what checks are required to ensure the seedlings are disease free?, is there a history of plant disease in the fields to be used?, can the plants in the margins of the field act as reservoirs of disease?. A HACCP approach needs to consider a very wide range of issues and expert advice may need to be developed and maintained by a multi-disciplinary team.

1.8 Microbiological Examination

One of the major technical challenges facing microbiologists is rapid detection of food poisoning organisms or plant pathogens at infectious dose levels in produce. Whilst modern molecular methods such as RT-PCR (e.g. Szabo et al. 2015; Zhang et al. 2011), or LAMP (e.g. D'Agostino et al. 2015; Wu et al. 2015) are sensitive and can detect for example *salmonella, listeria* and some *yersinia* spp. at levels likely to cause infection, this is not the case for all human pathogens. Some *E. coli* (Lynch et al. 2009; Friesema et al. 2008), norovirus (Cook et al. 2014), *Shigella* spp (Lewis et al. 2009) for example can have infective doses in food of the order of 10–100 colony forming units (cfu) which would be challenging to detect rapidly unless large sample volumes were taken for testing or if culturing was performed to grow microbes up to detectable levels. Inevitably culturing of bacteria takes time (e.g. 8–36 h) and with short shelf-life foods, this approach may only give a result after the product has been purchased and possibly consumed. Nevertheless it is clear that technologies are getting closer to being suitable for routine use in real time production system monitoring with increasing research consideration being given to development of field side testing capability, especially for plant pathogens. (Tomlinson et al. 2005).

An alternative approach to looking for target organisms that originate from faecal contamination is to look for other indicators of contamination (e.g. coliform markers rather than specific bio-threats) as these may be easier to find at higher concentrations that the biothreat agent (e.g. Harwood et al. 2014; Amoah et al. 2006). Optical detection of such contamination or disease on plants, e.g. using hyperspectral imaging (Bock et al. 2010) has been developed to the point where commercial food scanners are now becoming available. Test samples are irradiated with specific wavelengths of light and reflectance or fluorescence is used to detect surface anomalies where disease or faecal contamination may be present. Faecal material residues can be seen on plants using scanners at levels below that possible using the naked eye and are being evaluated for screening salad leaf crops and apples and isolate cultures. This is a rapidly evolving application (Pu et al. 2015) of established technology and scanners can range well beyond the visible spectrum on a production line.

Classical microbiological approaches to identifying plant disease or contamination are not discussed here but many of the "gold standard" methods available to laboratories require intensive and time consuming effort to develop, validate and obtain agreement that they are fit for purpose, examples being the many methods established as ISO standards (http://www.iso.org/iso/home.html). Even so, escalation of capacity to deal with an unexpected outbreak can be challenging if laboratories need to expand their scope or scale of operations, e.g. to develop high throughput methods (Adams et al. 2013) or consider unusual organisms. Many relevant laboratories have a portfolio of accredited methods or management systems (e.g. to ISO 17025:2005 (http://www.iso.org/iso/catalogue_detail.htm?csnumber = 39883) or ISO 9001:2015 (http://www.iso.org/iso/home/store/catalogue_tc/catalogue_detail.htm?csnumber=62085)) and thus can demonstrate a quality infrastructure around which extensions to scope or quality control of examinations can be based.

Networks of plant protection laboratories are key to safeguarding the plant food chain. Organisations such as the EPPO (http://www.eppo.int/), an intergovernmental organization which facilitates for European cooperation in plant health, develops international strategies to prevent the introduction and spread of dangerous pests and promote safe and effective control methods. The World Trade Organisation and the International Plant Protection Convention are also key drivers in this area (https://www.wto.org/english/thewto_e/coher_e/wto_ippc_e.htm). Human pathogens on plant issues are managed through a mixture of plant examination specialists and human health expertise, e.g. the European Centre for Disease Prevention and Control (http://ecdc.europa.eu/en/Pages/home.aspx) and their regional/national counterparts. The links between such control laboratories are vital in advising of outbreaks, novel developments (either in disease evolution or novel testing methods) and in managing cross-border issues.

Increasing surge capacity can be problematic in the midst of an outbreak where rapid screening is required and alternative examination methods may need to be considered, even if there are relatively large uncertainties associated with testing outcomes. As long as there is a low false negative testing rate and confirmatory methods are used to evaluate presumptive positive findings from screening, then less well defined methods may have utility if large numbers of samples are to be examined.

1.9 Other Critical Control Points

HACCP tends to be used close to production but more strategic considerations are also valuable, looking at where a wider food chain may be vulnerable or where a processing facility could be open to malicious abuse. Considering the latter point, threat assessment critical control point (TACCP) evaluations are a relatively new approach but work by the UK Food Standards Agency (https://www.food.gov.uk/) and Centre for Protection of the National Infrastructure (CPNI) (http://www.cpni.gov.uk/) has resulted in a helpful description of TACCP being published by the British Standards Institute under the reference PAS-96 (http://www.food.gov.uk/sites/default/files/pas96-2014-food-drink-protection-guide.pdf). In this type of control point assessment, a multi-disciplinary approach to protective security is applied to food production.

Once again a process flow is developed but in this case from the perspective of more than accidental or natural contamination. Experts from a number of disciplines (ingredient supply, security, personnel, engineering, marketing, distribution, production, packaging, etc.) consider the processes that go into getting a product to market and identify where there are weaknesses which could be exploited for financial or political gain. Many large scale suppliers, transportation companies and wholesale/retail outlets will routinely take steps to prevent such risks to their businesses in any event although there is a tendency to focus on fraud and similar criminal activity.

Having identified vulnerabilities in the process flow, a mitigation plan needs to be developed and decision makers have the options of:

- Treating the risk – taking action to remove the cause or take steps to prevent the risk from maturing. This could be as simple as locking up key ingredients when not in use to prevent loss or deliberate contamination or a more complex activity involving supplier audits and background checks on staff to increase trust in service provision.
- Tolerate the risk – the risk is accepted even though mitigating activities are not likely to be effective. In general, this categorisation would be for low probability events which cannot be managed. One example may be the risk of hurricane damage to a farm during the growing season where these were 1 in 1000 year events. Understanding the risk appetite of the stakeholders is critical in this option.
- Transfer the risk – this is where the risk is changed by moving it to another organisation. An example of this may be to move to planting seedlings rather than seed to reduce the risk of germination failure.
- Terminate the risk – use another process. The risk of deterioration of soil quality affecting production efficiency could be mitigated by switching to hydroponic methods or planting an alternative crop not susceptible to an endemic disease.

Simple TACCP mitigation actions may be as simple as knowing the staff on the farm, making sure they are appropriately supervised by trusted managers, opportunities for mischief are minimised (e.g. lock up cleaning materials and essential equipment when not in use by designated staff). However, with all of these assessments and recommended actions it is important to ensure that a proportionate response is maintained and that actions are prioritised appropriately.

Prioritisation of risk is essential. For example, although there are contingency plans within UK Government Departments to deal with risks ranging from extreme weather to a satellite falling to Earth and hitting the UK (https://www.gov.uk/government/uploads/system/uploads/attachment_data/file/61354/lead-government-department-march-2010.pdf), a structured approach to considering likelihood and impact is sometimes helpful in deciding how to best use limited resources.

1.10 Risk Prioritisation

There are a number of methods of risk ranking (e.g. as discussed by the European Food Safety Agency (EFSA) (van der Fels-Klerx et al. 2015)) ranging from a simple grid approach to quantitative assessments considering data uncertainty and detailed plant risk assessments for new species of plant/organism. A widespread, but simple, approach to prioritisation is to give the probability of a risk maturing a value from 1 (unlikely) to 5 (very probable) and an impact score from 1 (nothing appreciable) to 5 (major impact). Multiplying likelihood and impact gives a score which can be used to prioritise risks (Fig. 1.3)

This approach has the advantage of identifying those risks which can be tolerated (green scale), those that should be treated if cost effective (pink/amber) and those that must be treated or transferred (yellow/red).

Impact	5				Threat A	
	4		Threat C			
	3					Threat B
	2	Threat E				
	1			Threat D		
		1	2	3	4	5
	Likelihood					
Very high risk		Threat A				
High risk		Threat B				
Moderate risk		Threat C				
Low risk		Threat D				
Negligible risk		Threat E				

NOTE This is an example risk scoring matrix, organizations may choose different criteria for the different risk categories.

Fig. 1.3 Risk scoring matrix (PAS96:2014)

More complex quantitative processes and expert elicitation methods can be used to consider individual risks in significantly more detail which has benefits for policy makers to reduce risk profiles at national or international level.

One of the tools considered in PlantFoodSec was the so called PESTLE approach.

1.11 PESTLE

The so called PESTLE methodology was initially developed as a marketing tool but more recently has been expanded to assess impacts (with implicit risks) and to identify response triggers to complex scenario risks. The approach (or precursors) have been used since the late 1960s and provide a framework in which Political, Economic, Sociological, Technological, Legal and Environmental factors can be reviewed in a structured manner to produce a comparative framework that can be used to assess the relative management or impact priorities different scenarios. There are some differences to the simple 1–5 risk/impact method described above but once again a numerical score can be assigned for each PESTLE factor. The benefit of the approach is that it is flexible but because it is subjective a sense check should be included in the process to ensure it is not biased.

Experience has shown that one of the more effective approaches to scoring is to use:

- negligible outcomes score zero,
- low/very low outcomes score 1,
- a medium one gives a 3 and
- a significant/major outcome has a score of 9.

A non-linear approach encourages the assessor to differentiate which can be helpful in complex scenarios where key outcomes need to be identified.

The following describes a potential implementation of the approach. Examples are given for information only and do no describe a particular threat or vulnerability.

1.12 Political

There are several aspects to political risks and impacts ranging from whether national policies exist to support management of a given type of incident to considering the national and international impacts that might arise.

As an example, a malicious attack conducted using human pathogens on plants would require a multi-agency response with several agencies working in parallel. A clearly criminal act would require forensic investigation, affected persons would require treatment, guidance would need to be issued to consumers, epidemiology conducted to trace affected produce and steps taken to protect consumers and

Table 1.1 Political criteria and scores – an example

Criteria	Score
A minor incident dealt with by prompt action from local responders/officials/ landowner.	0
A minor event requiring formal action but with limited impact outside of emergency response community and key stakeholders/affected premises, easily managed.	1
Public is aware of the issue with Government(s) issuing targeted guidance to public and stakeholders. Appreciable public and media interest with local responders being at full stretch but coping.	3
National/international disquiet with significant public and media interest. Special control measures are required with use of emergency legislation or other special measures. A significant response to the situation is required (major resource utilisation) as local/regional response mechanisms are overwhelmed.	9

growers from further exposure. Coordinating activities would be necessary at local, regional, national and potentially international levels (Table 1.1). Understanding and managing these interfaces requires careful planning and very importantly exercising.

In general, existing food safety legislation, phytosanitary and health protection measures at local, national and international levels could work closely with counter-terrorism and criminal investigation authorities although detailed briefing may be needed to give context to any incident. Whilst there may be some debate as to which agencies would take the lead (unless agreed in advance), existing coordination mechanisms should enable decisions to be made promptly. The lead department may also change as the scenario develops from crisis to recovery to restoration, with the latter perhaps being a lengthy process of return to normality. Rating political factors could therefore focus on mitigation and consequences.

Where the incident is self-limiting and no special powers need to be enacted, the scores will be relatively low. However, as the *E.coli* 0104 outbreak in Germany demonstrated (Caprioli et al. 2012; Appel et al. 2012) a significant but local challenge can quickly build up into an outbreak with serious international consequences. One feature of the PESTLE approach is that regular reviews are required as incidents progress. This can provide evidence of when the different phases of incident can be considered as being over.

1.13 Economic

Economic impacts of an incident affecting food plants can result from direct consequences or appear in the form of collateral damage. Direct impacts could include losses due to seizure and destruction of infected crops or withdrawal of foods from sale and remedial action costs (land remediation, enhanced biosecurity, treatment costs, etc.). Collateral damage could include loss of reputation and therefore loss of

Table 1.2 Economic criteria and scores

Criteria	Score
No significant impact for a local incident beyond individual producers or small groups of enterprises. Financial losses < €10,000	0
Limited impact (albeit painful for those impacted upon) with losses of < €1 m expected	1
A major response is required with intervention affecting multiple stakeholders. Losses of < €10 m expected.	3
Major impact affecting the viability of the sector or sub-sector of the market. Losses in excess of €10 m likely with impacts beyond the food or agricultural sectors.	9

market share for affected foods or countries of origin or additional surveillance costs to allow positive release of fresh material. In positive release scenarios, the produce must be tested and shown to be contaminant free before release to wholesalers or retailers.

Fresh plant foods tend to have short shelf-lives with production chains being designed to allow a high turnaround of produce. Where the source of infection is unknown then collateral damage can be high because application of the precautionary principle will require intervention for more than the minimum number of products.

The numerical values in Table 1.2 are arbitrary and are for demonstration purposes only.

1.14 Sociological

The impact of a disease outbreak on society depends upon the societal groups involved or targeted, lifestyle choice adjustments either chosen or enforced because of the outbreak, and the impact of control measures at a cultural level.

Using the Rajneeshee cult attack on salad bars example described earlier, targeting salad items would have a disproportionate effect on those who choose not to eat meat if such items were removed from restaurants or the wider marketplace. It is likely that societal pressures would only ease with proactive measures, e.g. positive release of foods as being contamination free or re-certification of plant seedstock, etc to build stakeholder confidence.

Where food supplies are scarce, migration may be induced by poor crop yields with substantial social impacts, including potential unrest and cross-border issues (Table 1.3).

Perception of risk or hazard is an important aspect of managing vulnerabilities in the food chain and the social and natural sciences must work closer together to better manage incidents in future.

Table 1.3 Social criteria and scores

Criteria	Score
No significant impact on society at large (local impact only)	0
Impact is limited to individuals within no readily distinguishable socio-economic groups. Those affected appear random but small numbers only.	1
Impact felt by a small well defined socio-economic group (e.g. with specific dietary requirements) or ethnicity	3
Major impact with widespread fear leading to unrest and a lack of confidence that food is safe to eat.	9

1.15 Technological

In this proposed implementation of the PESTLE approach, some of the key considerations of a disease outbreak affecting humans from consumption of plants are given below. Note: Tables 1.4.1, 1.4.2, 1.4.3, 1.4.4, and 1.4.5 are guides and it is recognised that for each scenario the scores and criteria must be reviewed and revised to best suit particular circumstances.

1.15.1 Infectious Dose

If the disease is not infectious with little human to human transfer or inefficient transfer by known plant vectors, then it is unlikely to be a significant problem. However, some caution is needed as some groups within the population may be more sensitive than others (e.g., very young and very old or immune compromised individuals). Any assessments must therefore clearly identify the model boundaries to avoid mis-interpretation.

1.15.2 Incubation Period

There are scenarios where mass prophylaxis or intensified surveillance can reduce the impact of an outbreak. In general, the longer the incubation period, the more effective intervention can be at reducing exposure and infection but conversely time to impact will be longer and could result in wider distribution of the affected persons/foods.

1.15.3 Primary Route of Transmission

Once released, there are multiple routes by which the disease can move from host to host. This assessment considers disease vectors, e.g. in irrigation water, or via aerosol formation or manual inoculation.

Table 1.4.1 Infectious dose and scores

Criteria	Score
Not infectious – spread unlikely after initial inoculation or infection (self-limiting).	0
A reasonable amount of organism is required to cause infection e.g. >10,000 cfu per 25 g food or plant.	1
Infection is possible at moderately low levels, between 100 and 10,000 cfu	3
A very low exposure will give rise to infection (< 100 cfu).	9

Table 1.4.2 Incubation period

Criteria	Score
More than 5 days incubation period between infection and signs or symptoms becoming apparent.	1
Between 1 and 5 days	3
< 24 h before symptoms are apparent	9

Table 1.4.3 Transmission

Criteria	Score
Non-transmissible from infected host	0
Physical contact required between infected host and target	1
For human casualties, faecal-oral transmission or for plants, specific insect vector borne disease	3
Ingestion of contaminated foods or non-specific vectors required (e.g. wind, rain, etc.)	9

Table 1.4.4 Diagnostic method availability

Criteria	Score
Field test kits available and simple to use by growers or in other compartments of the food chain.	0
Test methods are routinely in use for the matrix/organism combination in many organisations (large surge capacity). Simple technologies are suitable.	1
Methods are in place for similar organisms or are used in specialist facilities	3
No routine methods available – research needed to develop/validate for a particular analyte/matrix combination.	9

Table 1.4.5 Health impacts

Criteria	Score
Infection causes mild discomfort in humans or affects cosmetic appearance in plants.	0
In humans, any illness is self limiting and mild lasting < 1 day. In plants yields are marginally affected only.	1
Disease has acute effects in humans or rapidly impacts on plant viability/yields. Product shortages may occur.	3
Fatalities, chronic effects in humans or mass crop failures are likely with long-term remediation strategies needed.	9

This is different to infectivity which relates to susceptibility of the host to infection. A disease can be highly infectious but if there are no transmission vectors then it may be of limited importance. As weather patterns change, there are concerns that changes in insect populations may impact on vectors of disease, positively or negatively and this has already been observed in the animal kingdom with the progression of bluetongue disease across Europe (e.g. Mehlhorn et al. 2009) being a well-publicised example.

1.15.4 Diagnostic Method Availability

Although modern molecular methods enable rapid assessment of large DNA sequences, simple screening tests to enable positive release of produce may not be widespread. The EHEC *E.coli* outbreak in Germany in 2011 arose from a genetic mutation that allowed *E.coli* 0104 to express the toxin normally associated with EHEC. The ECDC's Community Reference Laboratory and National Reference Laboratories were however, able to modify their procedures to detect this variant very quickly to support investigations.

1.15.5 Health Impacts

Human or plant health impacts are key to assessing the importance of a potential outbreak and the priority that is assigned for remediation. Some plant disease symptoms may only be a nuisance or affect the cosmetic appearance of the produce but others may be much more severe.

Health impacts of disease outbreaks are important to ensure optimal crop growth and yields, to allow efficient use as food/feedstock and ensure consumer safety.

1.15.6 Legal Consequences

There are two aspects to this part of the PESTLE analysis. One relates to whether there is legislation that can be used to support outbreak management and rapid epidemiological investigation. The other concerns whether or not that legislation has been applied effectively. The question of developing legislation has not been considered here as it is closely aligned to the Political issue review. Many countries have enabling legislation that allows a flexible response to an outbreak but that is not always the case (Table 1.5).

Table 1.5 Legal

Criteria	Score
No breach of legislation is likely	0
A minor infringement of laws may have occurred but handled at a local Government level	1
A major infringement has occurred requiring action by national authorities	3
International obligations have not been met with likely prosecution of national administrations.	9

Table 1.6 Environmental

Criteria	Score
Normal waste management/control protocols can be applied satisfactorily and safely. No real impact on the environment.	0
Special measures are required (e.g. increase in waste disposal capacity) with no discernible impact on environment in the longer term or beyond the immediate area. Reminders of good biosecurity measures issued.	1
Some wider environmental impact is inevitable with significant measures needed to enhance biosecurity in affected areas.	3
A major response is required to enhance biosecurity on a large scale including pro-active decontamination measures to make wastes safe before long terms storage or disposal	9

1.16 Environmental Considerations

The impact of a disease outbreak on the environment can be significant. Major remediation activities can produce waste and interventions to remove the threat to consumers can increase levels of waste requiring special treatment, e.g. incineration to destroy infected materials. An outbreak of Foot-and-mouth disease (FMD) in 2001 (The Comptroller and Auditor General 2002) and again in 2007 (Anderson 2008) in the UK highlighted the importance of environmental impact considerations in disease management strategies. Waste disposal in some FMD cases in 2001 had a major environmental impact where open ground incineration was used. Additional vehicular transport to move waste to authorised disposal sites added to local environmental disruption and extra landfill sites were required to deal with the 1.3 million animal carcases disposed of. It is to be hoped that a plant disease outbreak would not reach these levels of disruption but the need to monitor impacts on the wider environment remains a useful lesson. Should soil sterilisation be required then the environmental impact could be significant from use of gaseous fumigants or liquid disinfectants (Table 1.6).

1.16.1 PESTLE Outputs

Table 1.7 is an example of how the approach could have been used to model progression the German *E.coli* outbreak. As in any proposed use of the tool, it is subjective and in this example not subject to external review or challenge (which

Table 1.7 A retrospective example of the PESTLE approach

Factor\stage	Pre-incident		Early incident		Early investigations (cucumbers suspected)		Not cucumbers		Seeds as source	
Political	Probably national issue	3	Probably national	3	International	9	International	9	International	9
Economic	Significant	3	Significant	3	Major	9	Significant	3	Significant	3
Social	Limited population	1	Defined group	3	Significant fear factor	9	Cases reducing	3	Defined group	3
Technological	Highly infectious but tests available	9	Unknown cause	9	E Coli cause	9	E Coli cause	9	Screening	3
Legal	Local officials handling	1	Local officials handling	1	International incident	9	Eased international impact	3	Local officials handling	1
Environmental	Some special measures – minimal impact	1	Some special measures – minimal impact	1	Special waste management	9	Some impact	3	Some special measures – minimal impact	1
Score		18		20		54		30		20

should be part of any risk assessment undertaken in emergency response situations if time permits).

As shown in Table 1.7, as the incident develops, the PESTLE score changes and priorities alter, a situation experienced emergency planners will recognise. With finite resources, it is now clear where priorities lie although as this table was produced after the event it does not reflect the difficulty of standing back from an incident to develop enhanced situational awareness. This tool can also be used to inform development of a Commonly Recognised Information Picture (as used by UK authorities (https://www.gov.uk/government/uploads/system/uploads/attachment_data/file/192425/CONOPs_incl_revised_chapter_24_Apr-13.pdf) and others) to take a snapshot of a situation. Shared situational awareness and goals is vital if response to an incident is to be effective.

1.17 Plant Risk Assessments

Many countries undertake plant risk assessments and prioritise those organisms for which contingency planning is appropriate or equally those for which risks can be tolerated or treated. These assessments can be performed in advance of an outbreak and can support development of response plans.

Pest risk assessments (PRA) take several forms but contain similar information including:

- The name of the pest – phylogenetic grouping can and does change which can add confusion for the non-specialist so it is important to explain the exact nature of the pest under consideration. Similarly the names of the disease or diseases produced by the organisms should be described.
- The status of the pest in national and/or local legislation (including EPPO lists).
- Background to the reason for a PRA.
- Current geographical distribution.
- Is the pest established or transient?
- What are the pest's natural and experimental host plants?
- What is the economic and environmental importance of the host plants?
- Does the pest need a vector for transmission?
- Can the credible pathways upon which the pest may move be described?
- How likely is it that this pest can become established either indoors or outdoors?
- What geographical area is at risk?
- Is there evidence of economic, environmental or social impact where it is already found?
- Can the economic social or environmental impacts be estimated should the disease arrive here?

Outputs from this detailed assessment can be and are used to identify those organisms for which contingency plans are required.

Once again such assessments identify those organisms for which associated risks to plant heath can be treated, tolerated, terminated or transferred, e.g. by monitoring imports, increasing surveillance of growers, restricting use of high risk materials or using lower risk crops where this is an option.

1.18 PlantFoodSec

The PlantFoodSec project builds on existing national (e.g. https://www.gov.uk/guidance/plant-health-controls) and international arrangements (e.g. through the Food and Agriculture Organisation, World Health organisation IPPO and EPPO) to support growers to consumers in the "seed to salad" food chain.

This chapter has focussed on a variety of methods available to identify vulnerabilities and perhaps the main ones this project has sought to overcome are a lack of shared understanding across the various compartments of the food chain, a common lexicon to avoid misunderstanding, appreciation of technical limitations and future technologies and assessment and communication of risk.

There is no intention to repeat the detail found in other chapters of this book but when considering vulnerabilities and gaps, the following project activities to develop a toolbox that will be of value for many years to come, are of note.

Information gathering is key when considering vulnerabilities in a very complex area as prioritisation is only possible if based on evidence. Within this project substantial effort has gone into establishing a list of key crops of interest to EU consumers and the diseases that might impact upon them. With over 450 crops and 500 diseases being considered, this was a major undertaking and the tools developed to ensure a consistent approach to prioritisation fill an important capability gap.

Those diseases that impact on humans, the so-called Human Pathogens on Plants (HPOP), are used in examples given in this chapter but were part of a larger study conducted by specialists within the project. The PESTLE approach described above was one of the outputs of this work.

As discussed above, detection of when a disease is present, particularly in the pre-symptomatic phase is a challenge. Maintaining a consistent approach across Europe to both undertake presence/absence examination and more detailed contaminant species/toxin identification is not easy. Within this project, work has been undertaken to identify those laboratories with the necessary technical capability, to critically review methodologies and share understanding of performance criteria between national experts. In addition, expanding technical capability and capacity through formal training courses in both general and specialised test methods has boosted EU capability to respond to an incident. Extending analytical capability to forensically discriminate between strains is not trivial but is essential to support epidemiological investigations. The vast majority of food poisoning organisms occur naturally and, as the E Coli outbreak in Norther Europe has shown, misidentification of the causative agent can have disastrous consequences. Maintenance of a diagnostic network to support responses to future outbreaks and encourage efficient information exchange are important outcomes of this project.

It is because food borne and plant disease outbreaks do occur from time to time that tools are needed to help authorities decide if an outbreak is a natural or due to malicious intervention. Logic trees have been used for many years to support risk assessments but the tools developed in this project not only incorporate expert opinions but also consider uncertainty or lack of evidence. It would not be unusual for a new or poorly studied disease to have high levels of uncertainty over behaviour in the food chain although use of analogues could be helpful. Models incorporating uncertainty are important as they can provide more realistic indications of priority areas for research or intervention.

Much of the work in this area suffers from a lack of credible scenarios with many being protectively marked because of sensitivities over vulnerabilities that could be exploited for malicious purposes. Nevertheless, within this project over 100 scenarios have been developed and considered allowing the toolkit to be evaluated and benefits identified. This approach will have benefits beyond the plant food chain into wider emergency response.

The project team recognise that incidents will occur and it is important to be able to deal with them. Developing strategies of generic importance is a valuable result from this project, in particular emphasising the importance of taking a broader perspective. In one example considered by specialists in detail, endemic contamination of an onion crop by a fungal agent cannot be controlled by simple remediation methods, in part because of disease reservoirs in wild plants in the margins of grower's fields. Switching the type of product to a more resistant variety appears to be a pragmatic solution to restore production to previous levels and yields. Maintenance of a wide genetic pool of commercial crops is becoming increasingly important to mitigate outbreaks in high yield but susceptible crops. Ensuring national and international crop and pathogen collections are maintained and shared remains a key risk mitigation activity.

Within the Emergency Response community, significant effort is expended on dealing with the crisis phase of an incident and this phase is exercised at local, national and international levels. What is much more difficult to exercise is the recovery phase of an incident and for that reason the return to normality (whatever that may be) is less well practiced. Recently the UK published a recovery handbook for biological incidents which considers a wide range of contamination events, responses and recovery strategies (Pottage et al. 2015). Although not specifically aimed at plant disease outbreaks, there are lessons for those responding to HPOP issues and where food chain biosecurity needs are important. Some of the data contained on disease aetiology, survival in the environment and disinfection choice have wider utility.

With a project team encompassing major European countries and candidate EU member states, Israel and the USA, perhaps the key gap that this project has filled is in developing a shared understanding of the topic at an international scientific level. Further work is needed to reinforce the successes of this project and development of a virtual Centre of Competence of Plant and Food Biosecurity is one way of doing this.

1.19 Conclusions

Within this chapter some approaches to identifying vulnerabilities and their prioritisation for resolution have been considered. Of necessity, this chapter has drawn heavily on experience learnt from disease outbreaks that do not affect plants but the fundamental principles remain valid. The PlantFoodSec project is part of a wider programme of studies funded by the EU to enhance resilience to malicious, accidental and deliberate releases of infectious agents and there is little doubt that a number of capability and capacity gaps have been filled at its conclusion.

Acknowledgements The author would like to thank Audrey Harris, Christian Patermann, Fausto Pedrazzini and Paola Colla for their support and helpful contributions and suggestions.

References

Adams IP, Abidrabo P, Miano DW, Alicai T, Kinyua ZM, Clarke J, Macarthur R, Weekes R, Laurenson L, Hany U, Peters D, Potts M, Glover R, Boonham N, Smith J (2013) High throughput real-time RT-PCR assays for specific detection of cassava brown streak disease causal viruses, and their application to testing of planting material. J Plant Pathol 62:233–242. doi:10.1111/j.1365-3059.2012.02622.x

Amoah P, Drechsel P, Abaidoo RC, Ntow WJ (2006) Pesticide and pathogen contamination of vegetables in Ghana's urban markets. Arch Environ Contam Toxicol 50:1–6. doi:10.1007/s00244-004-0054-8

Anderson I (2008) Foot and mouth disease 2007: a review and lessons learned. Report HC312 2007–08 The Stationary Office. London

Appel B, Böi G-F, Greiner M, Larhssen-Wiederholt M, Hensel A (2012) EHEC Outbreak 2001. Investigation of the outbreak along the food chain. BfR. ISBN 3-938163-90-9

Bock CH, Poole GH, Parker PE, Gottwald TR (2010) Plant disease severity estimated visually, by digital photography and image analysis, and by hyperspectral imaging. Crit Rev Plant Sci 29(2):59–107. doi:10.1080/07352681003617285

Caporael LR (1976) Ergotism: the Satan loosed in Salem? Science 192(4234):21–26. doi:10.1126/science.769159

Caprioli S, Morabito S, De Smet K (2012) The role of European Union Reference Laboratories in food safety crisis: the experience of the EU-RL for Escherichia coli during the recent outbreak of E. coli O104:H4 infections. Euro Ref Winter 2012:2–4

Cook N, Bertrand I, Bosch A, Gantzer C, Maul A, Pinto R, van der Poel W, Stein J (2014) FSA Project FS101074: a critical review of the effect of heat, pH and water activity on the survival of Hepatitis A and E viruses. A Report to the United Kingdom Food Standards Agency July 2014

D'Agostino M, Diez-Valcarce M, Robles S, Losilla-Garcia B, Cook N (2015) A loop-mediated isothermal amplification-based method for analysing animal feed for the presence of Salmonella. Food Anal Methods 8(10):2409–2416. doi:10.1007/s12161-015-0148-0

European Food Safety Authority (2011) Shiga toxin-producing E. coli (STEC) O104:H4 2011 outbreaks in Europe: taking stock. EFSA J 9(10):2390. doi:10.2903/j.efsa.2011.2390

Friesema I, Sigmundsdottir G, van der Zwaluw K, Heuvelink A, Schimmer B, de Jager C, Rump B, Briem H, Hardardottir H, Atladottir A, Gudmundsdottir E, van Pelt W (2008) An international outbreak of shiga toxin-producing Escherichia coli O157 infection due to lettuce. Euro Surveillance 13(50):pii

Garcia RA, Hotchkiss JH, Steinkraus KH (1999) The effect of lipids on bongkrekic (Bongkrek) acid toxin production by Burkholderia cocovenenans in coconut media. Food Addit Contam 16(2):63–69

Grant J, Wendelboe AM, Wendel A, Jepson B, Torres P, Smelser C, Rolfs RT (2008) Spinach-associated Escherichia coli C157:H7 Outbreak, Utah and New Mexico, 2006. Emerg Infect Dis 14(10):1633–1636. doi:10.3201/eid1410.071341. PMCID: PMC2609868

Harwood VJ, Staley C, Badgley BD, Borges K, Korajkic A (2014) Microbial source tracking markers for detection of fecal contamination in environmental waters: relationships between pathogens and human health outcomes. FEMS Microbiol Rev 38(1):1–40. doi:10.1111/1574-6976.12031. Epub 2013 Aug 7

Jay MT, Cooley M, Carychao D, Wiscomb GW, Sweitzer RA, Crawford-Miksza L, Farrar JA, Lau DK, O'Connell J, Millington A, Asmundson RV, Atwill ER, Mandrell RE (2007) Escherichia coli O157:H7 in Feral Swine near Spinach Fields and Cattle, Central California Coast. Emerg Infect Dis 13(12):1908–1911. doi:10.3201/eid1312.070763

Johnson DA, Cummings TF (2015) Effect of powdery scab root galls on yield of potato. Plant Dis 99(10):1396–1403

Karch H, Denamur E, Dobrindt U, Finlay BB, Hengge R, Johanne L, Ron EZ, Tønjum T, Sansonetti PJ, Vicente M (2012) The enemy within us: lessons from the 2011 European Escherichia coli O104:H4 outbreak. EMBO Mol Med 4:841–848. doi:10.1002/emmm.201201662

Kelly S, Brereton P, Guillou C, Broll H, Laube I, Downey G, Rossman A, Hozl S, van der Veer G (2011) New approaches to determining the origin of food. In: Food chain integrity: a holistic approach to food traceability, safety, quality and authenticity. Woodhead Publishing in Food Science Technology and Nutrition, Oxford/Philadelphia, 238p

Lewis HC, Ethelberg S, Olsen KEP, Nielsen EM, Lisby M, Madsen SB, Boel J, Stafford R, Kirk M, Smith HV, Tikumrum S, Wisetrojana A, Bangtrakulnonth A, Vithayarungruangsri J, Siriarayaporn P, Ungchusak K, Bishop J, Mølbak K (2009) Outbreaks of Shigella sonnei infections in Denmark and Australia linked to consumption of imported raw baby corn. Epidemiol Infect 137:326–334

Lobb AE, Mazzocchi M, Traill WB (2007) Modelling risk perception and trust in food safety information within the theory of planned behaviour. Food Qual Prefer 18:384–395

Lynch MF, Tauxe RV, Hedberg CW (2009) The growing burden of foodborne outbreaks due to contaminated fresh produce: risks and opportunities. Epidemiol Infect 137:307–315

Mehlhorn H, Walldorf V, Klimpel S, Schaub G, Kiel E, Focke R, Liebisch G, Liebisch A, Werner D, Bauer C, Clausen H, Bauer B, Geier M, Hörbrand T, Bätza HJ, Conraths FJ, Hoffmann B, Beer M (2009) Bluetongue disease in Germany (2007–2008): monitoring of entomological aspects. Parasitol Res 105(2):313–319. doi:10.1007/s00436-009-1416-y, Epub 2009 Mar 26

Phelan S, Jonker D (2015) Applying DNA barcoding to the manufacturing supply chain, to reduce substitution and adulteration fraud. Genome 58(5):266–266

Pottage T, Goode E, Shieber C, S Wyke S, Speight S, Bennett AM (2015) UK recovery handbook for biological incidents 2015, Version 1. Public Health England, ISBN 978-0-85951-777-5

Pu YY, Feng YZ, Sun DW (2015) Recent progress of hyperspectral imaging on quality and safety inspection of fruits and vegetables: a review. Compr Rev Food Sci Food Saf 14(2):176–188. doi:10.1111/1541-4337.12123

Redmond EC, Griffith CJ (2007) Consumer perceptions of food safety risk, control and responsibility. (2004). Appetite 43(3):309–313. doi:10.1016/j.appet.2004.05.003

Scotter M, Robb P, Harris A (2015) Personal communication

Serra JA, Domenech E, Escriche I, Martorell S (1999) Risk assessment and critical control points from the production perspective. Int J Food Microbiol 46(1):9–26

Suffert F, Latxague E, Sache I (2009) Plant pathogens as agroterrorist weapons: assessment of the threat for European agriculture and forestry. Food Sec 1:221–232. doi:10.1007/s12571-009-0014-2

Szabo K, Trojnar E, Anheyer-Behmenburg H, Binder A, Schotte U, Ellerbroek L, Klein G, Johne R (2015) Detection of hepatitis E virus RNA in raw sausages and liver sausages from retail in

Germany using an optimized method. Int J Food Microbiol 215:149–156. doi:10.1016/j.ijfoodmicro.2015.09.013

The Comptroller and Auditor General (2002) The 2001 outbreak of foot and mouth disease. Report HC 939

Tomlinson JA, Boonham N, Hughes KJD, Griffin RL, Barker I (2005) On-site DNA extraction and real-time PCR for detection of Phytophthora ramorum in the field. Appl Environ Microbiol 71(11):6702–6710. doi:10.1128/AEM.71.11.6702-6710.2005

Torok TJ, Tauxe RV, Wise RP, Livengood JR, Sokolow R, Mauvais S, Birkness KA, Skeels MR, Horan JM, Foster LR (1997) A large community outbreak of Salmonellosis caused by intentional contamination of restaurant salad bars. JAMA 278(5):389–395

van der Fels-Klerx HJ, van Asselt ED, Raley M, Poulsen M, Korsgaard H, Bredsdorff L, Nauta M, Flari V, d'Agostino M, Coles D, Frewer L (2015) Critical review of methodology and application of risk ranking for prioritisation of food and feed related issues, on the basis of the size of anticipated health impact. EFSA supporting publication EN-710, 106pp

Vemireddy LR, Satyavathi VV, Siddiq EA, Nagaraju J (2015) Review of methods for the detection and quantification of adulteration of rice: basmati as a case study. J Food Sci Technol – Mysore 52(6):3187–3202. doi:10.1007/s13197-014-1579-0

Verbeke W, Frewer LJ, Scholderer J, De Brabander HF (2007) Why consumers behave as they do with respect to food safety and risk information. Anal Chim Acta 586:2–7

Wu GP, Chen SH, Levin RE (2015) Rapid real-time loop-mediated isothermal amplification combined with coated activated carbon for detection of low numbers of Salmonella enterica from lettuce without enrichment. Food Control 56:47–52. doi:10.1016/j.foodcont.2015.03.008

Zhang GD, Brown EW, Gonzalez-Escalona N (2011) Comparison of real-time PCR, reverse transcriptase Real-Time PCR, loop-mediated isothermal amplification, and the FDA conventional microbiological method for the detection of Salmonella spp. Produce Appl Environ Microbiol 77(18):6495–6501. doi:10.1128/AEM.00520-11

Chapter 2
Characterization of the Threat Resulting from Plant Pathogen Use as Anti-crop Bioweapons: An EU Perspective on Agroterrorism

Frédéric Suffert

Abstract This chapter provides an analysis specific for Europe of the risk of plant pathogens being used as anti-crop bioweapons, taking into account both the biological and human dimensions of the threat. An historical review of anti-crop bioweapons lays down the starting point of the characterization and contextualizes the threat in Europe. Four types of threat are developed and provide a structure for the analysis: (1) from military state programs to allegations of attacks; (2) from 'rogue state' hidden programs to claimed terror attacks; (3) biocrime, sabotage, private allegations and conspiracy theories on social media; (4) from the overzealous application of phytosanitary measures to the deliberate introduction of a regulated pest to justify trade protectionism. A database consisting of 21 important target crops and of 63 potentially dangerous pests (selected from a list of 570 pests) are combined with the development and categorization of 'scenarios'. This is proposed as a starting point of a prospective approach to quantify the risk of agroterrorism in Europe. Four challenges ('Convergence Tactics', 'Constraints', 'Climate', and 'Conspiracy') are suggested to be the most important determinants of the forthcoming evolution of the threat. The prospect for Europe to successfully confront the increasing risk and challenges for the next decade is discussed.

Keywords Agroterrorism • Anti-crop bioweapon • Biocrime • Biosecurity • Bioterrorism • Biowarfare • Epidemiology • Plant pathogen • Prospective scenario • Risk analysis • Target crops

---

F. Suffert (✉)
UMR BIOGER, INRA, AgroParisTech, Université Paris-Saclay, 78850 Thiverval-Grignon, France

Cropbiosecurity and Agroterrorism Watch, Paris, France
e-mail: frederic.suffert@inra.fr; cropsafexpertise@yahoo.fr; http://www.scoop.it/t/cropbiosecurity

M.L. Gullino et al. (eds.), *Practical Tools for Plant and Food Biosecurity*, Plant Pathology in the 21st Century 8, DOI 10.1007/978-3-319-46897-6_2

2.1 Introduction and General Concerns

The globalization of markets and social links poses new challenges for plant health, food safety and security. 'Crop biosecurity', defined by Brasier (2008) as "protecting a state from invasive plant pathogens", is usually ensured by plant health policies and regulatory measures imposed by the state, often by the national government. Maintaining biosecurity has become a subject of widespread concern, heightened by the recent focus on failures in biosecurity, such as disease emergence and pest introductions (Anderson et al. 2004) and by the world-wide increasing scrutiny of pest risk analysis (PRA) as the basis for commodity trade regulation (Schrader and Unger 2003). Europe has been concerned about biosecurity for some time, due to the specificities of its agriculture and its dominant commercial position in the international markets.

Several plant pests are perceived as serious threats to agricultural biosecurity and to agricultural industries and forestry in both developing and industrialized countries. The recent decades of booming trade in commodities and horticultural plants led to many new pest introductions (Waage and Mumford 2008; Sache et al. 2011). Some plant pests threaten natural ecosystems as well as managed ones. One of these introductions was, for example, the fungus *Phytophthora ramorum*, which threatens indigenous forest trees in the United Kingdom (Brasier et al. 2004). Another was the pine wood nematode *Bursaphelenchus xylophilus*, which currently is regarded as a major threat to French forests following its establishment in Portugal in 1999 (Mota et al. 1999). Another recent example of disease emergence is the bacteria *Xylella fastidiosa*, which was first recorded in Puglia (Italy) in 2013, where it causes serious damage to olive trees, and in Corsica (France) in 2015, where it affects the ornamental hosts *Polygala myrtifolia*.

The International Plant Protection Convention (IPPC) promotes global harmonization of phytosanitary measures that are imposed by the different national plant protection organizations to prevent accidental introductions of exotic pests through trade imports. Regional plant protection organizations, such as the European and Mediterranean Plant Protection Organisation (EPPO), improve the harmonization of plant health protocols on a regional level. Today in the European Union (EU) approximately 300 pests have been identified as quarantine pests, largely on the basis of EPPO's recommendations. In order to comply with the requirements of the new EU plant health regime (regulation EU 2016/2031 of the European Parliament of the Council of 26 October 2016 on protective measures against pests of plants), some European countries, such as the United Kingdom and France, have developed methodologies for prioritizing plant health risks from pests at the national level (Moignot and Reynaud 2013; Baker et al. 2014). These efforts have focused on conventional threats of exotic, invasive plant pests that historically have been either accidentally introduced through trade or passively spread, for example by wind currents. Until recently, little attention has been paid to the possible, deliberate misuse of plant pathogens as 'weapons' against agroecosystems.

As Josling et al. (2003) observe "since the terrorist attacks (...) on September 11th 2001 (...), biosecurity has taken on new dimensions and products that move across borders are treated more suspiciously, [creating] uncertainty and transaction costs that impinge particularly on trade that could put domestic animal, plant or human populations at risk". The term 'agroterrorism' corresponds to the deliberate misuse of biological agents against agriculture, including crops and cattle, by non-state actors, that is, a subset of 'bioterrorism'. 'Biocrime' and 'biowarfare', however, can be included in a general definition referring to the "intentional use, as well as the threat or simulation of use of plant pathogens by any individual or group in order to cause direct damage to crops or forests, or to indirectly affect the agricultural sector" (Latxague et al. 2007). The distinction between bioterrorism, biocrime and biowarfare was made for several transdisciplinary components, especially concerning the legal framework and risk assessment (Chap. 6). This distinction also acknowledges that each of these 'agro-risks' possesses a number of distinct characteristics, across a wide range of prospective scenarios. The economy of Europe is heavily dependent on its agricultural resources. Crops and forests cannot be entirely monitored and protected because they are grown on large and often patchy areas. Scientists and government stakeholders in several countries are reconsidering the vulnerability of agroecosystems to plant pests potentially used as bioweapons because of the socio-economical significance of crops and forests (Rogers et al. 1999; Foxwell 2001; Cochrane and Haslett 2002; Suffert 2003; Madden and Wheelis 2003; Khetarpal and Gupta 2007; Caldas and Perz 2013; Khalil and Shinwari 2014). The vulnerability of Europe is extremely difficult to assess, probably because the definition of the concept of the 'agroterrorist threat' is weak due to its dual nature: it has both a biological and a human dimension (Barbier 2008). This creates a paradoxical combination of science-based discourse about 'plant pathogens' or 'pests' (the weapons) and subjective views about 'perpetrators' (the human entity): Who are they, why are they acting, what are their capacities and knowledge? Understanding the ideologies and motivations that would direct a person, an organization or a state to attack the agricultural sector through biological means is important for understanding how better to assess the risk for Europe (Chap. 6). There are many ideological, economical and geopolitical interests that could lead to agroterrorism. Perpetrators can be motivated by a variety of objectives, including some specific to Europe. The tactics used to accomplish these objectives may be as varied as the motivations. The choice of attacking crops as a target could be aimed at a number of outcomes: inducing yield losses, undermining confidence in the agricultural sector, creating a profit-making opportunity, extorting money by threatening to introduce a pest, coertion or intimidation of a government, provocation of a response to support insurgent forces, etc. The risk assessment of such a scenario would be erroneous if it focuses only on a single type of act or perpetrator. This could result from the attention arising from the events or the topics reported by news media. On the other hand, the risk assessment also would fall short if it does not take into account the current context related to the human dimension.

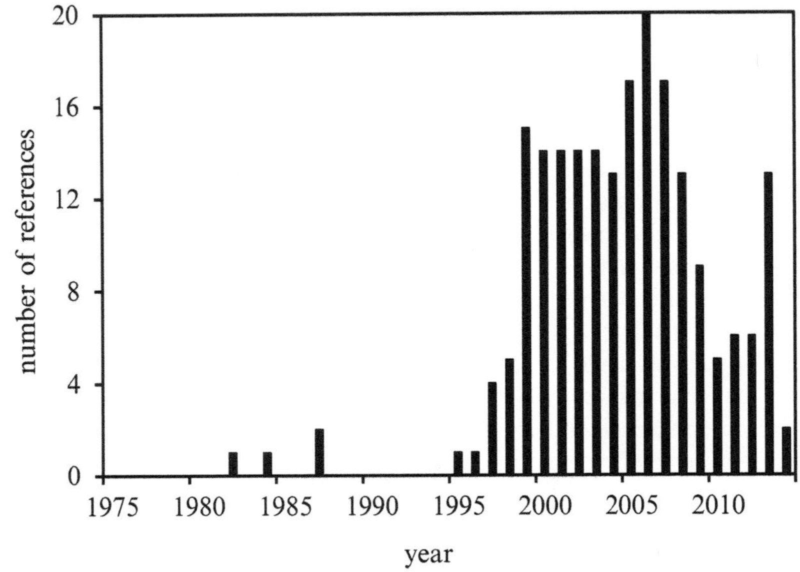

Fig. 2.1 Evolution of the number of books, scientific articles and public reports in English or French related specifically to anti-crop bioweapons and misuse of pest against plants, crops or agro-ecosystems, and the occurrence at least of one of the following keywords: agricultural biosecurity, crop biosecurity, agricultural terrorism, agroterrorism, anti-crop bioterrorism, environmental terrorism, ecoterrorism, rural crime, biowarfare, agro-warfare, anti-crop bioweapons (from Suffert et al. 2008, updated)

For the past two decades, agroterrorism has received increased attention (Fig. 2.1) and it has been subject to greater discussion within academic, media, and government circles, especially in the United States after the September 11, 2001 terrorist attacks and the subsequent anthrax infections. Studies around that period began arguing that agroterrorism represents a new and dire threat to national security (Casagrande 2000; Madden and Wheelis 2003; Cupp et al. 2004; Polyak 2004). Agroterrorism was framed as a specific issue of security research for crop protection, which contributed to the emergence of agricultural bioforensics (application of scientific methods to the investigation of possible violations of the law, where scientific knowledge and technology provide evidence in both criminal and civil matters) in the US during the 2000s (Budowle 2003; Murch 2003; Fletcher et al. 2006; Kamenidou et al. 2013). The vulnerability of the US agro-industrial sector was considered – rightly or wrongly – as high (Wheelis et al. 2002). Such a perception seems to be mainly based on the assessment of the human dimension of the threat, considering that the 'intentionality' correlated to the traumatic impact of terrorist attacks in the US. In reality, this intentionality is still very difficult to assess (Rohn and Erez 2013). The perception could be summarized by the motto "Because it's not a question of IF, but a question of WHEN" (Suffert et al. 2008), warning that agroterrorism is an imminent threat that should be taken seriously. In retrospect, the alarmist conclusions of some US reports were conjecture, based on worst case scenarios. This can be viewed by the other countries as too simplistic and, therefore, erroneous.

The asymmetry of knowledge between the biological and the human dimensions of the threat remains a key component of this issue. The lack of a common definition of agroterrorism, probably due to the recent more widespread interest in this topic, explains in part why the agroterrorist threat for European crops and forests had not yet been exhaustively assessed by appropriate methods. Unverified allegations (Table 2.1), alarmist reports (Rogers et al. 1999; Wheelis et al. 2002) and programs disclaimed for their cost (Schwägerl 2005) did not favor the recognition of agroterrorism in Europe as the real threat that the author believes it is. In this context the EU launched two successive research projects named CropBioterror (Gullino et al. 2006) and PlantFoodSec (Gullino et al. 2011). The goal was to build up expertise and develop awareness and preparedness concerning the risk of intentional threats against crops or the food chain, and to assess possible economic outcomes of such an attack in Europe. Those projects were complemented by a third, AniBioThreat, concerning the threat of agroterrorism against animals (Knutsson et al. 2013). The projects resulted in a scientifically-based framework, scientific knowledge and tools that can be used to delimit the scope of the issue and its associated narratives.

The goal of this chapter is to draw up an inventory and a specific analysis of agroterrorism risks for Europe based on both historical approaches and contextualization of the 'dual threat' (biological and human dimension). The chapter also attempts to describe and qualify the potential threat, before considering assessment of the overall risk (Chap. 6). The first problem with the term 'agroterrorism', as defined for example by Latxague et al. (2007), is that it refers to different types of acts related to the multiplicity of potential perpetrators, motivations, targets (crops) and agents (pests). In addition to the three main categories characterized by distinct objectives (biological warfare, bioterrorism, and biocrime; Latxague et al. 2007), a typology of consequences was proposed: impact on production (destruction of crops or reduced yields), impact on trade in agricultural products (due to prohibition or additional measures linked to the conditions caused by agroterrorism), impact on human or animal health, impact of an environmental and heritage nature, psychological impact on consumers, and social destabilization. This classification based on motivations and potential consequences was used to draw up and then analyze several prospective scenarios (Chap. 6).

2.2 Historical Review of Agroterrorism and Anti-crop Bioweapons: Starting Point of the Characterization and Contextualized of the Threat in Europe

The starting point of agroterrorism risk qualification is a global review of historical programs, allegations and acts. Analysis of such data is necessary to contextualize the assessment and to adapt it to the present and future European situation.

Table 2.1 Reports of 'agroterrorism' threats *sensu lato* against plant in different agro-ecosystems, from verified state programs to alleged accusations (after Suffert et al. 2009); human pathogens on plants (Fletcher et al. 2013) and case of malicious biocontamination of food (Elad 2005) not listed here (Chap. 5)

Pest	Target crop	Year	Target area	Origin[a]	Type of acts[b]	Threat veracity[c]	References[d]	
Puccinia triticina	Wheat	1950, 1970	USA	Soviet Union	BW2	++	A	Whitby (2002); Rimmington (2000)
		1950	Soviet Union	USA	BW2	++	A	Madden and Wheelis (2003)
		1990	India	Pakistan	BW2	+	A	Shoham (2014)
Puccinia graminis f. sp. *tritici*	Wheat	1950, 1970	USA	Soviet Union	BW2	++	A	Line and Griffith (2001); Rimmington (2000)
		1950	Korea	USA	BW2	++	A	Whitby (2002)
Puccinia striiformis	Wheat	1990	India	Pakistan	BW2	+	A	(Shoham 2014)
Tilletia tritici	Wheat	1980	Iran	Iraq	BW2	++	A	(Whitby 2002)
Tilletia laevis	Wheat	1980	Iran	Iraq	BW2	++	A	(Whitby 2002)
Tilletia indica	Wheat	2002	Afghanistan	India	BW2	–	A	Shoham (2014)
Aspergillus sp.(aflatoxin)	Wheat	1980	Iran	Iraq	BW2, BT1	+	A	Whitby (2002)
Cochliobolus miyabeanus	Rice	1940	Japan	USA	BW2	++	A	Madden and Wheelis (2003)
Magnaporthe grisea	Rice	1940	Japan	USA	BW2	++	A	Madden and Wheelis (2003)
		1940	China	Japan	BW2	++	A	Madden and Wheelis (2003)
		1950	USA	Soviet Union	BW2	+	A	Madden and Wheelis (2003); Rimmington (2000)

		1960	China, Korea	USA	BW2	++	A	Madden and Wheelis (2003)
Phytophthora infestans	Potato	1940	China, Southeast Asia	France	BW2	++	A	Madden and Wheelis (2003)
		1950		USA, Canada	BW2	++	A	Madden and Wheelis (2003)
		1950	USA	Soviet Union	BW2	++	A	Madden and Wheelis (2003)
Leptinotarsa decemlineata	Potato	1940	France	Germany	BW2	+	A	Madden and Wheelis (2003)
		1940	Germany	UK	BW2	+	A	Garrett (1996)
		1950	East Germany (GDR)	USA	BW2	–	G	Burns (2013)
Puccinia melanocephala	Sugarcane	1960	Cuba	USA	BW2	–	A	Zilinskas (1999)
Peronospora hyosciami f. sp. *tabacina*	Tobacco	1960	Cuba	USA	BW2	–	A	Zilinskas (1999)
Thrips plami	Several vegetables and ornamental plants	1960	Cuba	USA	BW2	–	A	Zilinskas (1999)
Hemileia vastatrix	Coffee	1950	Guatemala	USA	BW2	–	A	Suffert et al. (2008)
Crinipellis perniciosa	Cacao	1980	Brazil		BW2	–	A	Junior (2006); Caldas and Perz (2013)
Diabrotica virgifera	Maize	2000	Europe	Agrobiotech compagny	BC3	–	A	Suffert et al. (2008)
Pleospora papaveracea	Opium poppy	1990	Central Asia (Afghanistan)	UNDCP[e], USA, UK, Uzbekistan	BW3	++	A	O'Neill et al. (2000) and Jelsma (2001)

(continued)

Table 2.1 (continued)

Pest	Target crop	Year	Target area	Origin[a]	Type of acts[b]	Threat veracity[c]	References[d]	
Fusarium oxysporum f. sp. *erythroxyli*	Coca	2000	Andean Countries (Columbia)	US, Columbia	BW3	++	A	Connick et al. (1998); Jelsma (2001)
Fusarium oxysporum f. sp. *cannabis*	Cannabis	1980	USA, Europe		BW3	++	A	McCain and Noviello (1985)
Melampsora larici-populina	Larch	1990	France	France (eco-warriors)	BT2	–	A	Suffert et al. (2008)
Xylella fastidiosa	Olive tree	2013	Italy	Italy (property developers)	BC3	–	G	Simpson (2015)
Helicoverpa Armigera	Cotton, soybean	2013	Brazil			–	G	Anonymous (2013)
Pomacea canaliculata (snail)	Rice	2010	Spain			–	G	Pérez Pons (2012)
Glycaspis brimblecombei	Eucalyptus		USA			–	G	Corbyn (2012)
Gonipterus scutellatus	Eucalyptus		USA			–	G	Corbyn (2012)
Microcylus ulei	Rubber tree	1980	Sri Lanka	Sri Lanka (Tamil guerrillas)	BT1	–	G	Gurr and Cole (2000)

[a]Countries accused, suspected or known to have been involved as potential state-sponsors (or harboring perpetrators) in anti-crop programs or in agroterrorist acts planning
[b]*BW* biowarfare, *BT* bioterrorism, *BC* biocrime
[c]++ verified threat or program (considered as proved), + probable threat or program (with limited evidence), – unverified, alleged accusation (without significant evidence, very probably wrong)
[d]*A* for academic literature (scientific article, published report, book, etc.), *G* for grey literature (internet website, press article)
[e]United Nations International Drug Control Program

Indeed, the socio-economic and geopolitical situation in Europe is changing and is not always similar to other areas in the world.

2.2.1 First Type of Threat: From Military State Programs to Allegations of Attacks

The qualification of risk of agroterrorism was, and still is, strongly affected by the military dimension of the threat, particularly in reference to state biowarfare programs or 'state allegations'. The experience of the Second World War and the geographical and geopolitical nexus of the tensions between the US and the Union of Soviet Socialist Republics (USSR) during the Cold War has made Europe particularly concerned about this aspect. Biowarfare aimed at crops was state-sponsored in some European countries between 1920 and 1940. Around the Second World War, some countries developed research programs on anti-crop agents targeting staple crops, for instance potatoes (with late blight caused by the oomycete *Phytophthora infestans* and the Colorado potato beetle *Leptinotarsa decemlineata*; Table 2.1; Madden and Wheelis 2003; Suffert 2003). During the Cold War, the USSR Ministry of Agriculture was tasked with conducting a program codenamed 'Ekology' that aimed to develop biological weapons against animals and plants (Rimmington 2000; Alibek 1999); this anti-crop program was discontinued in the late 1980s. The US program of such research was the largest, until President Richard Nixon dismantled it in 1969 (Whitby 2002). In both the US and USSR, the most emblematic researched agents were probably *Puccinia graminis* f. sp. *tritici*, the cause of wheat stem rust (Table 2.1; Line and Griffith 2001) and *P. infestans* (Table 2.1; Madden and Wheelis 2003). In the German Democratic Republic (GDR), almost half of all potato fields were infested in the 1950s by the Colorado potato beetle, *Leptinotarsa decemlineata*, known as 'Amikäfer' (Yankee beetles; Table 2.1, Fig. 2.2) and the GDR government made the claim that the beetles were dropped by American planes. Similar allegations were made by Cuba, which accused the US of a biological attack with *Puccinia melanocephala* (sugarcane rust) and *Peronospora hyosciami* f. sp. *tabacina* (tobacco blue mold) in the 1960s (Table 2.1; Zilinskas 1999). During the same period, a wide range of plant pathogens, including *Magnaporthe grisea*, were the subject of research by Japan; the potential impact of these programs on Europe was low as they mainly concerned the rice crop.

2.2.2 Second Type of Threat: From 'Rogue State' Secret Programs to Emerging Terrorist Groups

While the states that signed the Biological and Toxin Weapons Convention (BTWC) in 1972 have officially renounced the development of biological warfare programs, a new cycle of concern over the possible use of anti-crop bioweapons began in the

Fig. 2.2 Leaflet taken from a GDR propaganda press campaign during the Cold War (1950) depicting the potato beetles (*Leptinotarsa decemlineata*) as tiny US soldiers (http://www.bbc.com/news/magazine-23929124)

late 1980s. This was based on the knowledge that several 'rogue states' (conducting their policy in a dangerously unpredictable way, disregarding international law or diplomacy) were trying to acquire this type of weapon. The 1980–2000 period, viewed as the transition from the Cold War to the globalization era, also raised concerns among several EU member states that some countries suspected of harboring potential anti-crop agents may be involved in developing them as weapons. More recently, evidence was purportedly found in caves in Afghanistan that suggested interest by Islamic militants in the weaponization of wheat rust (Fletcher et al. 2006). Following the First Gulf War, the United Nations Special Commission's inspections revealed that Iraq had expressed an interest in acquiring the military capacity to destroy Iranian crops and that progress had been made in research and development for the weaponization of wheat smut fungi (*Tilletia caries* and *T. tritici*) and aflatoxin-producing strains of the fungus *Aspergillus* (Table 2.1; Whitby 2002). The existence of an anti-crop state program was firmly established, while the supposed Iraqi stock of bioweapons subsequently used as pretext to start the Second Gulf War did not actually exist. The fact that some scientists involved in such research had studied in European universities raised the question of the tracking of students (both foreign and domestic) likely to be selected subsequently by malicious regimes for an anti-crop program. The emergence of the so-called Islamic State (ISIS), while not a state *sensu stricto*, has increased global concerns about terrorism and impacted a large part of the civilian population in Syria and northern Iraq. The terrorist attacks in Paris in 2015 and in Brussels in 2016 had a large impact on the European pschye and proved the motivation and the deployment capability of the organization.

Three elements should now suggest that the threat of agroterrorism for Europe coming from rogue states or terrorist organizations is not negligible. Firstly, the agriculture sector is strategic for Europe and also for isolated states or any organization attempting to hold large teritory (e.g. for ISIS, see Hansen-Lewis and Shapiro 2015). Secondly, anti-crop bioweapons programs have been developed in the past by states anticipating conflict. Thirdly, there is evidence of intentionality and the technical ingenuity – not yet related to agroterrorism capacity – among terrorist groups. Media reports (McElroy 2014) tell of a computer seized in 2014 from a Syrian rebel group contained a manual on how to turn bubonic plague into a bioweapon. These elements indicate that Europe has probably entered a new risk cycle in which the agroterrorism threat, possibly originating even from neighboring regions, has never been so high. Nevertheless, it is difficult to know if the risk level is unchangeable or how to reduce it.

2.2.3 *Third Type of Threat: Biocrime, Sabotage, Private Allegations and Conspiracy Theories on Social Media*

In the past there have been either false or unverified allegations that states or militant organizations have either used plant pathogens against crops or threatened to use them (Table 2.1; Junior 2006; Zilinskas 1999; Caldas and Perz 2013). An allegation of the deliberate introduction of the Western corn rootworm (*Diabrotica virgifera*) into European maize fields in the 1990s appeared on social media by internet; the fact that a population genetics study demonstrated the occurrence of multiple transatlantic introductions of the pest made it harder for the general public to reject the claims (Table 2.1). Scholarly publications are often ignored in this setting (Miller et al. 2005; Ciosi et al. 2008). One conspiracy scenario involved the deliberate release of the Western corn rootworm by a private company in order to sell biotech solutions in Europe, where the introduction of genetically modified organisms has been intensively debated. Now, after each new accidental introduction of a pest, allegations of deliberate introduction can be found on internet.

This was the case after the detection of *Xylella fastidiosa* in Italy in 2013. Accusations have ranged from a deliberate plot by a private company to introduce strains of olive trees that resist the bacteria to a mafia plot to force farmers to sell their land to land developers at low prices after the eradication olive trees. Much more seriously, in December 2015 nine scientists were investigated for a possible role in negligently enabling the disease outbreak by Italian prosecutors. They worried that *Xylella* strains may have been imported from California for a scientific training workshop in 2010, and may then have been released into the environment. Plant pathologists were officially suspected of "negligent spreading of the plant disease, presenting false information and materials to officials, environmental pollution and disfiguring natural beauty". Currently the truth of the matter is not established but this case illustrate that the consequences of allegations of deliberate introduction on the agricultural sector, from growers to scientists, are almost as high as from the introduction itself. Furthermore, the potential of intensified judicial involvement in a phytosanitary crisis will modify the posture of scientists and experts working in the field of plant protection.

Some plant pathogenic fungi that produce mycotoxins are already a recurrent cause of plant disease, such as *Fusarium graminearum* and *F. culmorum* on wheat or *Penicilium expansum* on apples (Russell and Paterson 2006). European assessments did not consider mycotoxin-producing fungi as serious anti-crop agents because of the low levels of mycotoxins and the availability of detection methods. However, based on biotechnical considerations and the fact that these may potentially affect human or cattle health, these pathogens might be reassessed. For example, the previous assessment disregarded the potential psychological effects of a malevolent contamination of food on the population. A deliberate introduction of a plant pathogen may cause significant public panic and a loss of confidence in a segment or the whole of the food chain, seriously affecting niche sectors of European agriculture (such as organic farming). Additionally, a perpetrator with limited technical and scientific skills would increase the potential impact by using simple intimidation or blackmail rather than actually attempting to contaminate the target: fear would have sufficient repercussions on trade and economy (Turvey et al. 2003, 2010; Waage and Mumford 2008).

2.2.4 Fourth Type of Threat: From the Application of Phytosanitary or Sanitary Measures in Response to Deliberate Introduction of a Regulated Pest to Justify Trade Protectionism

According to the Agreement on the Application of Sanitary and Phytosanitary Measures of the World Trade Organization (WTO 1995), every member country has the right to impose import restrictions to protect the health of crops and forests, or consumers in regard to food safety, as long as no unfair discrimination or hidden trade barriers are created. Import restrictions should be technically justified (so, for plant pests "justified on the basis of conclusions reached by using an appropriate pest risk analysis or, where applicable, another comparable examination and evaluation of available scientific information"; IPPC 2004; Heather and Hallman 2008). It is conceivable that a state or other actor could intentionally introduce a plant pest into an import consignment as a pretext to justify trade protectionism. The intention could be to preserve a domestic market, or disparage a competing supplying country. The objective of this kind of operation would not be to provoke direct damage to a crop, but to induce a false detection of a regulated pest or of a food hazard to cause the imposition of protectionist measures. However, trade disruption may not automatically follow a detection of a quarantine organism, unless there is an indication to the authorities of an ongoing unacceptable risk.

2.3 List of Targets Crops and Pests as Biological Data Base: Starting Point of a Prospective Approach to Quantify the Risk in Europe

Officially no act of agroterrorism has occurred in Europe in the past, excluding some criminal cases of human food poisoning. Programs existed, but none was applied. Yet the threat exists and European agriculture is a critical part of the regional economy. The combined agricultural and food sector forms an important part of the EU economy, accounting for 15 million jobs (8.3 % of total employment) and 4.4 % of the gross domestic product (GDP). The EU is the world's largest producer of food and beverages, with combined production estimated at 675 billion Euros (European Commission, Eurostat, November 2014). The self-sufficiency of the EU in basic agricultural products is vital, not only for the wellbeing of its citizens, but also for the geopolitical independence of its member states. The economic, social and political importance of agriculture is therefore much greater than its share in the GDP of the EU.

Crops and forests are vulnerable because they are grown over large ares, often with low levels of management. Although the opportunity for monitoring production areas in Europe is greater than in the rest of the world, these areas cannot be 'protected' from attack. A perpetrator may consider there is a low chance of being observed releasing plant pathogens in a field and there is little that can done initially to limit disease or pest spread (Madden and Wheelis 2003; Madden and van den Bosch 2002). In reality, results of risk assessments showed that, contrary to the assertion that agroterrorism is 'low tech, high impact' (Wheelis et al. 2002), deliberate contamination of plants in large forest areas, for example, are not technically easy to achieve (Suffert et al. 2009) and success of such an attack is not guaranteed. The misperception may result from an erroneous militarization of the threat. Lastly, while the probability that a given crop in a given European country will be a target for a given motivation by a given perpetrator is low, the overall probability that Europe will be concerned someday by an act of agroterrorism *sensu lato* is relatively high.

2.3.1 *Types of Scenarios, Human Dimension of the Threat*

The foresight approach developed in Chap. 6 is aimed at exploring the diversity of the potential scenarios (Table 2.2). They consist of a list of conditions and assumptions, pertaining to potential attacks, and a list of rules.

Table 2.2 Description of the nine types of agroterrorism scenarios

	Biowarfare
BW1	Attack by a country on the agricultural sector of another country. The aim of the attacker is to block commercial imports of the targeted products and prevent their entry into its national market or to enhance its own exports.
BW2	Attack by a country on the agricultural production of another country, in order to weaken the targeted country by reducing its domestic food supplies. This action could be undertaken before a military intervention or replace it.
BW3	Use of biological agents by a country to eradicate illicit crops in another country, such as drug cultivation.
	Bioterrorism
BT1	Terrorist attack targeting food crops. The use of the agent may have negative impacts on human or animal health.
BT2	Attack against planted trees or crops by ecowarriors who want to carry out a radical ecological action.
BT3	Terrorist attack aimed to damage a crop or a tree species that belongs to the patrimony of a country or a group of countries.
	Biocrime
BC1	Attack by activists or farmers groups against the production of a concurrent country.
BC2	Isolated attack by an individual working in the crop protection field, looking for revenge upon a colleague or an institution.
BC3	Deliberate use of a plant pathogen by a private company. The aim would be to render farmers dependant on specific cultivars or plant protection products.

From Latxague et al. (2007)

2.3.2 Target Crops, First Component of the Biological Dimension of the Threat

A comprehensive list of crops (cultivated plants or tree species) of economic or patrimonial interest was used as starting point. The listed species were cultivated or naturally present in Europe (27 EU member states, Switzerland, Norway, Iceland, Croatia, the Former Yugoslav Republic of Macedonia [FYROM], Montenegro, Bosnia-Herzegovina, Serbia, Albania, and Moldova, excluding outermost regions [OMR]). Species present in OMR[1] were excluded, while some crops that are not present in Europe but have a strategic importance for the European industry were taken into account (e.g. rubber plantations). The crops were organized in 11 groups: field crops, vineyards, orchards, vegetable crops, nursery and ornamental horticulture, medicinal and aromatic plants, forest production, beverage crops, straw, tree

[1] The most remote regions of the EU, known as the outermost regions are: Guadeloupe, French Guiana, Réunion, Martinique, Mayotte and Saint-Martin (France), the Azores and Madeira (Portugal), and the Canary Islands (Spain).

sap, seeds. In total 451 crops were inventoried and considered in the subsequent risk analysis. A first classification of the most important crops was established on the basis of the economic value of production (cultivated area × mean yield × mean price; data Eurostat). Crops were preliminarily selected when value of production exceeded 200 million Euros; 79 cultivated plants or tree species were concerned. For these 79 crops and tree, 17 criteria were filled. They were organized in 4 meta-criterias (MT1, 'economical importance'; MT2 'sociological importance'; MT3 'consumption importance'; MT4 'environmental importance') which were completed and assessed by as describe in Table 2.3.

Finally, a short prioritized list of 21 target crops strategic for Europe, chosen as important for socio-economic reasons, was established (Table 2.4).

Figure 2.3 illustrates that the use of a correction index modified the rank of only three crops (oilseed rape, oil olive and dessert apple) that did not appeared to be more important than others (e.g. sugar beet) if the "value of production" only was used for the ranking. The importance of tree species, such as scots pine, Norway spruce and oak, is probably underestimated because this considered only the annual wood production (in average, approximated by the annual increase in wood biomass). In this context, the importance of perennial crops (wine grape, oil olive, dessert apple, orange, peach) is probably also underestimated considering their replacement cost values (the actual cost to replace the crop to its pre-loss condition). This issue can be illustrated by the real socio-economical impact of extreme climatic events or epidemics that have destroyed plantations in the past, for example the *Phylloxera* which destroyed most of the European vineyards in the late nineteenth century, the consequence of the 1999 storm for forests along the Atlantic coasts, or more recently the French outbreak of *Ceratocystis platani* which led to the decision to cut down some plane trees along the Canal du Midi.

2.3.3 Pests Used as Bioweapons, Second Component of the Biological Dimension of the Threat

A non-prioritized comprehensive list of pests comprised of 570 pests of plant hosts cultivated or naturally present in Europe or having an high economic importance for some European countries was established based on historical lists of a similar nature (Table 2.5), and was completed by several experts.

Each pest that could have an impact on at least one of the 21 crops listed in Fig. 2.3 and which was listed in at least four historical reviews (Table 2.5) was added in the non-prioritized short list of pests (Table 2.6), then taken into account to assess the risk of agroterrorism for Europe. This short list was completed by adding a pest which was specifically used to elaborate the WP3 agroterrorist scenario (Table 2.3; Chap. 7).

Table 2.3 Criteria and metacriteria used to assess the importance of crops and forest trees for Europe

Criteria	Code	Unit	Data source	Definition	Coverage
Cultivated area	C1	ha	Crops: Eurostat (crops products; annual data; apro_cpp_crop)	Mean surface by country from 2005 to 2010 (sum EU27)	224 crops
			Forests: Köble and Seufert 2001		36 forest tree species
Mean yield	C2	t/ha/year	Crops: Eurostat (crops products; annual data; apro_cpp_crop)	Mean yield = production/surface (sum for EU27, from 2005 to 2010)	224 crop groups
		m3/ha/year	Forests: Köble and Seufert 2001		11 forest tree species groups
Mean price	C3	euro/t	Crops: EuroStat (selling prices of crop products; absolute annual prices 2000; apri_ap_crpouta)	Mean price (mean for EU27, from 2005 to 2010)	98 crop groups
Value of production	C4	euros	Calculated	Cultivated area × mean yield × mean price (C1 × C2 × C3)	98 crop groups
Volume of trade export	C5	t	Eurostat (international trade detailed data; EU27 trade since 1988 by CN8; DS_016890)	Mean export quantities from 2005 to 2010 (reporter EU27; partner extraEU27)	Over 10,000 products including agricultural and forestry products
Volume of trade import	C6	t	Eurostat (international trade detailed data; EU27 trade since 1988 by CN8; DS_016890)	Mean import quantities from 2005 to 2010 (reporter EU27; partner extraEU27)	Over 10,000 products including agricultural and forestry products
Agro-industrial integration	C7		Expert says	1 = null or low;	
				2 = moderate;	
				3 = high	
MT1, Economical importance					
Correction coefficient = mean(ln[(C1 × C2 + C5 + C6)/(C5 + C6) + exp(1)−1]; C7)					
States concerned by the crop	C8	number	Crops: Eurostat (crops products; annual data; apro_cpp_crop)	From 0 to 27 (number of states not concerned: with null mean surfaces or no data from 2005 to 2010)	224 crop groups

Farms concerned by the crop	C11	number	Number of farms and crop areas according to the farm size (SAU; area NUTS 2; ef_oluaareg)	Number of farms per country in 2007 (sum for EU27; max = 12000000)	76 crop groups
Territorial density	C12		Expert says	1 = diffuse production;	
				2 = large areas moderately specialized;	
				3 = highly specialized basins	
Patrimonial importance	C13		Expert says	1 = low;	
				2 = high	
MT2, Sociological importance					
Correction coefficient = mean(1 + C8/27;1 + 4 × C9/12000000; C10/2; C11)					
Importance in cooking traditions	C14		Expert says	1 = low or irrelevant;	
				2 = moderate;	
				3 = high	
Importance for feeding	C15		Expert says	1 = low or irrelevant;	
				2 = moderate;	
				3 = high	
MT3, Consumption importance					
Correction coefficient = mean(C12; C13)					
Significant presence in recreative areas	C16		Expert says	1 = yes;	
				2 = no	
Significant presence in protected areas	C17		Expert says	1 = yes;	
				2 = no	
MT4, Environmental importance					
Correction coefficient = mean(C14; C15)					

Table 2.4 Short prioritized list of 21 target crops strategic for Europe

Rank	Common name	Latin name
1	bread wheat	*Triticum aestivum*
2	potato	*Solanum tuberosum*
3	wine grape	*Vitis* spp.
4	maize	*Zea mays*
5	tomatoes	*Solanum lycopersicum*
6	barley	*Hordeum vulgare* subsp. *vulgare*
7	rape	*Brassica napus*
8	scots pine	*Pinus sylvestris*
9	sugar beet	*Beta vulgaris* subsp. *vulgaris*
10	oil olive	*Olea europaea*
11	dessert apple	*Malus domestica*
12	norway spruce	*Picea abies*
13	lettuce	*Lactuca sativa*
14	orange	*Citrus* spp.
15	bell pepper	*Capsicum annum*
16	peach	*Prunus persica*
17	alfalfa	*Medicago sativa*
18	strawberry	*Fragaria* spp.
19	carrot	*Daucus carotta*
20	sunflower	*Helianthus annuus*
21	oak	*Quercus* spp.

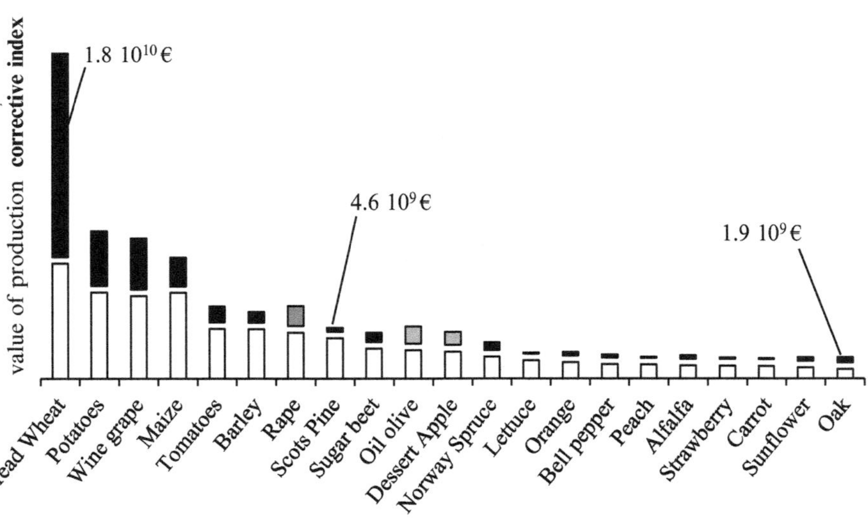

Fig. 2.3 Short list of the most important crops and forest tree species for Europe for socio-economical reasons. The classification was based on the value production, completed by a supplementary value obtained using a correction coefficient to take into account non-economic criteria (Table 2.3). Species whose rank was slightly modified after the use of this correction coefficient are indicated with *grey bars*

Table 2.5 Referenced lists of plant pathogens harmful to plants and plant products which could potentially be used in acts of agroterrorism in Europe (updated in September 2015)

Organization	List	Fungi[a]	Bacteria[b]	Viruses	Nematode
BTWC-SA	Plant pathogens important for the BTWC of the WP124 by South Africa[c]	13	6	1	0
BTWC-AHG	Ad Hoc Group 56/1 Procedural Report[d]	4	3	1	0
USDA-APS	List of biological agents and procedures for notification[e]	4	5	1	0
USDA-APHIS	Agricultural select agents and toxin list[f]	5	2	0	0
MRS	Microbial Rosetta Stone Central Agricultural Database[g]	65	16	12	5
UE-2000/29/CE	EU Plant Health Directive 2000/29/CE, Annex I[h]	18	3	34	5
UE-CBRN	EU list of high risk biological agents[i]	7	7	0	1
EPPO-A1	A1 list of pests recommended for regulation in Europe[j]	36	15	24	5
EPPO-A2	A2 list of pests recommended for regulation in Europe[k]	28	27	22	11
EPPO-Alert	Alert list of pests presenting a risk for Europe[l]	5	1	6	3
CNS	Select agent list of pathogens and toxins[m]	18	11	3	0
Australia Group	List of plant pathogens for export control by the countries member of the Australia Group[n]	11	5	2	0
ISSG-IUCN	100 of the World's worst invasive alien species[o]	3	0	1	0
ANSES	Prioritized list of pests by ANSES for the French Agricultural Ministry[p]	57	38	51	17
FR-31/07/2000	Additionnal list of pests regulated in metropolitan France[q]	3	7	4	3
INRA-CropBioterror	Candidate pathogens list of the UE CropBioterror project[r]	36	9	5	0
FERA-PlantFoodSec	FERA list of top pests of 21 major crops for the EU PlantFoodSec project[s]	38	13	5	10
Suffert	List compiled by Suffert et al. (2009)[t]	18	1	0	0

(continued)

Table 2.5 (continued)

[a]and oomycetes

[b]and phytoplasmas

[c]Established by the **Ad Hoc Group of the Biological and Toxin Weapons Convention (BTWC)** signed on April 10, 1972. T, (BWC/AD HOC GROUP/WP.124). The list, entitled "Plant pathogens important for the BWC", was drawn up at the 4th session of the Ad Hoc Group. It was re-evaluated and presented at the 6th session, held in Geneva **on 3–21 March 1997, in the Working Paper by South Africa**. The following criteria were used to develop the list

– agents known to have been developed, produced or used as weapons

– agents which have severe socio-economic and/or significant adverse human health impacts, due to their effect on staple crops

List: http://www.bradford.ac.uk/acad/sbtwc/ahg34wp/wp124.pdf

[d]Established by the **Ad Hoc Group of the Biological and Toxin Weapons Convention (BTWC)** signed on April 10, 1972. The list belongs to the **Procedural Report of the 23rd session**, held in Geneva, **23 April –11 May 2001** (BWC/AD HOC GROUP 56/1). "Each state party shall declare agents and toxins from the lists set out in Annex A, section I, in accordance with the format for declarations of facilities, activities and transfers referred to in Annex A, section IV". The following criteria were used to develop the list

– potential of individual agents and toxins to be used as weapons

– scientific and technological developments that may affect the potential of individual agents or toxins to be used as weapons

– effects of potential inclusion or exclusion of an agent or toxin in the list on scientific and technical research and development

List: http://www.bradford.ac.uk/acad/sbtwc/ahg56/doc56-1.pdf

[e]Established by the **American Phytopathological Society (APS)** and the **Animal and Plant Health Inspection Service (APHIS)** of the **United States Department of Agriculture (USDA)**. The list, published in the Federal Register of **August 12, 2002**,is displayed on the APS website, together with a paper that presents APS recommendations on countering agricultural bioterrorism with crop biosecurity practices. This list was prepared as part of the **Agricultural Bioterrorism Protection Act**, which was designed to "improve the ability of the United States to prevent, prepare for, and respond to bioterrorism and other public health emergencies that could threaten public health and safety or American agriculture". The following criteria were used to develop the list

– effect of an agent or toxin on animal or plant health or on products

– virulence of an agent or degree of toxicity of the toxin and the methods by which the agents or toxins are transferred to animals or plants

– availability and effectiveness of medicines and vaccines to treat and prevent any illness caused by an agent or toxin

– other criteria that the Secretary considers appropriate to protect animal or plant health, or animal or plant products

List: http://www.apsnet.org/publications/apsnetfeatures/Documents/2002/FEDREG8-12-02.pdf

[f]Established by the **Animal and Plant Health Inspection Service** of the **USDA**. In accordance with the **Agricultural Bioterrorism Protection Act**, implementing regulations detailing the requirements for possession, use, and transfer for select agents and toxins, this list was published by Heath and Human Services (HHS) and APHIS on **March 18, 2005**. The list was updated on **November 17, 2008**. It specifies select agents and toxins

List: http://www.selectagents.gov/SelectAgentsandToxinsList.html

[g]Established by Kamenidou et al. (2013) and published as special report in the APS Journal "Plant Disease". The list includes plant pathogens having significant potential for damage to US agricultural and natural ecosystems. Easily accessible informational resource tool was also developed to assist law enforcement personnel in the event of a disease investigation by providing key information on pathogens of concern

(continued)

Table 2.5 (continued)

List: http://apsjournals.apsnet.org/doi/pdf/10.1094/PDIS-03-12-0263-RE

[h]Established by the **European Union (EU)**. The Annex I of **Plant Health Directive 2000/29/EC** of **8 May 2000** on protective measures against the introduction into the EU of organisms harmful to plants or plant products and against their spread within the Community contain a list of quarantine pests. Published in the Official Journal L169, July 10, 2000

List: http://eur-lex.europa.eu/LexUriServ/LexUriServ.do?uri=OJ:L:2000:169:0001:0112:EN:PDF

[i]Established by the **European Union (EU) CBRN Action Plan**, adopted in **2009**. This plan aimed at strengthening CBRN security in the EU and reducing the threat and damage from CBRN incidents of accidental, natural and intentional origin. It is broadly based on an all-hazard approach, including terrorist threats, and contributes to the implementation of the EU Counter Terrorism Strategy. A EU list of high risk biological agents is under discussion since 2013

List: not public

[j]Established by the **European and Mediterranean Plant Protection Organization (EPPO)**, which is the regional plant protection organization for Europe, under the International Plant Protection Convention (IPPC). One of EPPO's aims is to inform its member countries about dangerous pests, thus helping them to prevent their entry or spread. The organization has therefore been given the task of identifying pests that may present a risk, and of making proposals on the phytosanitary measures that can be taken

The **EPPO A1 list** contains pests which have been evaluated as presenting a risk for Europe, which are absent from the EPPO region and which it recommends regulating as quarantine pests. The last version of this list, updated each year, was approved by EPPO Council in **September 2016**

List: http://www.eppo.int/QUARANTINE/listA1.htm

[k]Established by the **EPPO**, the **A2 list** contains pests which have been evaluated as presenting a risk for Europe and which are locally present in the EPPO region. The last version of this list, updated each year, was approved by EPPO Council in **September 2016**

List: http://www.eppo.int/QUARANTINE/listA2.htm

[l]Established by the **EPPO**, the main purpose of the **Alert List** is to draw the attention of EPPO member countries to certain pests possibly presenting a risk to them and achieve early warning. This list is updated each year

List: http://www.eppo.int/QUARANTINE/Alert_List/alert_list.htm

[m]Established by the **Center for Non-proliferation Studies (CNS)**, at the Monterey Institute of International Studies. It is the largest non-governmental organisation in the United States devoted exclusively to research and training on non-proliferation issues. It strives to combat the spread of weapons of mass destruction by training the next generation of non-proliferation specialists and disseminating timely information and analysis. The "select agent" list of pathogens and toxins was published in **November 2002** and compiles the data given by eight other biological agent lists. Authors: Croddy E. and Newhouse L

List: not public

[n]Established by the **Australia Group**, updated in **June 2012**. The Australia Group is an informal group with the aim of allowing exporting or transhipping countries to minimize the risk of assisting chemical and biological weapon proliferation. Participants in the Australia Group do not undertake any legally binding obligations: the effectiveness of their cooperation depends solely on a shared commitment to CBW non-proliferation goals and the strength of their respective national measures. All states participating in the Australia Group are parties to the Chemical Weapons Convention (CWC) and the Biological and Toxins Weapons Convention (BTWC). This group has established a list of plant pathogens for export control

List: http://www.australiagroup.net/en/plants.html

[o]Established by the **Invasive Species Specialist Group (ISSG)**, which is part of the **Species Survival Commission of the World Conservation Union (IUCN)**. The ISSG is an international group of scientific and policy experts on invasive species. It aims to reduce threats to natural ecosystems and the native species they contain by increasing awareness about invasive alien species, and defining ways to prevent, control or eradicate them. Species included in the list of "100 of the

(continued)

Table 2.5 (continued)

World's worst invasive alien species" were selected according to two criteria: their serious impact on biological diversity and/or human activities, and their illustration of important issues related to biological invasions. This list was updated in **2013**

List: http://www.issg.org/database/species/search.asp?st=100ss

[p]Established in **2013** by the **French Agency for Food, Environmental and Occupational Health & Safety** (ANSES) for the French Agricultural Ministry

List: not public

[q]Established by the **French Agricultural Ministry** in **Annex A of the 31 July 2000 decree**. This additionnal list of pests contain organisms harmful to plants and plant products subject to mandatory measures, which are not listed in the Annex I of EU Plant Health Directive 2000/29/CE

List: http://www.legifrance.gouv.fr/affichTexte.do?cidTexte=JORFTEXT000000584174

[r]Established by **INRA** in the **Workpackage 1 of the EU FP6 CropBioterror Project 2004–2007**. The list, delivered to the European Commission in 2005 consists of 50 candidate pathogens representing potential agroterrorist threats to the European agriculture and forests. It includes not only exotic and regulated pathogens, but also endemic pathogens with specific characteristics such as mycotoxinogenic ability, high potential of mutation and hybridization and records of highly pathogenic exotic strains

List: not public

[s]Established in **2012** by a group of experts of the **UKFood and Environment Research Agency** (FERA) coordinated by Dr. Christine Henry as part of the **EU FP7 PlantFoodSec Project 2011–2016**

List: not public

[t]Established by Suffert et al. (2009) and updated in **2016** (Table 2.1)

Table 2.6 Non-prioritized short list of pests considered a potential threat for Europe

Code	Name of the pest
1	Andean potato latent virus
2	*Anoplophora glabripennis*
3	*Aphelenchoides besseyi*
4	Beet leaf curl virus
5	*Bemisia tabaci*
6	*Bursaphelenchus xylophilus*
7	*Candidatus Liberibacter africanus*
8	*Candidatus Liberibacter americanus*
9	*Candidatus Liberibacter asiaticus*
10	*Candidatus Phytoplasma vitis (Grapevine flavescence dorée)*
11	*Ceratitis capitata*
12	*Ceratocystis fagacearum*
13	*Ceratocystis platani*
14	*Citrus tristeza virus*
15	*Clavibacter michiganensis subsp. sepedonicus*
16	*Diabrotica virgifera virgifera*
17	*Diaphorina citri*
18	*Ditylenchus dipsaci*
19	*Didymella exitialis*
20	*Endocronartium harknessii*

(continued)

Table 2.6 (continued)

Code	Name of the pest
21	*Erwinia amylovora*
22	*Fusarium graminearum (Gibberella zeae)*
23	*Globodera pallida*
24	*Globodera rostochiensis*
25	*Gymnosporangium yamadae*
26	*Leptinotarsa decemlineata*
27	*Leptosphaeria maculans*
28	*Meloidogyne chitwoodi*
29	*Meloidogyne fallax*
30	*Microcyclus ulei*
31	*Monilinia fructicola*
32	*Mycosphaerella pini*
33	*Mycosphaerella populorum*
34	*Paysandisia archon*
35	*Penicillium expansum*
36	*Pepino mosaic virus*
37	*Peronosclerospora philippinensis*
38	*Peronospora hyoscyami f.sp. tabacina*
39	*Phakopsora pachyrhizi*
40	*Phellinus weirii*
41	*Phoma tracheiphila (Deuterophoma tracheiphila)*
42	*Phytophthora infestans*
43	*Phytophthora ramorum*
44	*Pleospora papaveracea*
45	Plum pox potyvirus
46	Potato spindle tuber viroid
47	*Pseudomonas syringae pv. actinidiae*
48	*Pseudoperonospora cannabina*
49	*Puccinia graminis f. sp. graminis*
50	*Puccinia striiformis*
51	*Puccinia triticina*
52	*Ralstonia solanacearum race 3 biovar 2 (Pseudomonas solanacearum)*
53	*Sclerophthora rayssiae var. zeae*
54	*Synchytrium endobioticum*
55	*Tilletia indica*
56	*Tilletia laevis*
57	*Tomato spotted wilt virus*
58	*Tomato yellow leaf curl virus*
59	*Ustilago maydis*
60	*Xanthomonas axonopodis pv. citri*
61	*Xanthomonas campestris pv. campestris*
62	*Xylella fastidiosa (Pierce disease)*
63	*Xylophilus ampelinus*

2.4 Prospect for Risk Assessment and Management in the Next Decade

2.4.1 Evolution and Contextualization of the Threat in Europe

Anti-crop biowarfare was a relevant geopolitical and military issue until the 1980s. Awareness for biosecurity has increased from 1990 to 2000 owing to growing 'Trade', 'Travel', 'Transportation', and 'Tourism', summarized pertinently as the "four T's" components of globalization by Waage and Mumford (2008). While agroterrorism was a minor issue until the past two decades, it strongly emerged after 1997 (Suffert et al. 2008; Fig. 2.1). Subsequent general issues of agricultural biosecurity would be influenced during the next decade by a large set of different components. The nature of the changes was complex and it is necessary to identify in light of the current situation which modifications in the geopolitical and socio-economical context could transform the perception of the agroterrorist threat in Europe. After several years of in depth and, to the degree possible, neutral analysis, the threat of agroterrorism seems to fall into four categories by important determinants of the change. These can be identified presently as the "four C's" components 'Convergence Tactics', 'Constraints', 'Climate', 'Conspiracy'.

Convergence Tactics The different international terrorism activities, which have permanently altered the pschye of nations, from September 11th 2001 in New York to December 19th 2016 in Berlin, give evidence of intentionality, innovative strategies and motivation to look for novel technical means. Because of such undeniable motivation for novel action, the risk of bioterrorism in general, including agroterrorism, has significantly increased. The 'Convergence Tactics' of individual or small group actors who carry out guerilla style attacks may result from ideological or political motivations that differ greatly but all aim at vengeance or destruction of existing structures, systems, and states. The style of tactical convergence across borders initiated by ideologically motivated terrorism groups represents probably the most serious threat for the next decade.

Constraints The rise in research on potentially dangerous plant pests reflects the need to find solutions to new problems, but can also lead to problems if there are not adequate constraints on private commerce in such substances or illicit access by countries considered by Europe as 'rogue states' potentially involved in the development of bioweapons. Emerging capacity in biotechnology may allow intentional or unintentional proliferation of a wide range of dual-use technologies. The potential of future anti-crop biowarfare programs could rely on the effectiveness of constraints on both private and public research, in terms of preventing distribution of dangerous stock but also in terms of limiting access by students and researchers who are not in agreement with the principles of the BTWC and civil society. In this uncertain forthcoming context, international initiatives such as the 'Australia Group' could have a strategic importance for agricultural biosecurity in Europe and its actual exposure to agroterrorism. It is an informal forum of countries, including 30

European states, which use licensing measures to ensure that exports of certain chemicals, biological agents, and dual-use manufacturing facilities and equipment, do not contribute to the spread of chemical and biological weapons, including potential anti-crop agents (Table 2.4).

Climate When US Senator Bernie Sanders stated that "climate change is directly related to the growth of terrorism", he probably was thinking of the increase in drought and flooding and extreme weather disturbances as a result of climate change, and the added pressure and frustration that means to vulnerable people all over the world. While his statement was not substantiated, it raises the question of the relation between climate change, agricultural biosecurity and agroterrorism. This relationship must not be neglected under the pretext that it is complex, or not yet clearly established. During the next decade, climate change and a wide range of global and regional policies applied to minimize its adverse impacts will certainly modify the perception of the risk of agroterrorism in Europe. Furthermore, new forms of threat such as 'ecoterrorism' (Liddick 2006; Lodadenthal 2013) should be taken into consideration.

Conspiracy Several allegations about deliberate introduction of plant pests, viewed as the expression of a conspiracy theory, developed on the internet and social networks since the 2000s. In the past, allegations usually were state propaganda. Most of allegations are now 'civilian', in the sense that they are raised by private citizens or pressure groups, sometimes organized at an international level. Perpetrators or malicious whistleblowers can use social media as their modus operandi, while defenders, including organizations in charge of crop protection, can use it for peaceful purposes (i.e. for collecting valuable information and monitor social media before, during, and after an act of agroterrorism). The impact of this dual-use dilemma of social media in biopreparedness was analyzed by Sjöberg et al. (2013) in the case of an animal bioterrorism incident. Furthermore, Rohn and Erez (2013) asserted that early detection of 'data' enables preventive measures using overt data sources on internet is the best risk-management approach; however, to be efficient, this approach must allow to distinguish between between plausible and implausible allegation. In this context, the risk of 'false positive', such as the risk of considering that a pest introduction was deliberate while a natural or accidental cause was established, is as high as the risk of 'false negative', such as the risk of not being able to establish the deliberate nature of the pest introduction.

2.4.2 Current and Future Answers to Agroterrorism: Real or False Solutions?

The dual use potential of biotechnology research should be considered to pose a risk to crop biosecurity. For example, the United Nations Bioweapons Office has stated concern over the possibilities for weaponization of the 'gene drive' technology

(Begley 2015). There are some methods (e.g., CRISPR-Cas9 genome editing) that consists of designing a gene delivery system that will cause it to be inherited at greater than the usual inheritance rate, thereby possibly spreading into an entire insect or pathogen population in a relatively short time period. Although beneficial uses, such as control of disease vector, are under study, the possibilities for weaponizing gene drives range from suppressing populations of pollinators to giving innocuous insects the ability to carry plant diseases. This example raises the question of the need for surveillance of possible dual-use research, for example by the 'Australia Group'.

As Suffert et al. (2009) stated, the capacity of European countries to prevent an act of agroterrorism requires the involvement of all parties interested in crop biosecurity. They are expected to consider the multiplicity of threats and to collaborate to implement specific countermeasures. Regulation in terms of national biosecurity may not be a sufficient preventive approach to control intentional use of plant pathogens that have been found to fulfil the proposed criteria for a biological weapon. Furthermore, Pasquali (2006), Young et al. (2008) and Suffert et al. (2009) hold a view that European academic and scientific activities should not be inhibited by specific regulations (censure of scientific knowledge, restriction of exchanges of scientific material and movement of scientists, etc.). After the detection of a suspicious disease outbreak, in which a plant pathogen may have been used as an anti-crop weapon, an efficient response would require a collection of evidence that allows identification of the source as early as possible, as well as the method and timing of introduction, and of course the perpetrators (Schaad et al. 2003). In other words, such a situation would have a similar approach to a criminal investigation. To this end, the use of legal molecular-based detection technologies, summarized in the term 'bioforensic' (Fletcher et al. 2006), would be necessary to flag the occurrence of suspicious epidemics.

Biotechnology is only a tool, however, not the finality. The purpose of any investigation performed in a putative 'scene of agroterrorism' is to acquire epidemiological evidences, by both deductive and inductive reasoning and to gain knowledge of the events surrounding the alleged criminal act (Chap. 9). The main difference with a classical scene of crime is that the demonstration of the criminal nature of the contamination event (contrary to natural or accidental event) should be the first objective (Elbers and Knutsson 2013). It is also a real challenge. Bioforensic tools (Fletcher et al. 2006) and databases (Kamenidou et al. 2013) need to be coupled with classical epidemiological approach for assessing the likelihood that a plant disease outbreak may have been intentionally incited. One of the goals of the PlantFoodSec project was to produce scientific knowledge on the build-up, persistence and release of primary inoculum and the early stages of epidemics of selected plant pathogens to differentiate between the consequences of natural and deliberate field contamination. Would investigators be able to differentiate the deliberate introduction of a plant pathogen from an 'accidental' or 'natural' outbreak? In several cases the answer is probably no, because the main issue, "How does a natural epidemic start", is still a poorly resolved question in plant disease epidemiology. The concept of 'initial inoculum' persists as a black box. Two cases study of important pathogens of wheat, *Puccinia triticina* (the cause of leaf rust) and *Zymoseptoria tritici* (the cause of septoria leaf blotch) were developed by combining experimental and modeling approaches (Morais et al. 2015, 2016a, b; Soubeyrand et al., 2017) in order to track the early onset of epidemics. Despite the approach suggested above, countermeasures based exclusively on early detection would be ineffective in regard to the specific features of some prospective scenarios (Latxague et al. 2007; Suffert et al. 2009).

Despite the aforementioned challenges, the need for greater preparedness in Europe remains. The contributions of the PlantFoodSec project should improve the chances of 'getting it right' under the pressure of encountering possible agroterrorism in an increasingly uncertain world.

Acknowledgements This work was supported by the research project PlantFoodSec, which received funding from the European Union Seventh Framework Program (FP7/2007–2013) under Grant Agreement no. 261752. The views expressed in this chapter are exclusively those of the author and do not reflect the official policy of the INRA, the French Government, or the European Commission. The author thanks Bénédicte Moignot for her contribution to the development of the lists of target crops and pests, Ivan Sache, Corinne Le Fay-Souloy, Marc Barbier, Vincent Cardon, Julie Boumrar, John Mumford, John Holt and Adrian Leach for general discussion mixing themes of agroterrorism, epidemiology, sociology and risk analysis, and Maria Lodovica Gullino and Paola Colla for their substantial contribution to the management of the PlantFoodSec project. The author also thanks Megan Quilan and John Mumford for their help correcting the English and their editing work.

References

Alibek K (1999) The Soviet Union's anti-agricultural biological weapons. Ann N Y Acad Sci 894:18–19

Anderson PK, Cunningham AA, Patel NG, Morales FJ, Epstein PR, Daszak P (2004) Emerging infectious diseases of plants: pathogen pollution, climate change and agrotechnology drivers. Trends Ecol Evol 19:535–544

Anonymous (2013) Secretário da Agricultura confirma em nota hipótese de bio-terrorismo no caso da lagarta *Helicoverpa* https://jornaloexpresso.wordpress.com/2013/05/12/secretario-da-agricultura-confirma-em-nota-hipotese-de-bio-terrorismo-no-caso-da-lagarta/

Baker RHA, Anderson H, Bishop S, MacLeod A, Parkinson N, Tuffen MG (2014) The UK plant health risk register: a tool for prioritizing actions. EPPO Bull 44:187–194

Barbier M (2008) Expertise in the making about agroterrorism: issues and paradox of rationalizing hybrid threats. Presented at the Society for Social Studies of Science (4S) & European Association for the Study of Science and Technology (EASST), Rotterdam, The Netherlands

Begley S (2015) Why the FBI and Pentagon are afraid of this new genetic technology. https://www.statnews.com/2015/11/12/gene-drive-bioterror-risk/

Brasier CM (2008) The biosecurity threat to the UK and global environment from international trade in plants. Plant Pathol 57:792–808

Brasier CM, Denman S, Rose J, Kirk A, Hughes KJD, Griffin RL, Lane CR, Inman AJ, Webber JF (2004) First report of ramorum bleeding canker on Quercus falcata, caused by *Phytophthora ramorum*. Plant Pathol 53:804

Budowle B (2003) Public health: building microbial forensics as a response to bioterrorism. Science 301:1852–1853

Burns L (2013) The great Cold War potato beetle battle, BBC World Service. http://www.bbc.com/news/magazine-23929124
Caldas MM, Perz S (2013) Agro-terrorism? The causes and consequences of the appearance of witch's broom disease in cocoa plantations of southern Bahia, Brazil. Geoforum 47:147–157
Casagrande R (2000) Biological terrorism targeted at agriculture: the threat to US national security. Nonproliferation Rev 7:92–105
Ciosi M, Miller NJ, Kim KS, Giordano R, Estoup A, Guillemaud T (2008) Invasion of Europe by the western corn rootworm, *Diabrotica virgifera virgifera*: multiple transatlantic introductions with various reductions of genetic diversity. Mol Ecol 17:3614–3627
Cochrane H, Haslett D (2002) Deliberate release – what are the risks? N Z J For 47:16–17
Connick WJ, Daigle DJ, Pepperman AB, Hebbar KP, Lumsden RD, Anderson TW, Sands DC (1998) Preparation of stable, granular formulations containing *Fusarium oxysporum* pathogenic to narcotic plants. Biol Control 13:79–84
Corbyn Z (2012) California's bioterror mystery. http://failuremag.com/feature/article/californias_bioterror_mystery/
Cupp OS, Walker DE, Hillison J (2004) Agroterrorism in the US: key security challenge for the 21st Century. Biosecurity Bioterrorism 2:97–105
Elad D (2005) Risk assessment of malicious biocontamination of food. J Food Prot 68:1302–1305
Elbers A, Knutsson R (2013) Agroterrorism targeting livestock: a review with a focus on early detection systems. Biosecurity Bioterrorism: Biodefense Strateg Pract Sci 11:25–35
Fletcher J, Bender C, Budowle B, Cobb WT, Gold SE, Ishimaru CA, Luster D, Melcher U, Murch R, Scherm H, Seem RC, Sherwood JL, Sobral BW, Tolin SA (2006) Plant pathogen forensics: capabilities, needs, and recommendations. Microbiol Mol Biol Rev 70:450–471
Fletcher J, Leach JE, Eversole K, Tauxe R (2013) Human pathogens on plants: designing a multidisciplinary strategy for research. Phytopathology 103:306–315
Foxwell J (2001) Current trends in agroterrorism (antilivestock, anticrop, and antisoil bioagricultural terrorism) and their potential impact on food security. Stud Conflict Terrorism 24:107–129
Garrett BC (1996) The colorado potato beetle goes to war. Chem Weapons Control Bull 33:2–3
Gullino ML, Suffert F, Dehne H, Thomas J, Barker I, Gamliel A, Bonifert M, Stack J, Fletcher J, Abd-ElSalam K (2006) Crop and food biosecurity: first results of European research. Phytopathology 97:S44
Gullino ML, Thomas J, Henry C, Dehne H, Suffert F, Bonifert M, Mumford J, Alpas H, Bertin A, Marelli F, Gamliel A, Fletcher J, Stack J (2011) Plant and food biosecurity: a network of excellence funded by the European Union. Phytopathology 101:S65
Gurr N, Cole B (2000) New face of terrorism: threats from weapons of mass destruction. I.B. Tauris, London/New York, UK/USA
Hansen-Lewis J, Shapiro JN (2015) Understanding the Daesh economy. Perspect Terrorism 9:142–155
Heather NW, Hallman GJ (2008) Pest management and phytosanitary trade barriers. CABI, Wallingford
IPPC (2004) Pest risk analysis for quarantine pests, including analysis of environmental risks and living modified organisms. International Standard for Phytosanitary Measures no. 11
Jelsma M (2001) Vicious circle: the chemical and biological war on drugs. http://www.tni.org/archives/jelsma/viciouscircle-e.pdf
Josling T, Roberts D, Orden D (2003) Food regulation and trade. Towards a safe and open global system. Institute for International Economics, Washington DC, USA
Junior P (2006) Terrorismo biologico. Veja 1961:60–63
Kamenidou S, Jain R, Hari H, Robertson JM, Fletcher J (2013) The microbial rosetta stone central agricultural database: an information resource on high-consequence plant pathogens. Plant Dis 97:1097–1102
Khalil AT, Shinwari ZK (2014) Threats of agricultural bioterrorism to an agro dependent economy; what should be done? J Bioterrorism Biodefense 5:127

Khetarpal RK, Gupta K (2007) Plant biosecurity in India – status and strategy. Asian Biotechnol Dev Rev 9:83–107

Knutsson R, Skiby JE, Gronvall GK (2013) AniBioThreat: bioterrorism threats to animals, feed, and food. Biosecurity Bioterrorism suppl 1, 285 p

Köble R, Seufert G (2001) Novel maps for forest tree species in Europe. In: Proceedings conference on a changing atmosphere, 8th European symposium on the physicochemical behaviour of atmospheric pollutants, Torino, Italy

Latxague E, Sache I, Pinon J, Andrivon D, Barbier M, Suffert F (2007) A methodology for assessing the risk posed by the deliberate and harmful use of plant pathogens in Europe. EPPO Bull 37:427–435

Liddick D (2006) Eco-terrorism: radical environmental and animal liberation movements. Praeger Publishers, Westport, CT, USA

Line RF, Griffith CS (2001) Research on the epidemiology of stem rust of wheat during the Cold War. Stem rust of wheat: from ancient enemy to modern foe. Peterson, APS, St. Paul, MN, USA, pp 83–118

Lodadenthal M (2013) Deconstructing "eco-terrorism": rhetoric, framing and statecraft as seen through the insight approach. Crit Stud Terrorism 6:92–117

Madden LV, van den Bosch F (2002) A population-dynamics approach to assess the threat of plant pathogens as biological weapons against annual crops. Bioscience 52:65–74

Madden LV, Wheelis M (2003) The threat of plant pathogens as weapons against US crops. Annu Rev Phytopathol 41:155–176

McCain AH, Noviello C (1985) Biological control of *Cannabis sativa*. In: Proceedings of the VIth international symposium on biological control of weeds, 19–25 August 1984, Delfosse, pp 635–642

McElroy D (2014) Islamic State seeks to use bubonic plague as a weapon of war. The Telegraph. http://www.telegraph.co.uk/news/worldnews/middleeast/iraq/11064133/Islamic-State-seeks-to-use-bubonic-plague-as-a-weapon-of-war.html

Miller N, Estoup A, Toepfer S, Bourguet D, Lapchin L, Derridj S, Kim KS, Reynaud P, Furlan L, Guillemaud T (2005) Multiple transatlantic introductions of the western corn rootworm. Science 310:992

Moignot B, Reynaud P (2013) Developing a methodology for the prioritisation of pests in plant health. Euroreference 9:5–9

Morais D, Laval V, Sache I, Suffert F (2015) Comparative pathogenicity of sexual and asexual spores of *Zymoseptoria tritici* (Septoria tritici blotch) on wheat leaves. Plant Pathol 64:1429–1439

Morais D, Sache I, Suffert F, Laval V (2016a) Is onset of Septoria tritici blotch epidemics related to local availability of ascospores? Plant Pathol 65:250–260

Morais D, Laval V, Sache I, Suffert F (2016b) Inferring the origin of primary inoculum of *Zymoseptoria tritici* from differential adaptation of resident and immigrant populations to wheat cultivars. Eur J Plant Pathol 145(2):393–404

Mota MM, Braasch H, Bravo MA, Penas AC, Burgermeister W, Metge K, Sousa E (1999) First report of Bursaphelenchus xylophilus in Portugal and in Europe. Nematology 1:727–734

Murch RS (2003) Microbial forensics: building a national capacity to investigate bioterrorism. Biosecurity Bioterrorism 2:117–122

O'Neill NR, Jennings JC, Bailey A, Farr DF (2000) *Dendryphion penicillatum* and *Pleospora papaveracea*, destructive seedborne pathogens and potential mycoherbicides for *Papaver somniferum*. Phytopathology 90:691–698

Pasquali M (2006) Biosecurity research and agroterrorism: are there ethical issues at stake? In: Kaiser M, Lien ME (eds) Ethics and the politics of food. Wageningen Academic Publishers, Wageningen, The Netherlands, pp 203–206

Pérez Pons M (2012) El sabotaje del voraz caracol manzana. http://sociedad.elpais.com/sociedad/2012/10/12/actualidad/1350068798_751169.html

Polyak MG (2004) The threat of agroterrorism: economics of bioterrorism. Georgetown J Int Aff 5:31–38

Rimmington A (2000) Invisible weapons of mass destruction: the Soviet Union's BW programme and its implications for contemporary arms control. J Slavic Military Stud 3:1–46
Rogers P, Whitby S, Dando M (1999) Biological warfare against crops. Sci Am 280:70–75
Rohn E, Erez G (2013) A framework for agro-terrorism intentions detection using overt data sources. Technol Forecasting Soc Change 80:1877–1884
Russell R, Paterson M (2006) Fungi and fungal toxins as weapons. Mycol Res 110:1003–1010
Sache I, Roy AS, Suffert F, Desprez-Loustau ML (2011) Invasive plant pathogens in Europe. In: Pimentel D (ed) Biological invasions: economic and environmental costs of alien plant, animal, and microbe species. CRC Press, London, UK, pp 227–242
Schaad NW, Frederick RD, Shaw J, Schneider WL, Hickson R, Petrillo MD (2003) Advances in molecular-based diagnostics in meeting crop biosecurity and phytosanitary issues. Annu Rev Phytopathol 41:305–324
Schrader G, Unger JG (2003) Plant quarantine as a measure against invasive alien species: the framework of the International Plant Protection Convention and the plant health regulations in the European Union. Biol Invasions 5:357–364
Schwägerl C (2005) La recherche sur le bioterrorisme ne mérite pas tant d'argent. Courrier Int 752:56
Shoham D (2014) Pakistan and biological weapons. J Defence Stud 8:85–108
Simpson T (2015) Utopians v. Profiteers? Diffusion of *Xylella* in Italian olive trees. http://xylella-codiro.blogspot.fr/2015/04/utopians-v-profiteers.html
Sjöberg E, Barker GC, Landgren J, Griberg I, Skiby JE, Tubbin A, von Stapelmohr A, Härenstam M, Jansson M, Knutsson R (2013) Social media and its dual use in biopreparedness: communication and visualization tools in an animal bioterrorism incident. Biosec Bioterror Biodef Strat Prac August 11(S1):S264–S275
Soubeyrand S, Garreta V, Monteil C, Suffert F, Goyeau H, Berder J, Berge O, Moinard J, Fournier E, Tharreau D, Morris C, Sache I (2017) Testing differences between pathogen compositions with small samples and sparse data. Scient Rep, submitted
Suffert F (2003) L'utilisation volontaire d'agents phytopathogènes contre les cultures. L'agroterrorisme et ses conséquences sur notre approche de la lutte contre les maladies des plantes. Phytoma 563:8–12
Suffert F, Barbier M, Sache I, Latxague E (2008) Biosécurité des cultures et agroterrorisme. Une menace, des questions scientifiques et une opportunité: réactiver un dispositif d'épidémiovigilance. Le Courrier de l'Environnement de l'INRA 56:67–86
Suffert F, Latxague E, Sache I (2009) Plant pathogens as agroterrorist weapons: assessment of the threat for European agriculture and forestry. Food Secur 1:221–232
Turvey CG, Mafoua E, Schilling B, Onyango B (2003) Economics, hysteresis and agroterrorism. Canadian Agricultural Economics Society Annual Meeting, July 27–30, 2003, Montreal, Canada
Turvey CG, Onyango B, Hallman WK (2010) Risk, fear, bird flu and terrorists: a study of risk perceptions and economics. J Socio-Econ 39:1–10
Waage JK, Mumford JD (2008) Agricultural biosecurity. Philos Trans R Soc B 363:863–876
Wheelis M, Casagrande R, Madden LV (2002) Biological attack on agriculture: low-tech, high-impact bioterrorism. Bioscience 52:569–576
Whitby SM (2002) Biological warfare against crops, global issues. Palgrave, Basingstoke/New York, UK/USA
Young JM, Allen C, Coutinho T, Denny T, Elphinstone J, Fegan M, Gillings M, Gottwald TR, Graham JH, Iacobellis NS, Janse JD, Jacques MA, Lopez MM, Morris CE, Parkinson N, Prior P, Pruvost O, Neto JR, Scortichini M, Takikawa Y, Upper CD (2008) Plant-pathogenic bacteria as biological weapons – real threats? Phytopathology 98:1060–1065
Zilinskas RA (1999) Cuban allegations of biological warfare by the United States: assessing the evidence. Crit Rev Microbiol 25:173–227

Chapter 3
Vulnerabilities, Threats and Gaps in Food Biosecurity

Jacqueline Fletcher, Hami Alpas, Christine M. Henry, Edward Haynes, Heinz W. Dehne, Li Maria Ma, Yeşim Soyer, Filiz Yeni, Paola Colla, and Paul Robb

Abstract The food production system throughout the European Union, which includes farm production, harvesting, transport, processing, storage, marketing and consumption, is vast, complex and open. The high volume of trade in fresh vegetables and fruits contributes to the vulnerability to contamination, whether by accident or intent. Outbreak investigation is critical to understanding the sources of contamination and the steps required to minimize it. The fact that much of the trade in these commodities is international makes it critical that mediation efforts and cooperative research cross national barriers, just as the pathogens do. Enhancing the biosecurity of food production requires assessment of the following: how is the food production system currently organized, in what ways might it be vulnerable to contamination, either accidental or deliberate, what are the primary factors that would allow discrimination between deliberate vs. accidental outbreaks, how can the epidemiological and surveillance systems in Europe be strengthened to shorten outbreak response and mediation times, how can implicated fresh produce be traced to its source, and what forensically valid subtyping method(s) is/are available for detection and discrimination of associated foodborne pathogens.

J. Fletcher (✉) • L.M. Ma
National Institute for Microbial Forensics & Agricultural Biosecurity, Department of Entomology and Plant Pathology, Oklahoma State University, Stillwater, OK, USA
e-mail: jacqueline.fletcher@okstate.edu

H. Alpas • Y. Soyer • F. Yeni
Middle East Technical University, Ankara, Turkey

C.M. Henry • E. Haynes • P. Robb
Fera Science Ltd, York, UK

H.W. Dehne
Rheinische Friedrich-Wilhelms – Universitaet Bonn, Bonn, Germany

P. Colla
Universita Degli Studi di Torino, Torino, Italy

M.L. Gullino et al. (eds.), *Practical Tools for Plant and Food Biosecurity*, Plant Pathology in the 21st Century 8, DOI 10.1007/978-3-319-46897-6_3

Keywords Biofilm • *Campylobacter* • Centers for Disease Control and Prevention (CDC) • *Clostridium botulinum* • Contamination • Disease outbreak • Epiphyte • Enteric pathogens • Enterobacteriaceae • *Escherichia coli* • European and Mediterranean Plant Protection Organisation (EPPO) • European Centre for Disease Prevention and Control (ECDC) • European Food Safety Authority (EFSA) • European Union *E. coli* Reference Laboratory • Food and Agriculture Organization (FAO) • Food and Drug Administration (FDA) • FDA GenomeTrakr • Food Emergency Response Network (FERN) • Food safety • Food security • Food biosecurity • Foodborne illness • Global Food-borne Infections Network (GFN) • Human pathogens on plants (HPOP) • International Food Safety Authorities Network (INFOSAN) • *Listeria monocytogenes* • Phylloplane • Phytopathogenic bacteria • PulseNet • Mycotoxin • *Salmonella* • Shigatoxin • *Shigella* • Toxin • Traceback • World Health Organization (WHO)

3.1 Introduction

The European Union (EU) Networks of Excellence in Security project "Plant and Food Security" (PLANTFOODSEC) addressed biological threats to crops and food from production, through processing and marketing, to consumption. Unlike other EU initiatives in support of agricultural systems, PLANTFOODSEC focused on enhancing capabilities for prevention, detection, response, and recovery should crops or food be targeted with acts of bioterrorism or biocrime. Although most of the objectives focused on threatening plant pathogens and plant diseases, one work package focused on human pathogens, primarily enteropathogenic strains of *Escherichia coli* and *Salmonella* spp., that can colonize and contaminate plants and food products at any point along the production and distribution chains, creating the risk of outbreaks of foodborne illness.

EU member and associated nations have a culture of appreciation for diets featuring high quality, healthy fresh fruits and vegetables. Vigorous trade among EU nations, and between the EU and outside countries, contributes to the diversity, freshness, and quality of fresh produce that many Europeans enjoy and expect. Until recently outbreaks of foodborne illnesses caused by contamination of fresh produce with human pathogens such as *E. coli* O157:H7 and *Salmonella* spp. were reported in Europe relatively infrequently, and were generally of limited significance and confined to local areas. As a result, EU consumers generally ranked sensorial features of food of higher concern than issues related to food safety (TNS Opinion & Social 2012; Ventura-Lucas 2004; Hansstein 2014). Interestingly, young people in the EU are less likely than more mature members of the population to check food labels for point of origin and quality information (TNS Opinion & Social 2012).

In 2011an outbreak of unprecedented impact of serious, in some cases lethal, enterohemorrhagic illness struck Germany and France, with smaller case numbers reported in other countries (Buchholz et al. 2011; European Centre for Disease

Prevention and Control 2011a, b; Robert Koch Institute 2011). A series of unusual and complex disease features, including unexpected patient demographics (illnesses were more common in young, healthy females than in the elderly or immunocompromised), hindered early efforts to identify the causal agent and protect the public. Reliance on patient histories (patients' recollections of what they had eaten in the past) was hampered by the unusually long incubation period for disease development. Other complicating factors included the interconnected and international networks of fresh produce production and distribution, political and trade impacts, and public pressure. The alarm and uncertainty triggered pressure by patients and their concerned families for appropriate answers and response, by members of the press to capture the many complex elements of the story, and by national and EU-based health and food safety officials to contain the outbreak and minimize the human and commercial costs. Despite alarming rates of severe patient reactions, including hemolytic enteric uremia and death, confident association of the outbreak to a specific source of contaminated food proved to be difficult. At one point investigators suspected cucumbers that had been produced in Spain and then distributed to consumers within the EU; an announcement that led to significant economic losses to that agricultural sector. However, failure to confidently associate the pathogen, an unusual strain of *Escherichia coli*, with this source and the subsequent implication of fenugreek seeds, imported by EU countries from Egypt for salad sprout production, added to the general confusion. The case led many food safety officials throughout Europe to recognize a need for enhanced food safety awareness, new policies for food safety, international collaboration and coordination, and better scientific understanding of food safety issues. Although the stated research goals of Work Package 2, "Food Safety and Security," of the PLANTFOODSEC project were written well before the German *E. coli* outbreak, members of the project's Advisory Board encouraged project partners to adopt the recent outbreak as a case study, considering how each of the WP2 tasks, milestones and deliverables might be relevant to "lessons learned" assessments within the EU. Both national and EU-based initiatives prioritized the assessment and enhancement of EU capabilities to prevent, predict, identify, respond to, and recover from such high-impact incidents of foodborne illness. Significant progress has been made.

Human health impacts due to the consumption of contaminated plant foods are not limited to cases associated with enteric human pathogens, however. Some fungal plant pathogens produce mycotoxins that are highly toxic to humans. Despite the potential damage that can result from consumption of aflatoxins, fumonisins, and other dangerous mycotoxins, our knowledge of these compounds in food products is limited, and there are few regulations addressing minimal allowable limits of such toxins in food (Gamliel et al. 2015).

Another area that has received little attention within the EU, from either farmers or law enforcement personnel, is the possibility that food could be used as a vehicle for deliberate introduction of human pathogens or toxins. Individuals with nefarious intent, driven by motives such as economic competition, political conflict, or retaliation for perceived wrongs, could knowingly contaminate food items in order to cause illness, fear, and social unrest (World Health Organization 2002).

3.2 Sites Vulnerable to Microbial or Toxic Contamination

Fresh produce presents unique challenges for protection against microbial and toxic contamination. While plant-derived foods may have shells, waxy cuticles and low pH that help to exclude certain pathogens (Jay et al. 2005), other items having high moisture and nutrient content and natural openings (stomata, lenticels, hydathodes, etc.) present a suitable habitat for pathogens (Carlin 2007; Forsythe 2010). Microbial contamination can be facilitated by nutrients released during post-harvest processing of the fresh produce, such as cutting, slicing or peeling (Harris 2003). Moreover, since fresh produce is usually consumed raw, or is exposed to only minimal treatment for decontamination or pathogen inactivation, a focus on prevention of contamination, rather than reduction, is critical. If mycotoxins are present in food products, their complete elimination is impossible (Bennett 2003). However, various physical (brushing, hot water rinsing, irradiation), chemical (disinfectants) and biological (biocontrol agents) treatments for decontamination and reduction of pathogen numbers, as well as mycotoxin inactivation, are available (Drusch 2003; Parish 2003; Jay et al. 2005; Johnson 2007; Lampel 2007; Meng 2007; Swaminathan 2007; Forsythe 2010; Goodburn 2013).

Epidemiological and traceback investigation of recent foodborne illness outbreaks in Europe and elsewhere have illuminated the problem of microbial and toxic contamination of fresh produce. Assessments of the most common mycotoxins and human pathogens on plants (HPOPs), the most common plant hosts of these pathogens, common contamination pathways, illnesses that result from these agents, and populations to target in management efforts have been made (Harris et al. 2012; Yeni et al. 2015; Francischini 2013; Gamliel et al. 2015). Contamination of produce and/or grains with HPOPs or spores of mycotoxin-producing pathogens can occur at any point from farm to fork, including through contaminated seeds, water, soil, manure, dust, insects or cross contamination (Berger 2010). Consequences may be more severe if contamination occurs during postharvest handling and product development phases (Gorny 2006; Knutsson et al. 2011). Therefore, primary stages of the food supply chain should be considered as sites for preventative actions.

Assessment of the latest outbreaks of foodborne illnesses, in Europe and elsewhere, reveals that the most common HPOPs are pathogenic *Echerichia coli* strains (O157:H7, O26, O104 and O145), *Salmonella enterica* subp. *enterica* serotypes and *Listeria monocytogenes*. *Shigella sonnei* is associated less commonly with outbreaks, but still can pose a substantial health risk for humans (Yeni et al. 2015; Olaimat 2012; Fletcher et al. 2013; Martínez-Vaz 2014).

Many mycotoxin-producing pathogens of concern to human health, and their contamination pathways, are known. Most fungi that reside in or on plants do not produce mycotoxins; however, for those that do, environmental factors may trigger mycotoxin synthesis (Drusch 2003). Intoxication from inadvertent mycotoxin consumption can be prevented by controlling storage and harvest conditions and by routine screening in rigorous surveillance programs. In contrast, control of bacterial HPOPs is very challenging. Although routine diagnostic tests for commonly-found

HPOPs exist, these pathogens continue to be primary causative agents of deadly foodborne illness outbreaks. For example, a 2011 Listeriosis outbreak that spread ultimately to 28 states of the United States, was eventually associated with consumption of whole cantaloupes produced on a particular farm and then distributed throughout the country. In that outbreak, 147 persons were sickened with five outbreak-associated subtypes of *Listeria*, 33 persons died and one pregnant woman suffered a miscarriage (CDC 2012). The reputable company voluntarily recalled the related products but later filed for bankruptcy (CBS News 2013). In the *E. coli* O104:H4 outbreak in the EU area, the unusual bacterial pathogens that were implicated had undergone horizontal gene transfer, resulting in unusually high virulence. Interestingly, several years prior to this outbreak, it was reported that pathogenic *E. coli* might easily acquire virulence traits from other microbes in the population, thereby altering their disease-producing profiles (Meng 2007). However, this previous awareness did not eliminate the risk caused by such emerging strains. Today, among the greatest concerns of medical personnel are the increasing numbers of emerging multi-drug resistant strains (MDR) that fail to respond to therapeutic measures.

Such occurrences, in the EU and elsewhere, teach us the importance of understanding the routes by which HPOPs move to plants and then to humans, as well as the evolutionary pathways of emerging HPOPs, especially MDR strains. The establishment of timely and effective traceback and containment measures cannot occur without rapid, reliable, easily interpreted and standardized analytical methods to detect, identify and discriminate among similar strains of these pathogens.

3.3 Basic Biology and Agricultural Systems, as Well as Foodborne Illnesses, Associated with Fresh Produce Production and Distribution, and of Foodborne Illnesses in the EU and Associated Nations

Better understanding of the dissemination pathways of HPOPs and mycotoxin-producing fungi, and the ability to formulate enhanced strategies to manage the resulting risks to human health, will require significant new knowledge of microbial biology, ecology, environmental niches and community interactions (Fletcher et al. 2013). Most HPOPs are bacteria that normally reside in the guts of vertebrates, including cattle, swine, and poultry without causing disease in these hosts (Hancock et al. 1998). What factors allow such microbes to access, colonize, and sometimes even invade into the interiors of plants, and how do they interact with other members of the microbial community, as well as with the plant?

Interestingly, many HPOPs, including *E. coli* and *Salmonella*, are classified as Enterobacteriaceae, a bacterial family that also contains a number of important plant pathogens (*Enterobacter, Erwinia, Pantoea, Pectobacterium* and *Serratia*) causing serious blights, wilts, and soft rots in a wide variety of plant species

(Fletcher et al. 2013). The relatedness of these pathogen groups suggests interesting, researchable hypotheses related to possible niche competition or synergism on the phylloplane or within the plant, effects on the potential for horizontal gene exchange, and epigenetic influences on gene expression and niche adaptability.

Many phytopathogenic bacteria have an extended epiphytic phase during which they colonize plant surfaces and form biofilm-embedded, mixed-species microbial communities in which species-species communication and signalling (including quorum sensing and division of labor) occur (Barak 2004; Hao et al. 2012; Lapidot and Yaron 2009; Lapidot et al. 2006; Riedel et al. 2001). Nutrition is derived primarily from plant exudates, supplemented by insect honeydew, and decaying organic matter (Aruscavage et al. 2010, Teplitski et al. 2012). Only when niche conditions, including temperature, humidity, and possibly detection of plant signals from specific, susceptible host species, become favourable do these bacteria enter the plant through wounds or natural air exchange openings such as hydathodes and stomata (Brandl 2008; Melotto et al. 2006, 2008; Zeng et al. 2010). Sensor molecules in the host plant detect pathogens and set off signalling pathways that influence whether infection will proceed or whether plant defence responses will prevent pathogen establishment. Like many plant pathogens, HPOPs adhere to the surfaces of leaves, stems, flowers and fruits, and may undergo multiplication there. Some HPOPs enter into the tissues of leaves or fruits, followed in at least some cases by systemic movement into other plant parts (Berger 2010; Chitarra et al. 2014; Erickson 2012). Better understanding of the triggers for epiphyte-to-pathogen transformation and microbe-plant interactions of these human enteric microbes on plant surfaces could reveal targets for management opportunities.

Other community members add additional levels of complexity to the story. Many bacteria, viruses and fungi that are pathogenic to plants, animals and humans are disseminated by vectors, primarily arthropods but also, in some cases, nematodes or other small animals. Many such pathogens are carried on insect mouthparts, legs, and wings, while others may be ingested or otherwise internalized in the vector, later emerging by various means to infect new hosts (Brandl 2006; Doyle and Erickson 2012; Macovei et al. 2008; Talley et al. 2009).

Plant pathologists and crop specialists have developed a variety of plant disease management strategies based on elements of the concept of the "disease triangle," which maintains that disease initiation can occur only if there is a susceptible plant, a virulent pathogen, and a suitable environment (Agrios 1988). Management, then, is directed at (1) the host plant (enhancing plant tolerance or resistance), (2) the pathogen (reducing pathogen inoculum or virulence), or (3) the environment (cultural practices, including field site location, soil preparation and amendments, plant spacing, irrigation, etc.). Some of these strategies might be adapted to reduce the threat of foodborne human pathogens. A possible limitation to the translation of some of these methods to food safety applications is that, since human illness can be triggered by as few as 100 cells of pathogenic *E. coli*, pathogen reduction strategies must be even more effective than is the case for plant pathogens.

Work performed by both plant pathologists and food microbiologists has provided growing evidence that the ability of human enteric pathogens to persist on and within plants, as well as in the surrounding soil, is a natural part of their life cycle (Fletcher et al. 2013; Teplitski et al. 2009,). Thus, future research to explore how human pathogens interact with plants and what these processes share in common with those of plant pathogens will generate new clues for actively managing HPOPs.

Considering their use of common mechanisms for infection, colonization, and survival, strategies being tested or applied that interfere with these processes for controlling plant pathogens may also control HPOPs (Fletcher et al. 2013). Plants are protected from potential pathogens by physical surface structures such as cell walls, waxy cuticles and other architectural features, or by engaging plant innate immune responses (Abramovitch et al. 2006). Detection by the plant of pathogen-associated molecular patterns (PAMPs) (Ingle et al. 2006) using pattern recognition receptors (PRRs) triggers defensive response pathways that suppress microbial replication. A second, faster and more robust pathway known as effector triggered immunity (ETI) is activated by plant resistance (R) proteins that 'recognize' and inhibit the activity of introduced pathogen effector proteins, suppressing microbial multiplication and spread.

HPOPs encounter these same architectural barriers on the phylloplane, and communicate with plant cells, activating plant responses, in ways very similar to those used by plant pathogens (Schikora et al. 2011, 2012; Shirron and Yaron 2011) making these defence pathways potential targets for manipulation for management of both plant pathogens and HPOPs. For example, different tomato cultivars vary in their ability to support colonization by *S. enterica*, and these differences are correlated with variation in the types of leaf trichomes present (Barak et al. 2011). Variations in *E. coli* O157:H7 colonization patterns on different spinach cultivars were correlated with leaf surface topography (Mitra et al. 2009). Knowing that plant host variation is associated with resistance or differences in ability to support human pathogen populations is the first step in identifying the correlated heritable traits and integrating these into crop improvement strategies to reduce the risk of food contamination.

The examples above demonstrate why is important to gather more fundamental knowledge of the biology and genetic variability of HPOPs over time and by commodity and location, the differences in farming practices and other governmental requirements (e.g., conservation or environmental requirements), and, most importantly, the interactions between the human pathogens, plant associated microbes, the host plant, and the environment.

3.4 Needs for Traceback Capability

While steps are often taken to detect contaminants in food before they reach consumers, prevention is not always successful and food poisoning events still occur. Some incidents are natural (e.g. caused by evolving organisms), while others are accidental. Even if unintended, however, contamination due to negligence (e.g. poor handling, or failure to comply with food safety regulations or standards) may be considered criminal negligence and be subject to retribution. Some contaminations may be the result of malicious activities, requiring special measures for prevention and control (EFSA 2011).

As the detection and management of biological food contamination fall under the remit of food safety, human health specialist agencies within the European Centre for Disease Prevention and Control (ECDC 2011a, b) and the European Food Safety Authority (EFSA 2010, 2011) play key roles. These agencies are supported by networks of Community Reference Laboratories, National Reference Laboratories, Official Food Control Laboratories and other health monitoring governmental entities at national and local levels. If a plant-borne human pathogen is involved then a non-EU organisation, the European and Mediterranean Plant Protection Organisation (EPPO), and its members may have responsibilities in support of EFSA, which is responsible for community phytosanitary issues. Beyond Europe, the FAO and WHO are key high-level stakeholders.

Outbreaks of food poisoning caused by naturally or accidentally introduced biological agents are common. Malicious attacks against the food chain, while infrequent, could be undertaken by motivated individuals or groups using microbial agents to target foods as vectors to reach consumers or to reduce the availability of staple foods and cause secondary impacts on human wellbeing. A frequently cited example is the 1984 deliberate contamination of salad bar items with *Salmonella* in ten locations in the state of Oregon, USA, by members of the Rajneeshee cult to influence the outcome of area elections. In that case, over 750 people became ill, but there were no fatalities (Flynn 2006). The fact that this case was judged initially to have been due to accidental contamination reflects the challenges of recognizing signs of intentional, criminal activity. It was not until several years later that a cult member admitted responsibility and the case was prosecuted.

Malicious contamination may be covert (unannounced) or overt (declared and credible). Hoaxes, false claims of planned or accomplished intentional contamination events (credible but not realised), can be just as disruptive as actual events.

The covert attack can be recognised through casualties seeking medical relief. In such circumstances medical priorities will be treatment, new casualty identification and epidemiological investigation to identify the source of infection. In the absence of any intelligence to the contrary or the identification of an unusual or unexpected human pathogen, response will be as for a normal food poisoning outbreak. An overt attack will enable targeted investigations and intervention to embargo contaminated foods; food safety authorities can embargo crops, identify alternative food sources and initiate a positive release system to reassure consumers. Hoaxes are normally assessed for credibility ("could this be – or have been - done?") and for impact. However, the precautionary principle described by EFSA in the "Food Law General Principles" section of their "General Food Law" (http://ec.europa.eu/food/safety/general_food_law/principles/index_en.htm) would require action to protect consumers until the threat was found to be negligible.

In practice, it is not possible to develop specific and comprehensive response plans to deal with the number and range of plants managed through phytosanitary arrangements (>430) and the even larger number of known microorganisms (>525) that may be present on plants. Therefore, in preparing to respond to a potential malicious attack, it is important to prioritise organisms for consideration.

Molecular epidemiology, the use of molecular techniques to study the distribution and determinants of disease occurrence in populations, can be divided into two broad categories: (1) pathogen and toxin identification, and (2) strain typing. Identification is used to confirm the presence of a pathogen or toxin, or aid in disease diagnosis, both of which are especially important in a notifiable disease. Several of the most important human pathogen contaminants of fresh produce, including *Listeria monocytogenes*, *Salmonella enterica*, shigatoxin-producing *E. coli*, and others that are recognized as HPOPs or have caused recent outbreaks (*Clostridium botulinum*, *Shigella spp.*, *Campylobacter spp.*) (Harris et al. 2012) are also notifiable in various jurisdictions (e.g. The Health Protection (Notification) Regulations 2010 in the UK (2010)) or are monitored under the European Surveillance System (Bartels et al. 2014).

Rapid molecular, commercially-available methods are the usual choice for identifying HPOPs in emergency situations, while culture-enriched conventional analytical methods are preferred for confirmation because of their higher sensitivity. Immunological and HPLC-based methods are used for identification of mycotoxins (Yeni et al. 2014). In some cases, biosensors offer even greater speed and sensitivity than immunological assays, but more research and standardization is needed to improve the reliability of these newer tools (Yeni et al. 2014).

Strain typing can be thought of as a more discriminatory form of pathogen identification that identifies lineages within pathogen species. Typing allows identification of links between cases or disease outbreaks, to elucidate transmission routes or mechanisms and to allow epidemiological traceback. Pulsed field gel electrophoresis (PFGE) (Yeni et al. 2015), in which DNA fragments created by restriction enzyme digestion are separated in an oscillating electric field, has long been considered the 'gold standard' for typing human enteric pathogens, as the fragment banding pattern can be matched to existing databases (Yeni et al. 2015). PFGE is highly discriminating, and substantial experience and infrastructure (such as the CDC PulseNet database (Swaminathan et al. 2001)) exist for sharing gel images and metadata during an outbreak. However, PFGE has several limitations. Some pathogen subgroups cannot be discriminated or typed at all, and comparing data from different studies can be difficult due to differences in electrophoresis conditions (Gorman and Adley 2004). PFGE is not useful for determining evolutionary phylogenies (Gunel et al. 2015) and it is time consuming, requires a skilled technician, and presents challenges in the comparison of banding patterns (CDC 2013).

Today, PFGE is giving way to whole genome sequencing (WGS) of pathogens. In the United States, WGS has been instrumental in decision-making for recalls of dairy products (FDA 2014) and products made with plant materials (CDC 2014). It also has been used in source tracking and characterization in investigations of possible bioterrorism (Rasko et al. 2011). As WGS is based on the entire genome it can theoretically detect all single nucleotide polymorphisms (SNPs) present in an isolate, making it the most discriminatory typing method possible. The technology allows discrimination among foodborne pathogen isolates that share the same PFGE pattern (Allard et al. 2013) facilitating the identification of the outbreak strain (Lienau et al. 2011). Unlike PFGE banding patterns, DNA sequence information is easily comparable between investigations. Furthermore, WGS can provide specific information about genes responsible for important phenotypic traits such as virulence and antimicrobial resistance genes present in an organism. Importantly for the global implementation of WGS surveillance, the cost per base of high throughput sequencing continues to decline and there are global efforts to supply sequencers and expertise to developing nations (GMI 2014). In fact, the relative ease and diminishing cost of using modern sequencers and the availability of tools and online analyses may allow a step change in molecular epidemiology to take place in developing countries.

However, despite the genomics revolution, surveillance and communication systems continue to be weak, especially in developing nations (e.g. Gunel et al. 2015), and there is still a need for international food safety policies for global data and metadata sharing in real-time. Fortunately many data sharing and metasurveillance systems already exist, including those for WGS data such as the FDA GenomeTrakr hosted by NCBI (http://www.ncbi.nlm.nih.gov/bioproject/183844) and the BIGSDB database systems of pubMLST (Jolley and Maiden 2010). National surveillance efforts, including the International Food Safety Authorities Network (INFOSAN), the Global Food-borne Infections Network (GFN), and the Food Emergency Response Network (FERN) (Harris et al. 2012), can host and share PFGE data (e.g. the EURL *Lm* DB (Felix et al. 2014) and PulseNet/PulseNet International (Swaminathan et al. 2006; Yeni et al. 2015)), and such systems could be adaptable to WGS data. Furthermore, organizations such as GMI (2014) and EFSA (2014) are creating or encouraging new WGS databases. Before WGS will be adopted globally for foodborne illness investigation, international standards (Standard Operating Procedures and/or quality thresholds below which data is not acceptable) must be implemented. The GMI is currently working on proficiency testing for both data production and analysis (GMI 2014). Additional needs include bioinformatics training for researchers and government employees, and improved analytical tools. Individual organizations are producing such tools (Joensen et al. 2014; Soubeyrand et al. 2014), but all must be rigorously tested (Wyres et al. 2014).

3.5 Closing Comments

The inclusion of fresh vegetables and fruits in the diet is a priority to most nations in the EU, but increased awareness of the risk of microbial or toxin contamination and resulting foodborne illness has generated new interest in the development of enhanced systems and infrastructures throughout the EU for prevention, detection, response and management, communication and recovery from such incidents. Robust and well-funded programs in food safety research, education and public outreach are needed to provide a strong scientific framework on which to base new regulatory guidelines and policies. As the food supply chain for Europe and associated nations is complex, often crossing national boundaries, strengthened channels for international cooperation and coordination will facilitate effective action before, during and after outbreak incidents.

We reviewed outbreaks of foodborne illness that have occurred in the EU and associated nations, considered the pathogens of greatest impact in the EU and current methods for their detection. Project partners gathered data, reviewed published reports and communicated with EU agencies, professional societies, and university researchers to assess issues such as the most harmful HPOPs for Europe, methods for pathogen detection, strain discrimination and traceback (Alpas et al. 2012; Harris et al. 2012). Laboratory research in one project laboratory supported enhanced detection and strain discrimination of plant-resident human pathogens through the development of more effective PCR primers as well as a new molecular assay, based on a multi-locus variable repeat assay (MLVA) that can discriminate among strains of pathogenic *E. coli* (Timmons and Ma 2013; Timmons et al. 2013). The latter project was enhanced by collaboration with Stefano Morabito, of the European Union *E. coli* Reference Laboratory in Rome, Italy. This new research liaison was enhanced further when a project-funded Ph.D. student from the USA spent 9 weeks in the Rome laboratory, validating and optimizing the new MLVA assay on a collection of *E. coli* strains collected from all over the EU. Thus, PLANTFOODSEC served as a focal point and example for effective international collaboration in the food safety arena.

Much still remains to be learned about human pathogens that colonize the surfaces and interiors of plant species that we have come to consider part of a healthy diet. The EU has weathered, and learned much from, a serious outbreak of *E. coli* incited illnesses attributed finally to imported sprout seeds. The PLANTFOODSEC project has brought together researchers from the EU, the United States, Israel, and Turkey who, despite the project end, will continue as a trusted network, taking advantage of diverse funding opportunities to continue to address food safety issues of relevance to the European Union.

References

Abramovitch RB, Anderson JC, Martin GB (2006) Bacterial elicitation and evasion of plant innate immunity. Nat Rev Mol Cell Biol 7:601–611

Agrios G (1988) Plant pathology. Academic Press, Inc., San Diego

Allard MW, Luo Y, Strain E, Pettingill J, Timme R, Wang C, Li C, Keys CE, Zheng J, Tones R, Wilson MR, Musser SM, Brown EW (2013) On the evolutionary history, population genetics and diversity among isolates of *Salmonella Enteritidis* PFGE pattern JEGX01.0004. PLoS One 8:e55254

Alpas H, Soyer Y, Kiymaz T, Yeni F, Polat OG, Gunel E (2012) Review of extraction methods from exemplar food matrices and the analytical methods available for microbial or toxin contamination identification. Deliverable D 2.25, PLANTFOODSEC, EU Network of Excellence in Security

Aruscavage D, Phelan PL, Lee K, LeJeune JT (2010) Impact of changes in sugar exudate created by biological damage to tomato plants on the persistence of *Escherichia coli* O157:H7. J Food Sci 75:187–192
Barak J (2004) Biofilms and other strategies exploited by salmonella and friends on plants. Phytopathology 94:S128–S128
Barak JD, Kramer LC, Hao LY (2011) Colonization of tomato plants by *Salmonella enterica* is cultivar dependent, and type 1 trichomes are preferred colonization sites. Appl Environ Microbiol 77:498–504. doi:10.1128/AEM.01661-10
Bartels C, Beaute J, Fraser G, De Jong B, Martinez Urtaza J, Nichols G, Niskanen T, Palm D, Robeysn E, Severi E, Tavoschi L, Santos CV, Van Walle I, Warns-Petit E, Wetrell T, Whittaker R (2014) European centre for disease prevention and control. Annual epidemiological report – food- and waterborne diseases and zoonoses. ECDC, Stockholm
Bennett JK (2003) Mycotoxins Clin Microbiol Rev 16:497–516
Berger CS (2010) Fresh fruit and vegetables as vehicles for the transmission of human pathogens. Environ Microbiol 12:2385–2397
Brandl MT (2006) Fitness of human enteric pathogens on plants and implications for food safety. Annu Rev Phytopathol 44:367–392. doi:10.1146/annurev.phyto.44.070505.143359
Brandl MT (2008) Plant lesions promote the rapid multipication of *Escherichia coli* O157:H7 on postharvest lettuce. Appl Environ Microbiol 74:5285–5289
Buchholz U, Bernard H, Werber D, Bohmer MM, Remschmidt C, Wilking H, Delere Y, an der Heiden M, Adlhoch C, Dreesman J, Ehlers J, Ethelberg S, Faber M, Frank C, Fricke G, Greiner M, Hohle M, Ivarsson S, Jark U, Kirchner M, Koch J, Krause G, Luber P, Rosner B, Stark K, Kuhne M (2011) German outbreak of *Escherichia coli* O104:H4 associated with sprouts. N Engl J Med 365:1764–1770
Carlin F (2007) Fruits and vegetables. In: Doyle MP, Beuchat LR (eds) Food microbiology: fundamentals and frontiers, 3rd edn. ASM Press, Washington, DC, pp 157–170
CDC (2012) Multistate outbreak of listeriosis linked to whole cantaloupes from Jensen Farms, Colorado. Centers for Disease Control and Prevention
CDC (2013) Pulsed Field Gel Electrophoresis (PFGE) [Online]. http://www.cdc.gov/pulsenet/pathogens/pfge.html. Accessed 23 Mar 2013
CDC (2014) Multistate outbreak of *Salmonella* Braenderup infections linked to nut butter manufactured by nSpired Natural Foods, Inc. (Final Update) [Online]. Available: http://www.cdc.gov/salmonella/braenderup-08-14/index.html. Accessed 20 Mar 2015
Chitarra W, DeCastelli L, Garibaldi A, Gullino ML (2014) Potential uptake of *Escherichia coli* O157:H7 and *Listeria monocytogenes* from growth substrate into leaves of salad plants and basil grown in soil irrigated with contaminated water. Int J Food Microbiol 189:139–145
CBS News (2013) Feds charge Jensen Farms over 2011 listeria-cantaloupe outbreak. CBS News
Doyle MP, Erickson MC (2012) Opportunities for mitigating pathogen contamination during on-farm food production. Int J Food Microbiol 152:54–74
Drusch SR (2003) Mycotoxins in fruits, fruit juices, and dried fruits. J Food Prot 66:1514–1527
EFSA (2014) Use of whole genome sequencing (WGS) of food-borne pathogens for public health protection. EFSA Scientific Colloquium Summary Report, Parma, EFSA
Erickson M (2012) Internalization of fresh produce by foodborne pathogens. Annu Rev Food Sci Technol 3:283–310
European Centre for Disease Prevention and Control (2011a) Reporting on 2009 surveillance data and 2010 epidemic intelligence data. Annual Epidemiological Report. ISBN 978-92-9193-321-1
European Centre for Disease Prevention and Control (2011b) Understanding the 2011 EHEC/STEC outbreak in Germany. ECDC Director's Presentation, ICAAC conference, 17 September, 2011, Chicago, IL, USA. http://ecdc.europa.eu/en/aboutus/organisation/Director%20Speeches/201109_MarcSprenger_STEC_ICAAC.pdf
European Food Safety Authority (2010) Scientific colloquium on emerging risks in food: from identification to communication. Summary Rept. 15. Parma, Italy

European Food Safety Authority (2011) Annual Report. 2011ISBN: 978-92-9199-424-3. 2012

FDA (2014) FDA investigates presence of *Listeria* in some Hispanic-style cheeses [Online]. HYPERLINK http://www.fda.gov/Food/RecallsOutbreaksEmergencies/Outbreaks/ucm386726.htm http://www.fda.gov/Food/RecallsOutbreaksEmergencies/Outbreaks/ucm386726.htm. Accessed 20 Mar 2015

Felix B, Danan C, Van Walle I, Lailler R, Texier T, Lombard B, Brisabois A, Roussel S (2014) Building a molecular *Listeria monocytogenes* database to centralize and share PFGE typing data from food, environmental and animal strains throughout Europe. J Microb Methods 104:1–8

Fletcher J, Leach JE, Eversole K, Tauxe R (2013) Human pathogens on plants: designing a multidisciplinary strategy for research. Phytopathology 103:306–315

Flynn D (2006) Salmonella bioterrorism: 25 years later. Food Safety News. http://www.foodsafetynews.com/2009/10/for-the-first-12/#.Vo89UFq5f4E

Forsythe S (2010) The microbiology of safe food, 2nd edn. Wiley, Oxford

Francischini S (2013) Report on detection systems for mycotoxins in food. Deliverable D4.36, PLANTFOODSEC, EU Network of Excellence in Security

Gamliel A, Dehne HW, Karlovsky P, Fletcher J (2015) Minimum required performance of mycotoxin analysis for applications in a biosecurity context. Deliverable D 2.53, PLANTFOODSEC, EU Network of Excellence in Security

GMI (2014) Report of the 7th global microbial identifier meeting. FERA, York, United Kingdom. http://www.globalmicrobialidentifier.org/News-and-Events/Previous-meetings/7th-Meeting-on-GMI

Goodburn CW (2013) The microbiological efficacy of decontamination methodologies for fresh produce: a review. Food Control 32:418–427

Gorman R, Adley CC (2004) Characterization of *Salmonella enterica* serotype typhimurium isolates from human, food, and animal sources in the Republic of Ireland. J Clin Microbiol 42:2314–2316

Gorny J (2006) Microbial contamination of fresh fruits and vegetables. In: Sapers GM, Gorny JR, Yousef AE (eds) Microbiology of fruits and vegetables. CRC Press, Boca Raton

Gunel E, Polat Kilic G, Bulut E, Durul B, Acar S, Alpas H, Soyer Y (2015) *Salmonella* surveillance on fresh produce in retail in Turkey. Int J Food Microbiol 199:72–77

Hancock D, Besser T, Rice D (1998) Ecology of *Escherichia coli* O157:H7 in cattle and impact of management practices. In: Kaper J, O'Brien A (eds) *Escherichia coli* O157:H7 and other Shiga toxin-producing *E. coli* strains. ASM Press, Washington, DC, pp 85–91

Hansstein FV (2014) Consumer knowledge and attitudes towards food traceability: a comparison between the European Union, China and North America. 2014 International conference on food security and nutrition IPCBEE vol.67 (2014) © (2014) IACSIT Press, Singapore DOI: 10.7763/IPCBEE. 2014. V67. 22

Hao LY, Willis DK, Andrews-Polymenis H, McClelland M, Barak JD (2012) Requirement of siderophore biosynthesis for plant colonization by *Salmonella enterica*. Appl Environ Microbiol 78:4561–4570. doi:10.1128/AEM.07867-11

Harris LF (2003) Outbreaks associated with fresh produce: incidence, growth and survival of pathogens in fresh and fresh-cut produce. Compr Rev Food Sci Food Saf 2:141

Harris A, Henry C, Quill E, Robb P (2012) Prioritisation of human pathogens on plants in European Union countries. FERA, York

Health Protection (Notification) Regulations (2010) United Kingdom

Ingle RA, Carstens M, Denby KJ (2006) PAMP recognition and the plant-pathogen arms race. BioEssays 28:880–889

Jay JM, Loessner MJ, Golden DA (2005) Modern food microbiology, 7th edn. Springer, New York

Joensen KG, Scheutz F, Lund O, Hasman H, Kaas RS, Nielsen EM, Aarestrup FM (2014) Real-time whole-genome sequencing for routine typing, surveillance, and outbreak detection of verotoxigenic *Escherichia coli*. J Clin Microbiol 52:1501–1510

Johnson EA (2007) *Clostridium botulinum*. In: Doyle MP, Beuchat LR (eds) Food microbiology: fundamentals and frontiers, 3rd edn. ASM Press, Washington, DC

Jolley KA, Maiden MCJ (2010) BIGSdb: scalable analysis of bacterial genome variation at the population level. BMC Bioinf 11:595

Knutsson R, van Rotterdam B, Fach P, De Medici D, Frisker M, Löfström C, Ågren J, Segerman G, Wielenga P, Fenicia L, Skiby J, Schultz AC, Ehling-Schulz M (2011) Accidental and deliberate microbiological contamination in feed and food chains – how biotraceability may improve response to bioterrorism. Int J Food Microbiol 145:5123–5128

Lampel K (2007) Shigella species. In: Doyle MP, Beuchat LR (eds) Food microbiology: fundamentals and frontiers, 3rd edn. ASM Press, Washington, DC

Lapidot A, Yaron S (2009) Transfer of *Salmonella enterica* serovar Typhimurium from contaminated irrigation water to parsley is dependent on curli and cellulose, the biofilm matrix components. J Food Prot 72:618–623

Lapidot A, Romling U, Yaron S (2006) Biofilm formation and the survival of *Salmonella* Typhimurium on parsley. Int J Food Microbiol 109:229–233. doi:10.1016/j.ijfoodmicro.2006.01.012

Lienau EK, Strain E, Wang C, Zheng J, Ottesen AR, Keys CE, Hammack TS, Musser SM, Brown EW, Allard MW, Cao G, Meng J, Stones R (2011) Identification of a Salmonellosis outbreak by means of molecular sequencing. New Engl J Med 364:981–982

Macovei L, Miles B, Zurek L (2008) The potential of house flies to contaminate ready-to-eat food with antibiotic resistant enterococci. J Food Prot 71:432–439

Martínez-Vaz BF-G (2014) Enteric pathogen-plant interactions: molecular connections leading to colonization and growth and implications for food safety. Microbes Environ 29:123–135

Melotto M, Underwood W, Koczan J, Nomura K, He SY (2006) Plant stomata function in innate immunity against bacterial invasion. Cell 126:969–980

Melotto M, Underwood W, He SY (2008) Role of stomata in plant innate immunity and foliar bacterial diseases. Annu Rev Phytopathol 46:101–122

Meng JD (2007) Enterohemorrhagic *Esherichia coli*. In: Doyle MP, Beuchat LR (eds) Food microbiology: fundamentals and frontiers, 3rd edn. ASM Press, Washington, DC

Mitra R, Cuesta-Alonso E, Wayadande A, Talley J, Gilliland S, Fletcher J (2009) Effect of route of introduction and host cultivar on the colonization, internalization, and movement of the human pathogen *Escherichia coli* O157:H7 in spinach. J Food Prot 72:1521–1530

Olaimat AH (2012) Factors influencing the microbial safety of fresh produce: a review. Food Microbiol 32:1–19

Parish MB (2003) Methods to reduce/eliminate pathogens from fresh and fresh-cut produce. Compr Rev Food Sci Food Saf 2:161–173

Rasko DA, Worsham PL, Abshire TG, Stanley ST, Bannan JD, Wilson MR, Langham RJ, Decker RS, Jiang L, Read TD, Phillipy AM, Salzeberg SL, Pop M, Van Ert MN, Kenefic LJ, Keim PS, Fraser-Ligget CM, Ravel J (2011) *Bacillus anthracis* comparative genome analysis in support of the Amerithrax investigation. Proc Natl Acad Sci U S A 108:5027–5032

Riedel K, Hentzer M, Geisenberger O, Huber B, Steidle A, Wu H, Hoiby N, Givskov M, Molin S, Eberl L (2001) N-acylhomoserine-lactone-mediated communication between *Pseudomonas aeruginosa* and *Burkholderia cepacia* in mixed biofilms. Microbiology 147:3249–3263

Robert Koch Institute (2011) Report: final presentation and evaluation of epidemiological findings in the EHEC O104:H4 outbreak, Germany 2011. Berlin

Schikora A, Virlogeux-Payant I, Bueso E, Garcia AV, Nilau T, Charrier A, Pelletier S, Menanteau P, Baccarini M, Velge P, Hirt H (2011) Conservation of *Salmonella* infection mechanisms in plants and animals. PLoS One 6. doi:10.1371/journal.pone.0024112

Schikora A, Garcia AV, Hirt H (2012) Plants as alternative hosts for *Salmonella*. Trends Plant Sci 17:245–249. doi:10.1016/j.tplants.2012.03.007

Shirron N, Yaron S (2011) Active suppression of early immune response in tobacco by the human pathogen *Salmonella* Typhimurium. PLoS One 6, e18855. doi:10.1371/journal.pone.0018855

Soubeyrand S, Tollenaere C, Haon-Lasportes E, Laine A-L (2014) Regression-based ranking of pathogen strains with respect to their contribution to natural epidemics. PLoS One 9, e86591
Swaminathan BC (2007) Listeria monocytogenes. In: Doyle MP, Beuchat LR (eds) Food microbiology: fundamentals and frontiers, 3rd edn. ASM Press, Washington, DC
Swaminathan B, Barret TJ, Hunter SB, Tauxe RV, Pulsenet Taskforce CDC (2001) PulseNet: the molecular subtyping network for foodborne bacterial disease surveillance, United States. Emerg Infect Dis 7:382–389
Swaminathan B, Gerner-Smidt P, Ng L-K, Lukinmaa S, Kam K-M, Rolando S, Gutierrez EP, Binsztein N (2006) Building PulseNet international: an interconnected system of laboratory networks to facilitate timely public health recognition and response to foodborne disease outbreaks and emerging foodborne diseases. Foodborne Pathog Dis 3:36–50
Talley JL, Wayadande AC, Wasala LP, Gerry AC, Fletcher J, DeSilva U, Gilliland SE (2009) Association of *Escherichia coli* O157:H7 with filth flies (Muscidae and Calliphoridae) captured in leafy greens fields and experimental transmission of *E. coli* O157:H7 to spinach leaves by house flies (Diptera: Muscidae). J Food Prot 72:1547–1552
Teplitski M, Barak JD, Schneider KR (2009) Human enteric pathogens in produce: un-answered ecological questions with direct implications for food safety. Curr Opin Plant Biotechnol 20:166–171. doi:10.1016/j.copbio.2009.03.002
Teplitski M, Noel JT, Alagely A, Danyluk MD (2012) Functional genomics studies shed light on the nutrition and gene expression of non-typhoidal *Salmonella* and enterovirulent *E. coli* in produce. Food Res Int 45:576–586. doi:10.1016/j.foodres.2011.06.020
Timmons C, Ma LM (2013) Multi-locus variable-number tandem repeat analysis for strain discrimination of non-O157:H7 Shiga toxin-producing *Escherichia coli.* International Association Food Protection Annual Meeting Abstract. https://iafp.confex.com/iafp/2013/webprogram/Paper4691.html
Timmons C, Dobhal S, Fletcher J, Ma LM (2013) Primers with 5' flaps improve the efficiency and sensitivity of multiplex PCR assays for the detection of Salmonella spp. and Escherichia coli O157:H7. J. J Food Prot 76(4):668–673
TNS Opinion & Social (2012) Europeans' attitudes towards food security, food quality and the countryside. Special Eurobarometer 389. European Commission, DG Agric. and Rural Devel
Ventura-Lucas MR (2004) Consumer perceptions and attitudes towards food safety in Portugal. 84th EAAE Seminar, Food safety in a dynamic world, Zeist, The Netherlands, February 8–11, 2004
World Health Organization (2002) Terrorist threats to food: guidance for establishing and strengthening prevention and response systems. (Food safety issues). ISBN 92 4 154584 4
Wyres K, Conway T, Garg S, Queiroz C, Reumann M, Holt K, Rusu L (2014) WGS analysis and interpretation in clinical and public health microbiology laboratories: what are the requirements and how do existing tools compare? Pathogens 3:437–458
Yeni F, Acar S, Polat ÖG, Soyer Y, Alpas H (2014) Rapid and standardized methods for detection of foodborne pathogens and mycotoxins on fresh produce. Food Control 40:359–367
Yeni F, Yavas S, Alpas H, Soyer Y (2015) Most common foodborne pathogens and mycotoxins on fresh produce: a review of recent outbreaks. Crit Rev Food Sci Nutrit. doi:10.1080/10408398.2013.777021
Zeng W, Melotto M, He SY (2010) Plant stomata: a checkpoint of host immunity and pathogen virulence. Curr Opin Biotechnol 21:599–613

Chapter 4
Recent Outbreaks of Human Pathogens on Plants (HPOPs) on Fresh Produce – Lessons Learned from the Practice

Hami Alpas, Filiz Yeni, Yeşim Soyer, and Jacqueline Fletcher

Abstract Number of people being infected or intoxicated due to foodborne illness outbreaks reaches to millions annually. These outbreaks are also responsible from thousands of deaths and billions of dollars' worth of damage every year. However it is considered as an essential portion for a healthy diet, the fresh produce, which is contaminated with Human Pathogens on Plants (HPOPs), is one of the major food items causing this damage. *E.coli* O104:H4 outbreak, occurred in Germany in 2011, has attracted a great attention on foodborne outbreaks caused by contaminated fresh produce, and especially the vulnerabilities and gaps in the foodborne illness surveillance systems. In the frame of this chapter, we focused on the most common foodborne pathogens on fresh produce, epdimiological and traceback investigations of the outbreaks caused by these pathogens in the last 5 years (November 2010–December 2015) in all around the World.

Keywords Foodborne pathogens • Fresh produce • Epidemiology • Traceback • Notification • Early warning • RASFF • PulseNet

H. Alpas (✉) • F. Yeni
Department of Food Engineering, Middle East Technical University, 06800 Ankara, Turkey

Department of Earth System Science, Middle East Technical University, 06800 Ankara, Turkey
e-mail: imah@metu.edu.tr; filizyeni@gmail.com

Y. Soyer
Department of Food Engineering, Middle East Technical University, 06800 Ankara, Turkey
e-mail: ysoyer@metu.edu.tr

J. Fletcher
National Institute for Microbial Forensics and Agricultural Biosecurity, Department of Entomology and Plant Pathology, Oklahoma State University, 127 Noble Research Center, 74078 Stillwater, OK, USA
e-mail: jacqueline.fletcher@okstate.edu

M.L. Gullino et al. (eds.), *Practical Tools for Plant and Food Biosecurity*, Plant Pathology in the 21st Century 8, DOI 10.1007/978-3-319-46897-6_4

4.1 Introduction

Fresh produce is a group of products including fruits, vegetables, herbs, and seeds and nuts, which can be consumed in form of whole, prepared (pre-cut or reduced in size), ready to eat (requiring no preparation before consumption) and/or dressed (pH controlled or not) (Goldburn 2009). Fresh produce is an essential raw material for food manufacturers because of its nutritious content to meet the increasing damand for healthy food since 1980s (Carlin 2007; FAOSTAT 2012; Huang 2004). In accordance with the increasing consumption pattern, a considerable portion of recent foodborne outbreaks has been trace-backed to the contaminated fresh produce with pathogens in all around the world as a result of epidemiological studies (Gorny 2006). Although rarely causing foodborne outbreaks via fresh produce, mycotoxins are still a risk factor for foodstuff of plant origin, especially for seeds and nuts, in underdeveloped countries, too (Forsythe 2010).

4.2 The Most Common Foodborne Pathogens Found on Fresh Produce

Fresh produce can be contaminated with pathogens via environmental agents such as water, soil dust or insects during pre-harvest, and via contaminated water or cross contamination (equipment, surfaces, handlers, etc.) during the post-harvest (Adams and Moss 2006; Gorny 2006; Forsythe 2010; Jay et al. 2005). Since there is no inhibition step for pathogens (e.g. heat) before consumption (Ribot et al. 2008), it is crucially important to hinder pathogen contamination rather than reducing the pathogen load on the produce.

In humans, foodborne pathogen infections may cause complications ranging from mild ones such as diarrhea, vomiting, abdominal pain, fever, headaches and muscle aches (D'Aoust and Maurer 2007; McClane 2007; Seo and Bohach 2007) to severe complications such as enterotoxin poisoning, autoimmune complications, meningitis, septicemia, bloody diarrhea, hemorrhagic colitis, and hemolytic uremic syndrome (HUS), and also miscarriage in pregnant women (Adams and Moss 2006; Meng 2007; Robins-Browne 2007; Lampel 2007; Johnson 2007; Swaminathan 2007). Although the foodborne pathogens listed in this chapter do not have target populations, there are the risk groups which are primarily affected by these infections, namely, pregnant women, infants, elderly and immune-comprimised adults (Forsythe 2010).

In terms of fresh produce, *Salmonella* spp, pathogenic *Escherichia coli*, *Shigella* spp, *Yersinia* spp, *Listeria monocytogenes*, *Staphylococcus aureus*, *Clostridium* spp. are of fundamental importance (Table 4.1).

Table 4.1 List of Human Pathogens on Plants

Risk – Elimination/ Reduction	Illness	Contamination pathway	People at risk
Gram Negative Bacteria			
***Salmonella* spp**			
Very common in nature (Jay et al. 2005), over 2500 serotypes (Behling et al. 2010).	Nontyphoidal salmonellosis (diarrhea, fever, nausea, vomiting, and abdominal cramps)	Production phase: through insects, soil, water contaminated with faecal matter (Adams and Moss 2006).	People who have weak immune system
Reservoirs: humans and animals	Typhoid fever		
Low or high temperatures, extreme acidic environments (D'Aoust and Maurer 2007)			
Pathogenic *Escherichia coli*			
Enterotoxigenic *E. coli* (ETEC), Enteropathogenic *E. coli* (EPEC), Enterohemorrhagic *E. coli* (EHEC), Enteroinvasive *E. coli* (EIEC), Enteroaggregative *E. coli* (EAggEC), Diffusely adherent *E. coli* (DAEC) (Forsythe 2010).	Diarrhea, Bloddy Diarrhea, Hemolytic-Uremic Syndrome (Behling et al. 2010)	Production phase: through water contaminated with faecal matter (Forsythe 2010).	People who have weak immune system, children under 6, elderly (Weiss and Schmidt 2011).
Easily acquire virulence traits and cause wide range of diseases (Meng 2007).			
Reservoirs: ruminants			
Low pH, irridation (Meng 2007)			
***Shigella* spp**			
S. sonnei, *S. boydii*, and *S. flexneri*, *S. dysenteriae*	Diarrhea, bacillary dysentery, enterotoxin/shigatoxin related HUS with a low infetious dose (Forsythe 2010)	Production phase: through insects or water contaminated with human feces (Adams and Moss 2006)	Children younger than five
Tolerant to low pH and low temperatures but sensitive to ionizing, radiation and sodium hypochlorite (Lampel 2007)			People who have weak immune system
***Yersinia* spp**			

(continued)

Table 4.1 (continued)

Risk – Elimination/ Reduction	Illness	Contamination pathway	People at risk
Y. enterocolitica, Y. pseudotuberculosis	Abdominal pain, fever, bloody diarrhea (Jay et al. 2005).	Production phase: through contaminated water, soil or insects (Forsythe 2010)	Children, adolescents and people who have weak immune system are at more risk
Common in nature			People living in colder regions of the world
Since *Y. enterocolitica* is tolerant to low temperatures (below 4 °C) and alkaline conditions (pH = 4–10), it can contaminate frozen products (Robins-Browne 2007).			
It can be inactivated by ultraviolet irradiation (Butler et al. 1987) and aqueous ozone (Restaino et al. 1995).			
Gram Positive Bacteria			
Listeria monocytogenes			
Among 13 serotypes, serotypes 1/2a, 1/2b, and 4b that cause most of the foodborne outbreaks (Swaminathan 2007).	Muscle aches, diarrhea, meningitis, miscarriage (Behling et al. 2010).	Production phase: through soil, decaying vegetation or water contaminated with faecal matter (Adams and Moss 2006)	Pregnant women, newborns and the elderly, and people who have weak immune system are at risk (Forsythe 2010). Since mortality rate is high and infection can be transmitted from mother to placenta, listerosis may result in abortion, stillbirth and permature birth (Adams and Moss 2006).
Very common in nature			
Since it is tolerant to high salt concentrations and low temperatures (below 1 °C), it can contaminate frozen products (Behling et al. 2010).			
It can be eliminated when treated with acids (Swaminathan 2007).			
Staphylococcus aureus			

(continued)

Table 4.1 (continued)

Risk – Elimination/ Reduction	Illness	Contamination pathway	People at risk
13 staphylococcal enterotoxins were identified (Jay et al. 2005).	Abdominal pain, diarrhea (Forsythe 2010).	Production phase: through contaminated water, soil, dust (Behling et al. 2010).	??
Reservoirs: humans and animals (nose, throat, skin) (Behling et al. 2010)			
Being one of the most resistant non-spore forming foodborne pathogens to environmental conditions, It is tolerant to high salt concentrations and high temperatures (Seo and Bohach 2007). Contamination can be prevented by keeping foodstuff at either high or low temperatures (60 °C and above, 7.2 °C and below) before consumption (Forsythe 2010).			
Its enterotoxins are also tolerant to high temperatures so it can not be easily killed by cooking.			
***Clostridium* spp**			

(continued)

Table 4.1 (continued)

Risk – Elimination/ Reduction	Illness	Contamination pathway	People at risk
Spores are very common in nature	*C. perfringens*: milder symptoms such as diarrhea.	Production phase: through contaminated soil, water, and dust (Adams and Moss 2006)	The elderly and people who have weak immune system are at risk for *C. Perfringens* infection
The pathogen is tolerant to low oxygen levels and their spores are tolerant to high temperatures so it can survive in home-canned vegetables or the vegetables stored in oil (Johnson 2007).	*C. butyricum, C.baratii*, and especially *C. botulinum*: serious symptoms such as a neuroparalytic illness called botulism (Johnson 2007). As botulinum toxins (A, B, E and rarely F cause illnesses in humans) are the most toxic substances known and there are not much thing to do after the toxin is absorbed, early diagnosis is crucial (Adams and Moss 2006).		Everyone for botulin intoxication
However, its spores can be inactivated by ionizing irradiation, chlorine and hydrogen peroxide when thermal processing is not possible (Johnson 2007).			

4.3 The Most Common Mycotoxins Found on Fresh Produce

Mycotoxins are another major cause of foodborne illness outbreaks as secondary metabolites, which are produced by some fungi in the end of exponential growth phase (Jay et al 2005). Low pH content of fresh produce, especially fruits, is a disadvantage for bacterial pathogen contamination, however, it is an opportunity for fungi invasion (Moss 2008). On the contrary, relatively higher pH values in vegetables makes them more susceptible for bacterial pathogen contamination rather than mycotoxin invasion (Adams and Moss 2006).

Mycotoxins may contaminate the produce during pre-harvest phase through seeds, soil and air (Forsythe 2010) but presence of fungi on fresh produce does not always result in mycotoxin contamination. Environmental factors (e.g. inappropriate storage conditions) may trigger mycotoxin formation during post-harvest phase (Drusch and Ragab 2003). For instance, if pulses, nuts and oilseeds are not stored at appropriate temperatures, which keep these products at certain water content, fungi may produce mycotoxins (Adams and Moss 2006). If contamination occurs, it is not possible to completely eliminate mycotoxins from food products after this point (Bennett and Klich 2003). However, mycotoxins in food products can be partially degraded by physical and chemical methods as well as irradiation (Jay et al 2005).

For example, mycotoxin content of the fruits can be reduced with washing and sorting in the post-harvest phase (Drusch and Ragab 2003).

Mycotoxin uptake via food may results in mycotoxin poisoning which manifest it as symptoms including milder gastrointestinal problems such as diarrhea, abdominal pain, or more severe complications such as cancer (Adams and Moss 2006). The factors affecting susceptibility of humans against mycotoxins are the type of exposure (acute or chronic), and age, sex and overall health of the person exposed (Forsythe 2010). For instance, males are generally more susceptible than the female against aflatoxins (Adams and Moss 2006).

In terms of fresh produce, aflatoxin, ochratoxin A, citrinin and patulin are of major importance whereas zearalenone, fumonisins and deoxynivalenol cause mycotoxin-related problems in cereals (Table 4.2).

4.4 Epidemiological and Traceback Investigations of the Recent Outbreaks Caused by Contaminated Fresh Produce

Number of people being infected or intoxicated due to foodborne illness outbreaks reaches to millions annually. However, it is common for these outbreaks to stay unascertained or unreported by public health officials and uninvestigated by epidemiologists to trace back the source pathogen and/or food item consumed. Moreover, even if an outbreak is investigated, it is burden to reach outbreak investigation data for most of the countries in the world. There are a few countries which share the investigation data publicly via web pages of institutions or annual surveillance reports. The only way to reach out data about the rest of the countries is rare scientific publications and newspaper articles. Therefore, to obtain such data for every country in the world is almost impossible due to lack of a fully functional global disease surveillance system.

According to the published data on foodborne outbreaks, in all around the world in the last 5 years November 2010–December 2015, totally 8556 people were infected, 114 people died and 1 woman had a miscarriage due to HPOPs. Concerning these outbreaks, it is noteworthy that newborns, children, elderly and pregnant women are prone to contamination.

The most common food items associated with these outbreaks appeared as leafy greens, sprouts and fruits. The most common foodborne pathogens on fresh produce appeared to be pathogenic *E. coli* serotypes (mostly O157 – H4 and H7-, less commonly, O121, O26) and *Salmonella enterica* serotypes (Newport, Braenderup, Poona, Saintpaul, Montevideo, Mbandaka, Typhimurium, Strathcona, Agona, Panama) with 15 cases each (Yeni et al. 2015) (Table 4.3). Moreover, *Listeria monocytogenes*, *Clostridium* spp., and *Shigella sonnei* were found to be not common but still pose a significant threat to public health. Additionally, *Yersinia enterolitica* was found to be a very rare pathogen causing foodborne illness outbreaks via

Table 4.2 List of mycotoxins affecting the fresh produce

Mycotoxin	Risk	Illness
Aflatoxin		
	Produced by *Aspergillus* spp. (*A.flavus, A. parasiticus, A. nominus*).	Aflatoxins in groups B1, B2, G1, and G2 cause mild to severe complications such as liver failure (Hocking 2007; Forsythe 2010).
	More common in warmer regions of the world (tropics and subtropics) due to high temperatures and high humidity (Adams and Moss 2006).	
	Highly toxic and especially affects nuts, figs, dried fruits, cereals and oilseeds (Jay et al. 2005).	
Ochratoxin A (OTA)		
	Produced by *Aspergillus ochraceus* and *Penicillium verrucosum*.	As a nephrotoxin, OTA may cause mild to severe complications including various kidney problems (Bennett and Klich 2003).
	More common in temperate regions of the world (Moss 2008).	
	Especially affects products of tropical and subtropical origin including cereals (such as maize), cocoa, coffea and soy beans, but can also be present in spices, dried fruit, and nuts (Adams and Moss 2006).	
Citrinin		
	Produced by *Penicillium* spp. (*P. citrinum* and *P. verrucosum*) and *Aspergillus* spp. (*A. ochraceus*) (Adams and Moss 2006).	As a nephrotoxin, citrinin may cause mild to severe complications in humans including kidney problems and yellow rice fever (Bennett and Klich 2003)
	Affects fruits, herbs, beans, and spices and in other raw agricultural commodities such as cereals (EFSA 2012a, b).	
Patulin		
	Produced by *Penicillium* spp. (especially *P. expansum*), *Aspergillus* spp. and *Byssochlamys* spp. (Adams and Moss 2006).	Complications of exposure to patulin due to food products in humans is still unclear (Jackson and Dombrink-Kurtzman 2006).
	Especially affects fruits due to their low pH levels such as apples, pears, grapes,bananas, peaches and pineapples and it may also be present in seeds of plants (Jackson and Dombrink-Kurtzman 2006; Jay et al. 2005).	

fresh produce. Among these pathogens, *L. monocytogenes* and *Clostridium* spp. are of special importance because of their high mortality rates. Although they cause a limited number of outbreaks, in almost all cases some patients lose their lives. Although pathogenic *E. coli* and *Salmonella* species infected thousands of people and totally caused 71 deaths in thirty outbreaks, *L. monocytogenes* and *Clostridium* spp. caused 30 deaths in just one outbreak by infecting 146 people.

During investigations of these outbreaks, public health investigators used common methods in order to trace back the agent and contaminated source -food item-. They used web-based questionnaires and repeated interviews with patients, whose infection is confirmed in laboratory, in order to find the source of the outbreaks. In

Table 4.3 Recent outbreaks caused by foodborne pathogens on fresh produce

Date/Duration	Country	Pathogen – Disease	Source	Number of Case	Voluntary Recall
Oct–Nov 2015	Sweden	*S. sonnei* – Shigellosis	Fresh coriander	42 cases, no deaths	-
Samples from patients were analyzed with whole genome sequencing (WGS), pathogens could not be found on food samples (fresh coriander imported from Southeast Asia), however, bacterial strains were identical in the samples from patients.					
Ref: http://www.foodqualitynews.com/Food-Outbreaks/Sweden-finds-coriander-to-be-source-of-shigellosis					
March 2015	Canada	*E. coli* O157:H7	leafy greens (lettuces, kale, spinach, arugula or chard).	13 cases	-
E.coli from all samples had a matching genetic fingerprint. Although leafy greens were identified as a possible source of illness, a specific source of the outbreak could not be confirmed.					
Ref: http://www.phac-aspc.gc.ca/phn-asp/2015/ecoli-eng.php					
Jan 2014	Tajikistan	Botulinum toxins	Pickled tomatoes	18 cases	-
Ref: http://www.promedmail.org/direct.php?id=3135159					
Oct 2014-Jan 2015	USA	*L. monocytogenes*	Prepackaged Caramel Apples	35 cases, 34 hospitalizations, 7 deaths	+
Environmental testing revealed contamination with *L. monocytogenes* through PFGE at the firm's apple-packing facility and the pathogen was identified in the food samples through PFGE.					
Ref: http://www.cdc.gov/listeria/outbreaks/caramel-apples-12-14/index.html					
Dec 2014	Canada	*E. coli* O157:H7		14 sick, 2 HUS	-
Ref: http://www.marlerblog.com/case-news/e-coli-sources-in-quebec-windham-and-willimantic-connecticut-unnamed-14-sick/#.VpZmfvmLTDc					
Sept-Dec 2014	USA	*Salmonella* Enteritidis	Bean Sprouts	115 cases, 19 hospitalized, no deaths	+
Strains were identified through WGS and PFGE.					
Ref: http://www.cdc.gov/salmonella/enteritidis-11-14/index.html					
June-Aug 2014	USA	*Listeria monocytogenes*	mung bean sprouts	5 cases, 5 hospitalizations, 2 deaths	+

(continued)

Table 4.3 (continued)

Date/Duration	Country	Pathogen – Disease	Source	Number of Case	Voluntary Recall
Whole genome sequencing of the *Listeria* strains isolated from mug bean sprouts produced by Wholesome Soy Products, Inc. and environmental isolates collected at the production facility were found to be highly related to sequences of *Listeria* strains isolated from five people who became ill.					
Ref: http://www.cdc.gov/listeria/outbreaks/bean-sprouts-11-14/index.html					
July–Nov 2015	USA	*Salmonella* Poona	Cucumbers	838 cases, 165 hospitalizations, 4 deaths	+
It was identified that cucumbers imported from Mexico and distributed by one company are the likely source of the infections. The investigation into the source of these recent illnesses is ongoing.					
Ref: http://www.cdc.gov/salmonella/poona-09-15/index.html					
June 2014	Canada – USA	*Salmonella enterica* strains (Newport, Hartford, *Oranienburg, Saintpaul)*	sprouted chia seed powder	63 cases, 12 hospitalized Canada; 31 cases, 5 hospitalized in USA	-
It was confirmed that sprouted chia seed powder was the source of the illness. Strains were isolated from the patients and identified throught PFGE profiles.					
http://www.phac-aspc.gc.ca/phn-asp/2014/salmonella-nh-053114-eng.php					
http://www.cdc.gov/salmonella/newport-05-14/index.html					
May–Sept 2014	USA	*Salmonella* Newport	Cucumbers	275 cases, 48 hospitalized, 1 death	-
PFGE patterns and WGS was used to characterize the source strain.					
Ref: http://www.cdc.gov/mmwr/preview/mmwrhtml/mm6406a3.htm?s_cid=mm6406a3_e					
May 2014	USA	*Escherichia coli* O121 (STEC)	Raw clover sprouts	19 cases, 16 hospitalized, no deaths	-
It was indicated that contaminated raw clover sprouts from one producer was the likely source of this outbreak.					
Ref: http://www.cdc.gov/ecoli/2014/O121-05-14/index.html					
Nov–Dec 2013	USA	*Escherichia coli* O157:H7(STEC)	Ready-to-Eat Salads	33 case, 7 hospitalizations,2 HUS, no deaths	+
It was indicated that consumption of two ready-to-eat salads from one producer and sold at one grocery store locations was the likely source of this.					
Ref: http://www.cdc.gov/ecoli/2013/O157H7-11-13/index.html					

Sep 2013	UK	*E.coli* O157	pre-packed bagged salad products containing watercress	19 cases	+
Ref: https://www.food.gov.uk/enforcement/alerts/2013/6343/sainsburys-watercress-recall					
Jan–Apr 2013	USA	*Salmonella Saintpaul*	Cucumber	84 cases, 17 hospitalizations, no deaths	+
It was indicated that imported cucumbers supplied from Mexico were the source of this outbreak. The strain was identified through PFGE profiles.					
Ref: http://www.cdc.gov/salmonella/saintpaul-04-13/index.html					
Feb–May 2013	USA	*Salmonella Montevideo and Salmonella Mbandaka*	Tahini Sesame Paste	16 cases, 1 hospitalization, 1 death	+
Strains were isolated from the samples collected from the retail store and they were identified through PFGE profiles,					
http://www.cdc.gov/salmonella/montevideo-tahini-05-13/					
E.coli O157 has been detected at the watercress samples sold at Sainsbury's Supermarkets Ltd					
http://www.food.gov.uk/enforcement/alerts/2013/6343/sainsburys-watercress-recall					
Jan 2013	Canada	*E. coli O157:H7*	Shredded lettuce	30 cases, 13 hospitalizations, 1 HUS, no deaths	-
The investigation indicated that the most probable cause of the *E. coli* O157:H7 illnesses was shredded lettuce distributed by one company.					
Ref: http://www.phac-aspc.gc.ca/fs-sa/phn-asp/2013/ecoli-0113-eng.php					
Oct–Nov 2012	USA	*E.coli* (STEC) O157:H7	Pre-packaged organic spinach and spring mix	33 cases,13 hospitalized, 2 HUS	+
Ref: http://www.cdc.gov/ecoli/2012/O157H7-11-12/index.html					
July–Sep 2012	USA	*Salmonella* Braenderup	Mango	127 cases, 33 hospitalized	+
Ref: http://www.cdc.gov/salmonella/braenderup-08-12/index.html					
July–Sep 2012	USA	*Salmonella* Typhimurium *Salmonella* Newport	Cantaloupe	261 cases, 94 hospitalized, 3 deaths	+

(continued)

Table 4.3 (continued)

Date/Duration	Country	Pathogen – Disease	Source	Number of Case	Voluntary Recall
Ref: http://www.cdc.gov/salmonella/typhimurium-cantaloupe-08-12/index.html					
Aug 2012	Canada	*Salmonella* Braenderup	Mango	21 cases	-
It was confirmed that mangoes originating from Mexico were the source of this outbreak.					
http://www.phac-aspc.gc.ca/fs-sa/phn-asp/osm-esm-eng.php					
July–Aug 2012	Japan	*E. coli* O157	Pickled Chinese cabbage	110 cases, 7 deaths	-
Ref: http://www.yomiuri.co.jp/dy/national/T120821004270.htm					
http://www.bangkokpost.com/news/asia/308364/contaminated-pickles-kill-seven-in-japan					
Apr–June 2012	Canada	*E. coli* O157	Romaine lettuce	55 cases, 18 hospitalized	-
Ref: http://www.cbc.ca/news/canada/new-brunswick/story/2012/06/29/nb-e-coli-miramichi-lettuce-1033.html					
Dec 2011–Feb 2012	UK, Germany	*Salmonella* Enteritidis	Ready-to-eat sliced watermelon	50 cases, 1 death	-
Ref: http://www.dailymail.co.uk/health/article-2095387/Watermelons-linked-salmonella-outbreak-killed-left-50-ill.html					
Dec 2011–Mar 2012	USA	*E.coli* (STEC) O26	Raw clover sprouts	29 cases, 7 hospitalized	+
Ref: http://www.cdc.gov/ecoli/2012/O26-02-12/index.html					
Oct–Nov 2011	USA	*E.coli* (VTEC) O157:H7	Romaine lettuce	58 cases, 33 hospitalized, 3 HUS	+
Ref: http://www.cdc.gov/ecoli/2011/ecoliO157/romainelettuce/032312/index.html					
July–Oct 2011	France, Germany, Denmark	*E.coli* (EHEC) serotype O104:H4	Foodborne (unknown)	11 hospitalized (France 8, Germany 2, Denmark 1)	-
Ref: http://www.eurosurveillance.org/images/dynamic/EE/V17N04/art20065.pdf					
Sep–Oct 2011	Denmark, Germany, Austria	*Salmonella* Strathcona	Tomato	40 hospitalized (Denmark 40, Germany 14, Austria 1)	-
Ref: http://www.ssi.dk/English/News/News/2011/Salm%20imported%20tomatoes.aspx					

July–Oct 2011	USA	*L. monocytogenes*	Cantaloupe	146 cases (3 newborn, 4 pregnant), 142 hospitalized, 30 dead, 1 miscarriage	+
The pathogen was identified on whole cantaloupes collected from grocery stores and from ill persons' homes.					
Ref: http://www.cdc.gov/listeria/outbreaks/cantaloupes-jensen-farms/120811/index.html					
Oct2011	Norway	*Shigella sonnei*	Fresh basil	46 hospitalized	+
Ref: http://www.eurosurveillance.org/ViewArticle.aspx?ArticleId=20007					
Aug 2011	USA	*Salmonella* Enteritidis	Pine nuts	43 cases, 2 hospitalized	+
Ref: http://www.cdc.gov/salmonella/pinenuts-enteriditis/111711/index.html					
July 2011	Taiwan	Botulinum toxins	Homemade vegetable preserves	5 cases, 4 deaths	-
In the initial forensics tests, traces of botulism toxins were found in four of the bodies					
Ref: http://www.chinapost.com.tw/taiwan/local/nantou/2011/07/22/310649/Deaths-in.htm					
Jan–Aug 2011	USA	*Salmonella* Agona	Fresh papayas	106 cases, 10 hospitalized	+
Ref: http://www.cdc.gov/salmonella/agona-papayas/082911/index.html					
Dece 2010–July 2011	UK	*E.coli* serotype O157 (Phage type 8/PT8)	Leeks and potatoes	250 cases, 74 hospitalized (4 HUS), 1death	-
Ref: http://www.guardian.co.uk/world/2011/sep/30/ecoli-outbreak-uk-250-ill					
May–July 2011	EU/EEA (14 countries)	*E.coli* O104:H4	Sprout/Seed related	3842 cases (855 HUS, 2987 Non-HUS), 53 dead (35 HUS, 18 non-HUS)	-
Illnesses began on 1 May and ended on July 26th, 2011. After July 5th that the active phase of the outbreak ended, only sporadic cases of HUS occurred due to secondary transmission.					
Ref: http://ecdc.europa.eu/en/publications/Publications/1106_TER_EColi_joint_EFSA.pdf					
http://www.euro.who.int/__data/assets/pdf_file/0009/144981/EHEC_outbreak_10_June_2011.pdf					

(continued)

Table 4.3 (continued)

Date/Duration	Country	Pathogen – Disease	Source	Number of Case	Voluntary Recall
http://ecdc.europa.eu/en/aboutus/organisation/Director%20Speeches/201109_MarcSprenger_STEC_ICAAC.pdf					
http://www.rki.de/EN/Home/EHEC_final_report.pdf?__blob=publicationFile					
Apr–June 2011	USA	*Salmonella* Enteritidis	Alfalfa sprouts and spicy sprouts	21 cases, 3 hospitalized	+
Ref: http://www.cdc.gov/salmonella/sprouts-enteritidis0611/062611/index.html					
Feb–Apr 2011	USA	*Salmonella* Panama	Cantaloupe	20 cases, 3 hospitalized	+
Ref: http://www.cdc.gov/salmonella/panama0311/062311/index.html					
Feb 2011	Japan	*Salmonella* spp.	Broccoli salad	1500 cases	-
Broccoli salads provided for school lunches were the cause of massive food poisonings that occurred at nine elementary and junior high schools. An investigation has found that broccoli salads cooked at the city's joint school meal cooking center for lunches.					
Ref: http://barfblog.foodsafety.ksu.edu/blog/146846/11/02/24/salmonella-broccoli-salad-sickens-1500-japanese-students					
Dec 2010–Feb 2011	USA	*E.coli* O157:H7	In-shell hazelnuts	8 cases, 4 hospitalized	+
Ref: http://www.cdc.gov/ecoli/2011/hazelnuts0157/index.html					
Dec 2011	Armenia	Botulinum toxins	Homemade souses of red pepper, cucumber, tomato, caviar of eggplant, canned bean	19 cases	-
Ref: http://www.aysor.am/en/news/2010/12/29/botulism-health/234100					
Sep 2011	France	Clostridium botulinum – Type A	Homemade green olive paste	15 cases	-

Serum, gastric juice and stool samples from the suspected cases and samples recovered from leftover foods from the meals were analyzed and presence of botulinum toxin type A was confirmed by the mouse lethality test, by sero-protection with specific antibodies, and presence of *C. botulinum* was investigated by real-time PCR amplification of the type A, B and E neurotoxin genes, and strain isolation and characterization.					
Ref: http://www.eurosurveillance.org/ViewArticle.aspx?ArticleId=20035					
Apr 2011	Norway	*Yersinia enterocolitica* O:9	Pre-packaged lettuce mixes	20 cases	-
Ref: http://regionalnews.safefoodinternational.org/page/Europe%3A+Food%2FWaterborne+Illness+Outbreaks+2011					
Nov 2010–Feb 2011	USA	*Human Salmonella* serotype I 4,[5],12:i:-	Alfalfa/Spicy sprouts	140 cases, 34 hospitalized	+
Ref: http://www.cdc.gov/salmonella/i4512i-/021011/index.html					

some cases, electronic purchase records of patients were also used with their permission. After the same pathogen is confirmed in the suspected food source from questionnaires, the source pathogen is announced to the public via public health institutions. Besides the biochemical analysis and phenotypic characterization, subtyping methods are commonly used to detect the agent of outbreak from the suspected source. Although different subtyping methods are used to find the contaminated produce, pulsed-field gel electrophoresis (PFGE) is the gold standard molecular method for subtyping foodborne pathogens. Whole genome sequencing (WGS) is also started to be used more frequently in the last couple of years in order to find a relation between the pathogenic strains isolated from the patients and the food items.

In all of the outbreaks occurred in the USA, after the contamination pathway is determined, the responsible company voluntarily recalled the product from the market. Apart from the USA, voluntary recall of the contaminated product was seen in some outbreaks outside the USA such as Norway and UK.

There has been no new outbreaks occurred in the recent years due to mycotoxin contamination because strict limits were set by both national and international legislation with the efforts of European Commission and institutions of United Nations like Food and Agricultural Organization and World Health Organization (Adams and Moss 2006). It can be deduced that mycotoxins are not of major importance as causative agents of foodborne outbreak anymore. However, it should be noted that mycotoxins have a high mortality rate and they still pose a substantial threat to public health in the underdeveloped countries due to malfunctioning of surveillance systems and lack of resources, technology and infrastructure (Alves et al. 2010; Bhatnagar et al. 2008; Strosnider et al. 2006). For instance, in the two outbreaks occurred in Kenya (in 2004) and Brazil (in 2006–2008), aflatoxin and citreoviridin appeared to be responsible from 837 cases and 157 deaths through contaminated cereals (rice and maize) due to inappropriate storage conditions.

Among these outbreaks occurred in the last 5 years, *E.coli* O104:H4 outbreak in 2011 has a specific importance because it not only revealed the vulnerabilities and gaps in the EU foodborne illness surveillance system, but also stressed the importance of timely and accurate communication between countries during an international outbreak. Throughout the outbreak, all the institutions related to food safety under European Commission (EC) and World Health Organization (WHO) worked closely to investigate the case. In order of occurrence, Germany notified the World of this outbreak through EU Early Warning and Response System (EWRS) (WHO 2011). A team sent to Berlin including experts from European Center for Disease Prevention and Control (ECDC), European Food Safety Authority (EFSA) and European Commission (ECDC 2011). EC temporarily banned the import of the source food item (fenugreek seeds) on the same day that source was declared (European Commission 2011). World Health Organization denoted that this outbreak was the biggest ever occurred in Europe, the second biggest ever seen on worldwide, and the most deadly EHEC outbreak ever reported due to its size and virulence (WHO 2011). Finally, RKI published a final report clarifying the details of the outbreak (RKI 2011) as: more than 50 % of the cases were notified to the RKI within 2 days and 75 % were notified within 4 days after receipt of the report at the local health authority and incubation period of the infection was calculated as 8 days on average (RKI 2011).

However there seems to be no problem in the flow of actions taken by these institutions, malfunctions and delays in the detection and confirmation of the source food item and pathogen caused thousands of people to become ill, dozens of people to die, and unaccountable economic loss (0.5–3.5 billion US dollars) (Food Safety News 2011). Moreover, the contamination pathway of the source is still unclear (BfR 2011). Since the outbreak strain (*E.coli* STEC O104:H4) is very rare, has unusual features (highly virulent and affecting mostly adults), and has never been reported in food before (ECDC-EFSA 2011); standard methods to test for STEC used in most EU laboratories could not detect this rare serotype (ECDC 2011) in a timely manner. Therefore, the outbreak strain could not be detected in any of the food samples of plant origin (Winter 2012). Because of the problem in diagnosis, the first source food item was announced was Spanish cucumbers (ECDC 2011). However, after a small outbreak in France, where the cases had no recent travel history to Germany, EFSA declared the source as a specific lot of fenugreek seeds imported from Egypt, which were subsequently used for sprout production both by a horticultural farm in Lower Saxony and by private individuals (EFSA 2012a). After this outbreak, EU has been restructured its Early Warning and Response System (EWRS) in order to increase effectiveness and *E.coli* O104:H4 strain has started to be counted among the most common HPOPs.

4.5 Final Remarks

Consumption of fresh produce has been increasing since 1980s on year-by-year basis, since there is no kill step for pathogens (e.g. heat) before consumption, outbreaks caused by contaminated fresh produce have been in an upward trend. However, except a few countries, reaching global outbreak investigation data is almost impossible because these outbreaks to stay unascertained, unreported by public health officials and uninvestigated by epidemiologists to trace back the source pathogen and/or food item consumed, or the public health agencies do not share the data with the public.

From the available data, the most common foodborne pathogens on fresh produce are pathogenic *E. coli* serotypes (mostly O157 – H4 and H7-, less commonly, O121, O26) and Salmonella enterica serotypes. However *Listeria monocytogenes* and *Clostridium* spp. are not common but still pose a significant threat to public health with their high mortality rates.

Epidemiological and trace-back investigations of outbreaks revealed vulnerability and gaps in the national and global surveillance systems. Basic weakness can be listed as incapabilities and delays in detecting emerging pathogens and real-time data sharing for the countries having an in-place surveillance system. However, the slight increase in the voluntary recalls of the suspected products from the market outside the USA look promising in order to improve the public-private partnership in the realm of food safety. In order to fill these gaps and establish a fully functional global surveillance system including rapid early warning and detection networks for both known and emerging pathogens on food products, improving available regional and global networks seem to be crucial.

References

Adams M, Moss M (2006) Food microbiology, 2nd edn. The Royal Society of Chemistry, Cambridge

Alves HC, Lima V, Porto EAS, Marins JRP, Alves RM, Machado RR, Braga KNL, De Paiva FB, Carmo GMI, Santelli E, ACFS, Sobel J (2010) Outbreak of beriberi in the state of Maranha‹ o, Brazil: revisiting the mycotoxin aetiologic hypothesis. Trop Doct 40:95–97

Behling RG, Eifert J, Erickson MC, Gurtler JB, Kornacki JL, Line E, Radcliff R, Ryser ET, Stawick B, Yan Z (2010) Selected pathogens of concern to industrial food processors: infectious, toxigenic, toxico-infectious, selected emerging pathogenic bacteria. In: Kornacki JL (ed) Principles of microbiological troubleshooting in the industrial food processing environment. Springer, New York, pp 5–61

Bennett JK, Klich M (2003) Mycotoxins Clin Microbiol Rev 16:497–516

Bfr (2011) EHEC outbreak 2011: updated analysis as a basis for recommended measures. http://www.bfr.bund.de/cm/349/ehec-outbreak-2011-updated-analysis-as-a-basis-for-recommended-measures.pdf

Bhatnagar D, Rajasekaran K, Brown R, Cary JW, Yu J, Cleveland TE (2008) Genetic and biochemical control of aflatoxigenic fungi. In: Wilson CL (ed) Microbial food contamination. CRC Press, Boca Raton, pp 395–425

Butler RC, Lund V, Carlson DA (1987) Susceptibility of Campylobacter jejuni and Yersinia enterocolitica to UV radiation. Appl Environ Microbiol 53(2):375

Carlin F (2007) Fruits and vegetables. In: Doyle MP, Beuchat LR (eds) Food microbiology: fundamentals and frontiers, 3rd edn. ASM Press, Washington, DC, pp 157–170

D'Aoust JY, Maurer J (2007) Salmonella species. In: Doyle MP, Beuchat LR (eds) Food microbiology: fundamentals and frontiers, 3rd edn. ASM Press, Washington, DC, pp 187–236

Drusch S, Ragab W (2003) Mycotoxins in fruits, fruit juices, and dried fruits. J Food Prot 66:1514–1527

European Commission (2011) E. coli outbreak: EU withdraws Egyptian seeds from the market and temporarily bans their import. European Commission Press Release. IP/11/831. http://europa.eu/rapid/press-release_IP-11-831_en.htm

European Centre for Disease Prevention (2011) Understanding the 2011 EHEC/STEC outbreak in Germany. ICAAC conference, Chicago, USA2011

European Centre for Disease Prevention and Control and European Food Safety Authority (2011) Shiga toxin/verotoxin-producing Escherichia coli in humans, food and animals in the EU/EEA, with special reference to the German outbreak strain STEC O104, Stockholm. http://ecdc.europa.eu/en/publications/Publications/1106_TER_EColi_joint_EFSA.pdf

European Food Safety Authority (2012) E.coli: rapid response in a crisis. http://www.efsa.europa.eu/en/press/news/120711.htm

European Food Safety Authority (2012) Scientific opinion on the risks for public and animal health related to the presence of citrinin in food and feed. EFSA J 10(3):2605. http://www.efsa.europa.eu/en/efsajournal/doc/2605.pdf

Food Safety News (2011) Germany's E. coli outbreak most costly in history. http://www.foodsafetynews.com/2011/06/europes-o104-outbreak---most-costly-in-history/

Forsythe SJ (2010) The microbiology of safe food, 2nd edn. Wiley, Oxford

Goldburn K (2009) Management of E coli O157 in fresh fruit and vegetables: how can we be safe? Chilled Food Association. Conference an E. coli: cases, controls and common sense. London, UK

Gorny J (2006) Microbial contamination of fresh fruits and vegetables. In: Sapers GM, Gorny JR, Yousef AE (eds) Microbiology of fruits and vegetables. CRC Press, Boca Raton

Hocking AD (2007) Toxigenic Aspergillus species. In: Doyle MP, Beuchat LR (eds) Food microbiology: fundamentals and frontiers, 3rd edn. ASM Press, Washington, DC, pp 187–236

http://www.aysor.am/en/news/2010/12/29/botulism-health/234100

http://www.bangkokpost.com/news/asia/308364/contaminated-pickles-kill-seven-in-japan

http://barfblog.foodsafety.ksu.edu/blog/146846/11/02/24/salmonella-broccoli-salad-sickens-1500-japanese-students

http://www.cbc.ca/news/canada/new-brunswick/story/2012/06/29/nb-e-coli-miramichi-lettuce-1033.html

http://www.cdc.gov/listeria/outbreaks/caramel-apples-12-14/index.html

http://www.cdc.gov/salmonella/enteritidis-11-14/index.html

http://www.cdc.gov/listeria/outbreaks/bean-sprouts-11-14/index.html

http://www.cdc.gov/salmonella/poona-09-15/index.html

http://www.cdc.gov/salmonella/i4512i-/021011/index.html

http://www.cdc.gov/ecoli/2011/hazelnuts0157/index.html

http://www.cdc.gov/salmonella/panama0311/062311/index.html

http://www.cdc.gov/salmonella/sprouts-enteritidis0611/062611/index.html

http://www.cdc.gov/salmonella/agona-papayas/082911/index.html

http://www.cdc.gov/salmonella/pinenuts-enteriditis/111711/index.html

http://www.cdc.gov/listeria/outbreaks/cantaloupes-jensen-farms/120811/index.html

http://www.cdc.gov/ecoli/2011/ecoliO157/romainelettuce/032312/index.html

http://www.cdc.gov/ecoli/2012/O26-02-12/index.html

http://www.cdc.gov/salmonella/typhimurium-cantaloupe-08-12/index.html

http://www.cdc.gov/salmonella/braenderup-08-12/index.html

http://www.cdc.gov/ecoli/2012/O157H7-11-12/index.html

http://www.cdc.gov/salmonella/montevideo-tahini-05-13/

http://www.cdc.gov/salmonella/saintpaul-04-13/index.html

http://www.cdc.gov/salmonella/newport-05-14/index.html

http://www.cdc.gov/ecoli/2013/O157H7-11-13/index.html

http://www.cdc.gov/ecoli/2014/O121-05-14/index.html

http://www.cdc.gov/mmwr/preview/mmwrhtml/mm6406a3.htm?s_cid=mm6406a3_e

http://www.chinapost.com.tw/taiwan/local/nantou/2011/07/22/310649/Deaths-in.htm

http://www.dailymail.co.uk/health/article-2095387/Watermelons-linked-salmonella-outbreak-killed-left-50-ill.html

http://ecdc.europa.eu/en/publications/Publications/1106_TER_EColi_joint_EFSA.pdf

http://ecdc.europa.eu/en/aboutus/organisation/Director%20Speeches/201109_MarcSprenger_STEC_ICAAC.pdf

http://www.euro.who.int/__data/assets/pdf_file/0009/144981/EHEC_outbreak_10_June_2011.pdf

http://www.eurosurveillance.org/images/dynamic/EE/V17N04/art20065.pdf

http://www.eurosurveillance.org/ViewArticle.aspx?ArticleId=20007

http://www.eurosurveillance.org/ViewArticle.aspx?ArticleId=20035

https://www.food.gov.uk/enforcement/alerts/2013/6343/sainsburys-watercress-recall

http://www.foodqualitynews.com/Food-Outbreaks/Sweden-finds-coriander-to-be-source-of-shigellosis

http://www.guardian.co.uk/world/2011/sep/30/ecoli-outbreak-uk-250-ill

http://www.marlerblog.com/case-news/e-coli-sources-in-quebec-windham-and-willimantic-connecticut-unnamed-14-sick/#.VpZmfvmLTDc

http://www.phac-aspc.gc.ca/fs-sa/phn-asp/2013/ecoli-0113-eng.php

http://www.phac-aspc.gc.ca/phn-asp/2014/salmonella-nh-053114-eng.php
http://www.phac-aspc.gc.ca/phn-asp/2015/ecoli-eng.php
http://www.phac-aspc.gc.ca/fs-sa/phn-asp/osm-esm-eng.php
http://www.promedmail.org/direct.php?id=3135159
http://regionalnews.safefoodinternational.org/page/Europe%3A+Food%2FWaterborne+Illness+Outbreaks+2011
http://www.rki.de/EN/Home/EHEC_final_report.pdf?__blob=publicationFile
http://www.ssi.dk/English/News/News/2011/Salm%20imported%20tomatoes.aspx
http://www.yomiuri.co.jp/dy/national/T120821004270.ht
Huang SW (2004) Global trade patterns in fruits and vegetables. U.S. Department of Agriculture (USDA) – Economic Research Service (ERS). http://www.ers.usda.gov/media/320504/wrs0406_1_.pdf
Jackson L, Dombrink-Kurtzman MA (2006) Patulin. In: Sapers GM, Gorny JR, Yousef AE (eds) Microbiology of fruits and vegetables. CRC Press, Boca Raton, pp 281–311
Jay JM, Loessner MJ, Golden DA (2005) Modern food microbiology, 7th edn. Springer, New York
Johnson EA (2007) Clostridium botulinum. In: Doyle MP, Beuchat LR (eds) Food microbiology: fundamentals and frontiers, 3rd edn. ASM Press, Washington, DC
Lampel K (2007) Shigella species. In: Doyle MP, Beuchat LR (eds) Food microbiology: fundamentals and frontiers, 3rd edn. ASM Press, Washington, DC
McClane BA (2007) Clostridium perfringens. In: Doyle MP, Beuchat LR (eds) Food microbiology: fundamentals and frontiers, 3rd edn. ASM Press, Washington, DC
Meng JD (2007) Enterohemorrhagic Esherichia coli. In: Doyle MP, Beuchat LR (eds) Food microbiology: fundamentals and frontiers, 3rd edn. ASM Press, Washington, DC
Moss MO (2008) Fungi, quality and safety issues in fresh fruits and vegetables. J Appl Microbiol 104:1239–1243
Restaino L, Frampton EW, Hemphill JB, Palnikar P (1995) Efficacy of ozonated water against various food-related microorganisms. Appl Environ Microbiol 61(9):3471
Ribot EM, Hyytia-Trees E, Cooper K (2008) PulseNet and emerging foodborne pathogens. In: Wilson CL (ed) Microbial food contamination, 2nd edn. CRC Press, Boca Raton
RKI (2011) Final presentation and evaluation of the epidemiological findings in the EHEC O104:H4 outbreak, Germany 2011 Berlin, Germany, Robert Koch Institute, 2011
Robins-Browne RM (2007) Yersinia enterocolitica. In: Doyle MP, Beuchat LR (eds) Food microbiology: fundamentals and frontiers, 3rd edn. ASM Press, Washington, DC
Seo KS, Bohach GA (2007) Staphylococcus aureus. In: Doyle MP, Beuchat LR (eds) Food microbiology: fundamentals and frontiers, 3rd edn. ASM Press, Washington, DC
Strosnider H et al (2006) Workgroup report: public health strategies for reducing aflatoxin exposure in developing countries. Environ Health Perspect 114(12):1898–1903
Swaminathan BC (2007) Listeria monocytogenes. In: Doyle MP, Beuchat LR (eds) Food microbiology: fundamentals and frontiers, 3rd edn. ASM Press, Washington, DC
The Statistics Division of the FAO (FAOSTAT) (2012) http://faostat.fao.org. Accessed 19 Dec 2012
Weiss A, Schmidt H (2011) Mechanisms of enterohemorrhagic Escherichia coli spread along the food-chain and precautionary measures. J Verbrauch Lebensm 6:503–510
WHO (2011) Public health review of the enterohaemorrhagic Escherichia coli outbreak in Germany. The Regional Office for Europe of the World Health Organization, 2011
Winter M (2012) Lessons learned from the E. coli outbreak. Summer School & Training RASFF, Cremona: The Federal Ministry for Food, Agriculture and Consumer Protection; 2012
Yeni F, Yavas S, Alpas H, Soyer Y (2015) Most common foodborne pathogens and mycotoxins on fresh produce: a review of recent outbreaks. Crit Rev Food Sci Nutrit. doi:10.1080/10408398.2013.777021

Chapter 5
A Risk Management Framework for Plant Biosecurity

Abraham Gamliel, James P. Stack, and John D. Mumford

Abstract Plant and food bisecurity threats in the context of this book target the agricultural industry or the food supply chain by accidental or deliberate introduction of a plant pathogen or insect pest, hence, damaging crops, food and feed. It is therefore a challenges for the stability of a society, and its economy, since, agriculture and the food industry are a primary sector of any nation's economy. There are many windows for threats from the introduction of pest or disease agents. Therefore, a risk management program, an interdisciplinary set of actions before and following the introduction of harmful organisms, should be carefully prepared and effectively executed. The risk management framework comprises a sequence of activities with structured responsibilities among regulatory authorities, scientists, extension professionals and farmers. Risk management includes the activities for pest risk assessment, preparedness, detection, diagnosis, procedures of containment, eradication and management, and finally a recovery plan. The risk management steps should be coordinated within a country and at a regional level in order to ensure that a disease or insect pest outbreak is successfully managed and does not establish in the area or spread to neighboring countries.

Keywords Emerging infectious pests • Prevention • Pest risk assessment • Containment • Eradication • Management recovery plan

---

A. Gamliel (✉)
Laboratory for Pest Management Research, Institute for Agricultural Engineering, ARO Volcani Center, P.O. Box 6, 50250 Bet Dagan, Israel
e-mail: agamliel@volcani.agri.gov.il

J.P. Stack
Department of Plant Pathology, Kansas State University, Manhattan, KS, USA

J.D. Mumford
Centre for Environmental Policy, Imperial College London, Silwood Park Campus, Ascot SL5 7PY, UK

M.L. Gullino et al. (eds.), *Practical Tools for Plant and Food Biosecurity*, Plant Pathology in the 21st Century 8, DOI 10.1007/978-3-319-46897-6_5

5.1 Introduction

5.1.1 Vulnerability of Agriculture to Disease Outbreak

Agricultural crops are vulnerable to attack by a wide spectrum of insects and plant pathogens. The deliberate introduction of a new plant pest or pathogen into an agricultural area could have a serious impact on crop yield and on the cost of management over the short and long-term. Introduction of new pests and pathogens can disrupt the trade from a country or region, resulting in lost markets. Moreover, it may impose negative attributes on the quality of food and feed and may lead to unavailability of certain foods.

Historically, anti-crop weapons based on plant pathogens or insects have been developed in parallel with the development of chemical and biological anti-human weapons (Wheelis et al. 2002; Whitby 2001). Anti-crop weapons have always been of lower priority compared with anti-human ones and in most cases less successful. However, this threat should not be overlooked, since certain terrorist groups and some countries are suspected of actively developing biological weapons (Myerson and Reaser 2002; Cook and Proctor 2007; Desprez-Loustau et al. 2007) (see Chap. 4). Therefore, agricultural systems are exposed to a real risk from anti-crop weapons and the potential consequences should be measured (Frischknecht 2003).

Plant pathogens and insect pests are globally distributed. However, the source of most plant pathogens and insect pests, especially the crop-specific ones, is in most cases linked with the geographic origin of the host plant (Stukenbrock and McDonald 2008). At the crop origin many highly contagious organisms may be endemic as part of an ecological equilibrium with the host plant and the biological environment (Anderson et al. 2004). The global movement of crops (especially, staple and food crops) outside their natural environment enhances their vulnerability to pathogens or insects in their new habitats, and might result in severe economic consequences (Foxell 2001). Outbreaks of plant diseases have been associated throughout history with crops outside their origin, resulting in a wide-scale famine (such as late blight of potatoes in Ireland). The significance of an outbreak within only one region can be significant even on a global perspective, if that region is a major world supplier for one of the food staples. For example, Fusarium head blight of wheat and barley has affected several successive harvests in several states of the USA between 1993 and 1998 (Stack 2000). Hence, if such outbreaks are deliberately initiated and further stimulated, they can lead to a global food shortage. Therefore, any country should protect its agricultural production and set up a contingency plan to mitigate these threats. The objectives of this chapter are to define the risk management framework for a pest or disease outbreak and outline the related procedures, before, during and after its introduction.

5.2 Risk Management in the Context of Plant and Food Biosecurity

An outbreak of a nonnative pest is an unpredicted course of events initiated by the transfer of a pathogen or insect from a remote habitat through various possible routes, using physical or biological vectors. The survival and establishment of the harmful population after its introduction will govern the scale and magnitude of the outbreak. The goal of a risk management framework is to create a mechanism that will allow society within the outbreak area to mitigate this risk and counteract its potential impact. An integrated risk management framework must evaluate, control, and monitor the assessed risks as a combination of the probability or frequency of an event and its consequences (see Chap. 9). Uncertainty, by definition, always accompanies the process, hence, various scenarios and preventive activities are important components of a successful risk management process. Risk management has first evolved as a formal process in the financial context after World War II, and has long been associated with the use of market insurance to protect individuals and companies from various losses associated with accidents. It is defined as "a coordinated set of decision tools, activities and methods that is used to direct an organization and to mitigate the many risks that can affect its ability to achieve objectives" (Kaplan and Garrick 1981; Anon 2009). The phytosanitary terms for pest management are described in ISPM 5 (IPPC 2015a).

The first step of risk management is risk assessment which combines several steps. Risk identification, thinking about what could go wrong, with regard to an outbreak of a new pest or disease no matter how obvious or simple it is. Risk characterisation, which follows, is the approach to describe the harm from an outbreak of a given pest or pathogen, evaluate the impact it will have on agriculture, the economy, food security and food safety, and calculate the probability that such scenarios might occur. All the risk assessment steps are usually combined to give a total risk ranking (Mumford et al. 2013). An effective management process is based on understanding the system and its interactions with its environment. This understanding enables risk analysts to model various scenarios of outbreak progression under different conditions (see Chap. 9). Effective risk management minimizes the impact of unexpected changes which can dramatically shake a society. The Irish famine in 1845, as a result of an unexpected and prolonged outbreak of potato late blight is a significant example of a change that affected the economy and the entire Irish society. The impact of the late blight was extended over 6 years, thus exacerbating the inability of the Irish society to manage the crisis and quickly recover. Hence, identifying risks of unexpected negative changes in a production system should be done in advance and risk management frameworks should be established to ensure there is adequate capacity to cope with its risks. This is already done in most countries for conventional plant health problems (Mumford 2013), and should also include risks from deliberate introductions.

5.3 Risk Management from a Geographical Perspective

The risk management framework is a national responsibility, however, it is practically applied in either of three major geographical scales: a province, a country, or an international region. The risk management framework can be scaled up or down with respect to the commonality of areas with regard to food safety, ecology, demography, trade issues and political concerns (Gamliel et al. 2008).

- **Province** – a confined and relatively small area (state, county, municipality, geographical region) with defined borders and specific topographical contours and climatic conditions. In most cases a region is located in proximity to other agricultural areas and also in the vicinity of populated urban areas. The crop diversity is usually narrow, especially with intensive agriculture, and the products are intended for specific markets. It also involves specific procedures which are relevant to the risk management process, such as intensive import of propagation material or the use of reused boxes and packing containers. There are, however, provinces which are characterized by a diversity of crops (as in developing countries, in which a local community supplies most of their own needs). Threatening pathogens or insects should include not only exotic quarantine pests, but also harmful organisms which exist in other provinces of the country. The local crop protection and phytosanitary rules and procedures are usually part of the national regulations, but in certain cases there may also be additional specific local regulation, such as for island regions.
- **Country** – a country, a state, or a big province which is usually diverse in its topographical contours and climatic conditions, and includes many agricultural areas. There are various crops and productions systems intended for local and export markets. The agriculture may be vulnerable to many threats to crop production and requires a wide range of mitigation components in a risk management framework. The highest priority is given to the pathogens or insects threatening major crops. These usually form the national list of quarantine pests, or other list of plant health risks (Mumford et al. 2013). Risk management at the national level is carried out and regulated by a national plant protection organisation.
- **Geographical region** – refers to a group of countries which are geographically related (such as the British Isles, the Balkan peninsula) or share mutual climatic characteristics (such as the Mediterranean basin). A region often has open borders for free movement of people, plants and plant products. Crop production systems are diverse and the threats are divided into two main groups. The first group includes pathogens or pests which do not appear in any of the countries within the region. The second includes pathogens or pests which appear in some countries but not in other. The threats from the second group are more imminent, especially if the boundaries between countries are close. International regions may have formal regional plant protection organisations, such as the European and Mediterranean Plant Protection Organization (EPPO). Such organisations usually work in harmony with the national authorities and with the central

federation (USA or EU) to form the frameworks and regulation for quarantine, prevention of pest or pathogen invasion and for response. The EU comprises countries from various geographical, climatic conditions and agricultural sectors, (the Mediterranean Sea, North, Central and Eastern Europe). Europe is characterized by crop diversity and a long list of targets for plant pathogens and pests (see Chap. 4). The staple food crops, such as wheat and potatoes, are of mutual interest to most of the countries in Europe. Potatoes are very vulnerable to pathogens which do not exist in Europe or which exist in some countries but not in others. For example, the Mediterranean countries are still free from the pathogens *Synchytrium endobioticum,* the causal agent of potato wart, and *Clavibacter michiganensis* subsp. *Sepedonicus,* the causal agent of potato ring rot (EPPO 2016). On the other hand, intensive vegetable production characterizes southern Europe and the Mediterranean countries, where threats from organisms in Africa or Asia occur. Also, a wide spectrum of new pathogens pose a threat to the main staple crops in Europe. Examples include the Ug99 strain of wheat stem rust and Karnal bunt of wheat *(Tilletia indica).* The agricultural nature and commercial aspiration of different countries in the EU are different and derive from specific threats of each sub-region (the Mediterranean countries, Central Europe, and northern countries). This results in different crop biosecurity and risk management priorities for the different countries. The formal organization for plant protection is EPPO. Other organizations such as Interpol, and the European law enforcement agency Europol, also have roles in enforcement across borders and the prevention of environmental crime, such as the destruction of natural forests or other habitats. Other EU organizations include Frontex which liaises closely with other EU partners responsible for the security of the external borders, customs cooperation and the cooperation on phytosanitary and veterinary controls.

5.4 Risk Management Framework – Temporal Activities and Responsibilities

The risk management process is a hierarchical structure of a sequential set of procedures and activities which are executed in order to mitigate an outbreak which was initiated by a nonnative insect or pathogen. Moreover, it defines the responsible unit for performing and controlling each procedure and the interaction with the other units involved in each process. The sequential set of the procedures within the risk management framework is presented in Table 5.1.

- **Prevention.** The first set of precautionary measures, aiming at preventing a new plant health outbreak. Prevention deals first with priorities, such as assessing the probability and possible impact of a threat and taking precautionary measures prior to its outbreak (Gamliel 2008). The prevention process includes the formation of a list of quarantine pests, followed by contingency plans to mitigate an

Table 5.1 Flowchart of iterative actions involved in a risk management framework of an outbreak of a new pest, and the sequential activities which are carried out at each stage. Details for the specific activities are given in the text

		Iterative steps	Sequential measures to be taken		
Risk management processs	1	**Prevention**	Quarantine pest list	Pathway analysis	Contingency plans
	2	**Detection**	Locating the outbreak area	Surveillance of the region	Delineation of a quarantine area
	3	**Diagnosis**	Pest isolation and identification	Validation of identification	Strain discrimination
	4	**Risk assessment**	Epidemiological aspects	Possible other hosts	Possible vectors
			Survival in soil	Weather forecast	
	5	**Containment**	Applying quarantine measures	Sanitation in	Foliar and soil treatments
	6	**Eradication**	Sanitation	Crop distraction	Foliar and soil treatments
	7	**Management**	Sanitation	Foliar and soil treatments	Resistant cultivars
	8	**Recovery**	Resistant cultivars	Alternative crops	Cultural practices

outbreak. Prevention requires the establishment of a special infrastructure of institutions, trained personnel and emergency regulatory actions. Prevention for deliberate introduction of harmful plant pests should also be directed at the motivation of potential perpetrators (see Chaps. 7 and 9).

- **Detection**. When a potential threat becomes evident in the form an outbreak, the activities include its detection, identification, and delineating the area in which the outbreak is evident. An important part of this process is reporting the outbreak and establishing communications strategies, and a network of responsible agencies which will coordinate the next steps.
- **Diagnosis**. After a new outbreak is detected, the diagnosis process consists of identifying the organism, including finer taxonomic detail where relevant, such as a pathogen strain. The diagnosis process should be rapid in order to provide relevant information which can be used for containment and eradication. It also provides the initial information for forensic work, to assess if the outbreak was deliberately initiated.
- **Pest risk assessment.** The major task in the risk assessment process, after pest introduction, is to evaluate probability and possible ways of further spread and distribution of the introduced pest. At this point it is important to identify weak points in order to minimize the pest impact. Pest risk assessment considers relevant aspects of the pathogen or insect biology and epidemiology, host range, survival in soil, dissemination in seeds, etc. The assessment also considers climatic factors and other issues which can favor spread and challenge the success of the containment and eradication process.

- **Containment.** Involves the establishment of a quarantine zone around the outbreak area, and the application of phytosanitary measures within this area in order to restrict further distribution of the introduced organism from the affected area to new locations.
- **Eradication.** Applied following and in parallel to the containment process, eradication involves a range of phytosanitary measures to eliminate the pathogen or pest from the outbreak area. Eradication aims at the elimination of all the existing inoculum from a contained area, including potential vectors, other hosts and various physical reservoirs.
- **Management.** A process to confirm the elimination of the organism from the outbreak area, including, pesticide application, other pest management approaches, and introduction of resistant cultivars. The management process may need to be applied for an extended period for final confirmation of eradication.
- **Recovery plan**. The ultimate goal is to provide the area with alternative options for crop production. Hence, the recovery plan is about the introduction of new crops, resistant cultivars and the adoption of cultural practices which are suppressive to future reestablishment of the pathogen or pest.

5.5 Prevention

The prevention process begins with assessment of the probability of the introduction of a new organism together with the analysis of the potential impact to the local agro-ecosystem and the motivation of any potential intentional introduction. The outcome of this assessment is a set of priorities in the shape of a list of quarantine or high risk pests. Additionally, contingency plans are prepared including training of special units for early detection, establishing a reporting network and enacting laws, regulations or cooperative processes for effective implementation. Prevention requires a special infrastructure in order to achieve these goals. A successful prevention process means no outbreak. However, this can be confusing, since the absence of an outbreak can mean either a lack of introduction challenge or a successful strategy of control before the introduction. The first scenario can often lead to a debate about the necessity of such infrastructure and the allocation of funds for its activities. Nonetheless, the history of many new outbreaks of nonnative organisms emphasizes the need for additional effective preparedness programs after an outbreak has occurred. Specific processes related to prevention are described in further detail.

5.5.1 High Consequence Pathogens

Quarantine lists for non-indigenous and high consequence pathogens are specified in each country and are aimed at minimizing the invasion of the specified pest in order to prevent the potential damage to the local agriculture and economy. The most common approach to define and classify high consequence pathogens as quarantine pests is based primarily on their potential impact on the crop or to the food supply and safety. Anderson et al. (2004) suggested classification of emerging pathogens into four major groups: (1) pathogens of the four staples (wheat, rice, corn and potato); (2) pathogens of cash crops and secondary food crops; (3) pathogens of non-food crops; (4) pathogens of wild plants. This could also apply to high risk insect pests. However, such classification often overlooks significant local aspects of crop systems which vary geographically and commercially among different countries and regions. The list for quarantine may be small or large. For example the US Animal and Plant Health Inspection Service (APHIS) identifies a list of 8 most threatening plant pathogens (USDA 2016), while the EU list is much longer. The European Plant Protection Organization (EPPO) has listed two select harmful organism lists: A1 – which lists pests which are absent from the EPPO region; A2 – which lists pest which are already locally present in the EPPO region. In addition, EPPO has an alert list which aims to draw the attention of Member Countries to pests which have either recently been added to A1 or A2 lists, or present an urgent phytosanitary concern (EPPO 2016). The list of quarantine pests is reviewed periodically.

A list of high consequence pathogens or pests is the first step for prevention in order to form a quick and comprehensive response in case of a disease outbreak (Gamliel 2008). However, this list is not necessarily identical with the most threatening plant health organisms which can be used as anti-crop weapons. Hence it is important to have an additional list of pathogens which can target crops as biological weapons (see Chap. 4). Formal lists of related organisms are available for human and animal weapon pathogens in various countries, but not for plant pathogens. Several lists were voluntarily generated and published for various purposes, such as restriction of international trade. However in most countries the only formal list is the quarantine list.

High consequence pathogens with regard to their potential as weapons can be classified according to factors which reflect their threat to agriculture (Schaad et al 1999):

- **The type of threat and the circle of its impact**. These include threats to food and feed, trade, crop and yield loss, and threat to biodiversity.
- **The target crop**. Staple food crops, other large scale crops, non-food crops.
- **Classification of pathogen type**. The nature of the organism and its potential ability to establish in a new area after its introduction.
- **Pathogen epidemiology and pest spread**. The nature and rate of pest and disease development and spread govern the threat potential of an introduced organism.
- **Available control measures**. The ability to reverse the impact of an introduced organism and later on to contain and eradicate it from the introduced area, depends upon the available knowledge, means and technologies to maintain, eradicate and control the invading agent.

5.6 Pathway Risk Analysis

Pathway risk analysis is the process of identifying possible pathways of entry and assigning the probability for introduction, establishment and impact. The identification of potential pest arrival hotspots involves a cluster of uncertainties and is a key to assumptions in modeling (Yang et al. 1991; Yemshanov et al 2013). Pest pathway maps can serve as effective tools to describe potential arrival, establishment and spread. Such maps can support pest management decisions, including restrictions on international trade, domestic quarantine practices and response programs (Venette et al. 2010). Pathway risk analysis maps consider the following possible routes and measures:

- **Global movement of plants and plant products**. The global trade of agricultural products includes extensive trade in fresh fruits, vegetables and grains which can be infested with pathogens and insects. Trade in live plants has been recognized worldwide as an important invasion pathway for nonnative plant pests (Seebens et al 2015). Such organisms can eventually reach agricultural field by many routes resulting in an outbreak.
- **Trade in propagation material**. Seeds and other propagation material are globally traded and transported for many purposes. Potato seed tubers which are imported to Israel for the spring crop from European countries are the prime source for many soilborne pathogens (Tsror et al. 1999). The bacterium *Ralstonia solanacearum* race 3 biovar 2, that causes wilt diseases on eggplant, geraniums, potatoes, and tomatoes, was imported into the United States in infected geranium plants from Kenya and Guatemala; *Ascochyta rabiei*, the causal agent of chickpea blight on infected seed (Liebhold et al. 2012; Strange and Scott 2005), may be spread by transplants and other infected propagating material, such as cuttings. Another example is the Bayoud disease of date palms which is caused by *Fusarium oxysporum f. sp. Albedinis,* which currently is confined to North West Africa. The pathogen can spread by all plant parts (de la Perriere et al. 1995). Additionally, the transport of seeds by passengers for private purposes, by tourists looking for new exotic plants in their back yard, or immigrants seeking to maintain their traditional food in their new country. All these should be considered as potential routes for pathogen or pest entry.
- **Cargo and shipping containers**. Cargo mobility is massive and increased exponentially since 1970 (Hulme 2009). Cargo shipping across countries and continents has important consequences for the spread of pests and their vectors over multiple scales. For example, studies have shown that the volume of imports of fire wood and logs has been positively correlated with the number of invasive species that have established (Pujadas 2001; Tkacz 2002; Westphal et al. 2008).
- **Natural aerial dispersal.** The spread of plant pathogens over long distance is an important factor in the entry of invasive species. The ability of pathogen propagules to survive the atmosphere and reach new areas is well documented and modeled (Aylor 2003; Main et al 2001; Marshall et al. 2003).

- **Climate change and drastic weather events.** Predicting the consequences of climate change for the spread of infectious disease or pests is a challenge (Altizer et al 2013). However, evidence exists for the contribution of drastic weather events to spread of pathogens and pathogen vectors. Soybean rust is thought to have moved from Brazil or Venezuela into North America for the first time in the fall of 2004 with Hurricane Ivan. Other drastic weather events (hurricanes, floods, tsunamis) may have impacted other crop plants with bacterial, fungal and viral diseases (Wolken et al 2003). Since such events are nearly inevitable, pathway analysis should assess this route of pathogen entry.
- **Intentional introduction.** There may be motivation and capacity for intentional introductions (see Chaps. 4, 7, and 9) which should also be included in pathway analyses to ensure all risks and management options are covered.

5.7 Contingency Plans (Preparedness)

While prevention is the art of handling "how to keep the door closed for invasion", contingency plans are all about "what is needed to be ready when ultimately the outbreak occurs" (EPPO 2014). Preparedness and contingency plans are crucial to ensure a rapid, skilled and responsive infrastructure. Effective contingency plans should consist of the additional following elements.

- Training of scouts for early detection of the pest or disease
- Development of early-warning systems (Roberts et al 2006)
- Establishing a rapid and central reporting network
- Setting sentinel plots for high consequence pathogens in sensitive areas
- Tools and equipment for rapid diagnosis of the pathogens, including specific primers and microbial isolates
- Outlining an outbreak response plan including the team structure and responsibility

An effective responsive set of actions can and should be improved by frequent exercises. These in turn target the weak links in the chain and suggest options for improvement. Such practices are already executed in many countries. Otherwise the effectiveness of responsive action can be judged only after the occurrence of a real outbreak. However, the agroindustry cannot afford costly mistakes in case the contingency plan did not stand up in the reality test.

5.8 Detection

The occurrence of a quarantine pest creates a new scenario in which the threat is now real and evident in the area. Occurrence does not necessarily mean outbreak, but may also be an appearance on a small scale, at an early stage (IPPC 2015c). Once the organism is evident, the main goal shifts from prevention towards minimizing the impact of the pathogen or pest on the immediate cropping systems. Detecting the pathogen or pest is an important element to delineate the area of outbreak in order to later preform containment and eradication procedures. The initial detection of quarantine pests can be achieved either in a proactive approach or a responsive one. The proactive approach involves a set of sentinel field plots, in which a crop susceptible to the pathogen or pest is planted in a sensitive area. Frequent monitoring of the crop is performed during the entire crop season in order to detect the first appearance of the organism. A good example for this approach is monitoring the possible appearances of soybean rust in the USA. The responsive approach involves routine scouting and monitoring for new insects and diseases in commercial fields, which is done as part of the pest management program of each agriculture industry. It is, however, important that the scouts who perform monitoring are trained to detect the nonnative pathogen or pest and report immediately anyunusual presence.

Once a new outbreak is detected and reported, the following steps are performed:

- Prompt confirmation of the detection should be done by a reliable and rapid diagnostic tool.
- Surveillance, taking of official samples. Data gathered at the site of detection includes geographical information, hosts infested at the site, extent and impact of damage and level of pest incidence. Scene assessment may also include taking an inventory of the main plants in the affected area and the quarantine area to support the later risk assessment.
- Delineating the area in which an outbreak is detected for the purpose of setting the quarantine area.
- Setting a quarantine area. The main objectives are to prevent further spread and to protect special locations. Thus, prevention applies to the zones around the affected area or to protected objects, such as nurseries and other propagation fields.
- Establishing a responsible quarantine team, and assigning a person responsible for managing the quarantine area.
- Tracking back for possible sources of the insect or pathogen. This is especially important since an infested seed batch may spread a pathogen over a vast area and in many locations. Hence, such information will direct surveillance in wider circles than just the outbreak area.
- Reporting and establishing a national communication network. It is also essential to inform neighboring countries and coordinate international responses.

The responsibility of handling and managing the outbreak after it is detected goes to the official regional or national plant protection authority. At this stage the contingency plans should serve as a work plan for the delineation of the outbreak area, performing risk assessment and sorting out the available measures for response.

The activities and responsibilities are summarized in Table 5.2. A rapid detection of an outbreak and a hierarchical plan will usually lead to the delineation of a more confined area to be treated. Eventually it will increase the probability for successful containment and eradication process.

Table 5.2 Activities and responsibilities for the detection process following an outbreak of a new plant disease

Activity	Relevant area	Responsibility
Identification of a new plant disease outbreak	Infected area and around	Local grower extension service
Confirmation of intelligence and involvement of low enforcement	Infected area and around	Police/grower/extension service
Surveillance, taking of official samples, and delineating the area (IPPC 2015b)	Infected area and around	NPPO
Tracking back the source of propagation material	Seed companies, nurseries, contractors, machinery, etc.	NPPO
Setting the quarantine area	Infected area and around	NPPO
Reporting	Neighboring, farms, regions and countries	NPPO
Establishing communication network	Infected area, neighboring, farms, regions and countries	NPPO

5.9 Diagnosis

Once an outbreak is detected it is essential to rapidly identify the causal organism, and to verify its responsibility for the outbreak (Stack et al 2006). A rapid, accurate and reliable identification is the key for effective implementation of containment and eradication procedures. The diagnosis process involved the following important aspects:

- Rapid isolation and culturing of the pathogen, which is important for morphological and molecular diagnosis and for additional tests, such as pathogenicity test, resistance to pesticides and other biological traits. For insects, specimens for taxonomic classification should be collected.
- Application of rapid and accurate diagnostic tools,such as molecular primers.
- Identification of the pathogen and the relevant strain/serovar/pathovar. This is an important step for selecting the appropriate management measures.
- Validation of the identification results with other laboratories and other identification methods.

The activities and responsibilities are listed in Table 5.3. An absolute positive identification of the causal organism provides important information for an effective control process thereafter and increases the chances for successful mitigation. Rapid identification also provides the initial information for forensic investigation, to assess if the outbreak was deliberately initiated and by whom. Wrong identification of the organism may, however, lead to the use of ineffective measures and the expansion of the outbreak. This is also relevant when the strain is misidentified, leading to the use of pesticides which are not effective against the pest or pathogen.

Table 5.3 Temporal and spatial activities and responsibilities for the diagnosis process following an outbreak of a new plant disease or insect pest (applied to the infected area)

Activity	Responsibility
Isolation and identification of the pathogen	National/regional Plant protection and regulation authority
Identifying the relevant strain/serovar/pathovar	National/regional Plant protection and regulation authority

5.10 Pest Risk Assessment

Following the detection of outbreak, the focus of pest risk assessment (PRA) changes from "what can go wrong", to "how bad it can develop from this point". Hence, the goal of the PRA process it to provide a detailed risk map which will enable actions to minimize the impact of the outbreak with regard to the immediate damage and long-term consequences. The output of an effective PRA is the choice of appropriate counter measures. PRA involves addressing the relevant key points with regard to the pathogen or pest:

- **Disease or life cycle**. Plant disease epidemics consist of repeated cycles of pathogen development in association with the plant host and as influenced by the environmental conditions, as do insect pest infestations. Monocyclic pathogens produce only one cycle of development (one infection cycle) per crop cycle, while polycyclic pathogens can produce many infection cycles per crop cycle. Thus, the damaging impact from a polycyclic pathogen is potentially higher. This is also true of multi-voltine insect pests.
- **Relevance of a vector**. The involvement of a vector in the life cycle of the pathogen and its existence in the area of invasion, amplify the threat. The existence of the vector in or around a diseased location favours a rapid spread of infective inoculum, and rapid distribution to new infected loci. Concommitantly, the presence of a vector reduces the prospect for effective containment and eradication. One of the effective spread factors of Citrus canker disease in Florida is the Asian leaf miner, which is not reported in other infected countries such as Brazil and Latin America (Gottwald et al 2002).
- **Survival in the invaded area**. Most (if not all) plant pathogens can infect other host plants, even without disease symptoms. The host range of a pathogen can include plants from the close botanical family, or from various cultured plants and wild plants and weeds. The host range for some pathogens and insects (including threatening ones) can be very wide. For example the host range of *Phytophthora ramorum,* the causal agent of sudden oak death, covers 137 different plant hosts (APHIS 2016).
- **Dissemination**. A wide array of plant pathogens and insects can be present in propagation material, mainly the seeds. The main impact of seed infection is the pathogen ability to spread over a vast area and in many locations within a short period. Numerous examples of disease outbreak as a result of infected seeds have been documented. Moreover, the transmission of a pathogen through the seeds

Table 5.4 Activities and responsibilities for the pest risk assessment process following an outbreak of a new plant disease

Activity	Relevant area	Responsibility
Addressing the relevant key-point with regard to the pathogen's epidemiology,	N/A	National/regional Plant protection and regulation authority
Examine adjacent susceptible crops	Affected and surrounding area	Local grower/extension service
Look for relevant vectors	Affected and surrounding area	Local grower/extension service
Check for infection or infestation of relevant wild plants	Affected and surrounding area	Local grower/extension service
Check for infection or infestation of relevant weeds	Affected and surrounding area	Local grower/extension service
Test the possible survival of the pathogen in soil	Affected and surrounding area	Local grower/extension service
Examine dissemination in seeds	N/A	NPPO
Check for pathogen spread in water	Affected and surrounding area	NPPO
Assess weather forecast (rain wind)	N/A	NPPO
Test vitro sensitivity to pesticides	N/A	NAPPO, regional Plant protection authority

will not necessarily express disease symptoms during the first crop generation. However it will enable the pathogen to establish in a new area long before it might be detected. Therefore, it is important to address this point and perform surveillance in all the fields which originated from the same seed batch.

Effective and detailed PRA should also consider other aspects which are not directly related to pests, pathogens or the disease type. These include:

- The significance of a non-agriculture area, such as proximity to urban and populated areas. Such areas embrace a wide variety of plants, such as small vegetable gardens,fruit trees and ornamentals. These can serve as hosts to the pathogen or insect and may be also affected. Hence assessment of the potential of a reservoir in such areas should be considered in any scenario of a responding program. A significant example is the outbreak of the citrus canker in Florida in two highly populated counties with many backyard citrus trees (Gottwald et al 2002)
- Available measures for the control of invasive species. The ability to reverse the impact of an introduced pathogen or insect and later on to contain and eradicate it from the introduced area depends upon the available knowledge, means and technologies to maintain, eradicate and control the invading agent.

A detailed and informative PRA provides a powerful visual communication tool to describe the potential impact and address the need of additional tools, such as the further design of pest surveys around and outside the outbreak area. It also provides the guidelines for general and specific pest containment and eradication decisions. A summary of the temporal and spatial activities and responsibilities regarding the PRA process are listed in Table 5.4.

5.11 Containment, Eradication and Management

Eradication is the ultimate goal when an outbreak from non-native organism is detected. Successful eradication of invasive organisms involves a concerted complex of actions, including the adoption of the appropriate strategy and careful execution of all the control procedures (Gamliel and Fletcher 2008; IPPC 2015d). A quantitative assessment of all the factors that influence the eradication process can lead to the adoption of an appropriate eradication approach and strategy. The following aspects should serve as the guidelines for assessing the appropriate response. Most of these aspects were described in the previous section while referring to the organism. However, below these factors are discussed with relevance to the potential response measures.

- **Type of threat and possible impact.** Four types of threats exist: food safety, trade, economic damage to product yield and quality, and loss of consumer confidence. A threat to biodiversity in a forest or other natural habitat may not be considered to be as critical as a threat from food toxicity. In the latter situation, the response should be swift, as human and animal health is at the top of the priority ladder. Higher vulnerability is a factor in the motivation of intentional introductions.
- **The pathogen or pest**. The systematic grouping of pathogens (*i.e.* viruses, bacteria, fungi) or insects sets not only the type and magnitude of the agricultural threat, but also the type of response. Since viruses cannot be controlled directly, they often can be managed effectively by vector eradication and destruction of infected plants. In contrast, the strategy to eradicate myctoxin-producing fungi involves activities both targeting the fungus directly and managing the contaminated products.
- **Pathogen biology and disease epidemiology**. The ultimate objective of containment is to suppress new infectious inoculum. To apply appropriate countermeasures and accomplish this goal requires knowledge of the pathogen and host biology, life cycle and disease progress. Soilborne fungi have a relatively slow, special distribution pattern. Their containment can be accomplished if the inoculum is suppressed. In contrast it is much difficult to contain foliar disease, such as Karnal bunt (*Tilletia indica),* in which large masses of spores are produced.
- **Vectors.** The involvement of vectors in the disease cycle increases the potency of an invading pathogen, as it favours rapid spread within and beyond the infection site. Therefore, it is critical to prevent the entry of the vector to the infected area, or to eradicate if it is already there. *Xylella fastidiosa,* the bacterium that causes Pierce's disease of grapevine, was reported recently in olive trees in southern Italy. Numerous species of *Cicadellidae* and *Cercopidae* known to be vectors of *X. fastidiosa*, (Hopkins and Purcell 2002) reside in Europe or the Mediterranean. Hence, vector management should play an important role in the infected areas in Italy and in any prevention and eradication program in other neighbouring European countries.

- **Other hosts**. Pathogens can infect, survive and spread on hosts other than the main crop. The range of pathogen hosts can include cultured or wild plants that are botanically close (or not), and a wide spectrum of weeds. Failure to identify and eradicate all host species from the invaded area can result in failure of the overall eradication process. Containment and eradication must include all the possible local hosts in the eradication.
- **Forecast of unusual climatic events.** Unusual climatic conditions can induce, spread or suppress epidemics. For example, citrus canker was further spread in Florida by hurricanes.
- **Size and location of the affected area**. The success of containment and eradication measures is inversely correlated with the size of the infected area. When a pathogen is detected in a small and confined area, a rapid response could be successful. However, a wide area of infection, or multiple infection sites, implies that distribution of the pathogen occurred beyond the detected location. In the second scenario the chances for success of containment and eradication are lower. Deliberate introduction of a pathogen into an urban area or forest could be much more difficult to handle than one in a field setting, since other factors may dominate the response approach. For example, the fact that the Florida citrus canker outbreak was within an urban area with many back yard citrus trees was one of the main reasons for the eradication failure during 1995–2001 (Gottwald et al. 2002).
- **Available measures and time of response.** The two critical processes of and outbreak are the pathogen or pest establishment in a new area and its spread to other areas. Because preventing these events is time dependent, the success or failure will depend on the measures taken and the rapidity of the response.
- **The time passed from introduction to detection**. – Early detection and accurate diagnosis are crucial to prevent the establishment and dispersal of introduced pests and pathogens to minimize subsequent impact. Once an invading species becomes established in an area it can be difficult or impossible to eradicate. A good example of effective and quick detection is the case of pathogens in propagation material which are detected before their introduction into the soil. In contrast, symptoms of citrus greening (caused by *Liberobacter* sp.) were expressed in a period of 2.5–3.5 months after leaves emerged from buds on diseased trees. Furthermore, detection of citrus greening pathogens in asymptomatic tissue is inconsistent by any known method. Molecular detection assays may be complicated, and results are not always reliable. The incubation period (i.e. the time from infection to disease), and the latent period (the time from infection to production of an infectious propagule) further extends the time from invasion, possibly beyond the threshold timing for effective containment and eradication.
- **Forensic and legal action.** If introduction is intentional it is important that a rapid forensic investigation leads to the identification and apprehension of the perpetrators to ensure that containment measures will not be undermined by further releases.

A management team should be established before the containment-eradication process is carried out, in order to effectively coordinate and control the eradication process (IPPC 2015d). The management team has the responsibility for:

- Ensuring that the eradication program meets the agreed criteria for successful eradication
- Formulating, implementing, and modifying as necessary an eradication plan
- Ensuring program operators have appropriate authority and training to undertake their duties
- Financial and resource management
- Appointing and defining duties of operators, ensuring operators understand their responsibilities, and documenting their activities
- Managing communication, including a public relations program
- Communicating with affected parties, e.g. growers, traders, other government departments and non-governmental organizations
- Implementing an information management system, including program documentation and appropriate record-keeping
- Daily management of the program
- Continuous monitoring and evaluation of critical elements
- Periodic overall program review

5.12 Containment

Containment steps are taken within the infected area and the close outskirts around it (defined together as the quarantine zone). The ultimate objectives of containment are to detain the pest within the infected area and block its egress. The first step in containment is geographical delineation of the infected area and the buffer zone to be treated. The containment procedures in the outlined area should cover the agricultural, rural and the urban sectors, to reach and eliminate all the existing inoculum. The procedures for successful containment include:

- **Quarantine**. The main focus is to block every possible escape of the pathogen from the contained area. Quarantine measures are aimed at restriction of entry and exit of machinery, equipment, farm materials and products. Furthermore, it is essential to clean and disinfect vehicles, machinery, commodities and any products that can potentially carry contaminants.
- **Sanitation**. As a supplement to quarantine, sanitation aims to reduce and suppress the spread of the organism and to prevent infection in pathogen or pest free locations within the affected zone. Sanitation includes disinfection of equipment, machinery and working tools after leaving one area and entering another.
- **Physical barriers**. Creating barriers to contain the inoculum within the infected area is especially important with soilborne pathogens, which can spread by root to root contact.

- **Vector control.** Intensive insect control and monitoring is directed at eliminating any vector that transmits the pathogen, and preventing further infection within and outside the infected area. Vector elimination is important especially with insect transmitted viruses, phytoplasmas and spiroplasmas, and fastidious walled bacteria
- **Destruction and removal of infected plants**. Infected and diseased plants that could serve as inoculum sources should be eradicated to prevent further infection and possible spread outside the infected area. Therefore, destruction and removal of infected plants is crucial for the success of the eradication.
- **Intensive pesticide application program (perennial crops).** There are cases in which whole plants are not removed for many reasons. Thus containment efforts include continuous efforts are made to suppress the internal inoculum, and prevent further infection and spread. Such approaches may be successful when an affected area is small and spread is limited, however may lead to a failure in the entire eradication chain of actions.
- **Eradication of weeds**. Elimination and eradication of weeds and other plants that can serve as volunteer hosts may be very helpful in managing pests and diseases.

An effective containment procedure results in keeping the affected outbreak area small enough to give some chance ofsuccess for the eradication process to follow. Preventing further accidental or intentional introductions is critical to containment success, so good knowledge of the causes of outbreaks are needed at the earliest stage.

5.13 Eradication

Eradication is a key link in the chain of the process toward the termination of the outbreak. Eradication procedures are performed both within the affected area and the surrounding area. In practice, containment and eradication overlap significantly, and the measures described earlier as part of a containment strategy may serve also as initial steps for eradication. However, eradication can only be achieved if it follows effective and rigorous containment activities, and prevention of further accidental or deliberate introductions. The following are additional measures contributing to eradication.

- **Removal and destruction of plants or plant parts**. Destruction and removal of the infected or infested plants are required to eliminate the major inoculum source. This process can start by applying the appropriate pesticide to minimize the pest or pathogen spread during plant removal, followed by the destruction of the affected plants.
- **Soil disinfestation**. Eradication of soilborne pathogens from the soil requires a robust treatment such as soil disinfestation using highly toxic soil fumigants with non-selective activity. Effective soil disinfestation depends upon establishing

proper application conditions and efficient application. The treatment should be effective down to deep soil layers and should be repeated.

A successful eradication program may require repeated treatment in order to suppress residual surviving inoculum. Additionally, the principles of quarantine and sanitation must continue within the eradicated area, and a pesticide application program should be maintained to suppress any inoculum left on plant debris, wild weeds and other wild host plants. Monitoring for newly symptomatic plants should be performed routinely, and such plants removed and destroyed. In addition, shoots emerging from remaining plant roots must be suppressed, often by repeat applications of soil fumigants or herbicides. It can be effective also to control weeds and other plants that can serve as alternate hosts for the pathogen.

5.14 Post-outbreak Management

Post-outbreak management serves as the third step, following the execution of a successful containment and eradication program in the outbreak area. The management program prevents the emergence of new inoculum by maintaining conditions unsupportive of an epidemic. Management strategies, applied subsequent to eradication, also provide crop protection tools for a disease recovery plan.

An effective and successful post-outbreak management program should maintain all the quarantine practices as listed above under containment. Sanitation should include disinfestation of tools, equipment, and machinery. Measures performed during the eradication process, including removal of infected plants, pesticide applications against the pathogens and/or their possible vectors, and destruction of weeds and wild hosts, should continue. Intensive pesticide applications are most important in tree crops, where trees were not removed during the eradication process. Additionally management includes:

- **Use of pathogen-free propagation material**. Only certified propagation material should be allowed into the area. Such material should be disinfected by chemicals, thermal treatment or a combination of approaches.
- **Soil treatment**. Soil fumigants or herbicides can be used to destroy the host plant root system, eliminating existing pathogen inoculum and suppressing the formation of new inoculum. It is also recommended to combine cultural practices and suppressive measures such as compost amendments.
- **Cultural practices**. The cropping system should be modified to maintain conditions that do not favour re-emergence of the pathogen. To suppress new infections of *Erwinia amylovora* (the causal agent of fire blight of pome fruit trees), recommendations include reducing fertilization to slow the growth rate of the trees, withholding irrigation water, nitrogen fertilizer, and cultivation (Koseoglu et al 1996). Other cultural procedures may include changes in planting dates and the establishment of wind breakers as mechanical barriers to pathogen movement.

Table 5.5 Temporal and spatial activities and responsibilities related to the management process

Activity	Relevant area	Responsibility
Use of clean propagation material	Entire area	Grower, or extension service
Use of resistant cultivars	Entire area	Local extension service
Sanitation and disinfection of any tool, equipment, machinery	Entire area	Grower
Further Removal and destruction of plants or plant parts	Entire area	Local extension service
Intensive pesticide application program (perennial crops),	Entire area	Local extension service
Intensive vector monitoring and control	Entire area	Local extension service
Destruction of new emerging plants from a treated area	Entire area	Local extension service
Eradication of volunteer cultivated hosts	Entire area	Local extension service
Eradication of wild weeds.	Entire area	Local extension service
Adoption of appropriate cultivation practices	Entire area	Local grower extension service
Leaving land fallow	Quarantine area	NPPO
Host-free periods	Quarantine area	NPPO
Restriction of subsequent cropping	Quarantine area	NPPO
Use of cultivars that suppress or eliminate pest populations	Quarantine area	NPPO

- **Use of resistant cultivars**. Planting resistant varieties of crops is particularly effective in suppressing initial infections, promoting healthy crop production.

The activities and responsibilities are listed in Table 5.5. The management actions may need to be repeated over more than one crop season, in order to accomplish and confirm the elimination of the pathogen from the outbreak area. Post-outbreak management should also address the causes of unintentional or deliberate introductions to ensure there is no further problem from that source.

5.15 Recovery Plan

Final confirmation of eradication may take some time, even years to be assured no further risk is present. During that period a plan enabling sustainable crop production in the affected area is needed. It may be necessary to switch to resistant cultivars. However, resistant cultivars do not necessarily imply the absence of the pathogen or insect pest. Therefore, adoption of cultural practices which are suppressive to the organism should be maintained. A radical change in the introduction in the area may involve the need to grow alternative crops. Such a shift in production requires a well prepared recovery plan which include rapid introduction of the new crop together with the knowledge and technology for production. Activities and responsibilities for the recovery plan following eradication of an outbreak are listed in Table 5.6.

Table 5.6 Activities and responsibilities for the recovery plan following eradication of an outbreak

Activity	Relevant area	Responsibility
Host-free periods and restriction of subsequent cropping	Quarantine area	NPPO
Adoption of appropriate cultivation practices	Entire area	Local grower extension service
Use of cultivars that suppress or eliminate pest populations	Quarantine area	NPPO
Introduction of new crops	Quarantine area	NPPO

5.16 Conclusion

A risk management framework is the key for rapid and effective counteracting measures, aiming at reducing the likelihood or nullifying the impact of an outbreak from a nonnative pathogen or pest. The potential impact of such a pathogen or pest may be devastating, especially if targets are the major staple food crop in a country. Risk management is relevant for both an accidental or deliberately introduced pathogen since the significance of both is in many cases similar. The activities which comprise the risk management framework should, above all, be structured and coordinated with responsibilities and hierarchical authorities. The margins for error in the process are narrow since the time and space to correct failure in the management of an outbreak are minimal. While risk management in the event of an intentionally caused outbreak is very similar to unintentional outbreaks, additional preventive measures directed at the motivation of perpetrators and their prosecution should be taken.

References

Altizer S, Ostfeld RS, Johnson PTJ, Kutz S, Harvell CD (2013) Climate change and infectious diseases: from evidence to a predictive framework. Science 341:514–519

Anderson PK, Cunningham AA, Patel NG, Morales FJ, Epstein PR, Daszak P (2004) Emerging infectious disease of plants: pathogen pollution, climate change and agrotechnology drivers. Trends Ecol Evol 19:535–544

Anon (2009) Risk management – principles and guidelines. ISO – International Organization of Standardization, Geneva, Switzerland, ISO 31000:2009

APHIS (2016) Phytophthora ramorum. United States Department of Agriculture, Animal and Plant Health Inspection Service. https://www.aphis.usda.gov/aphis/ourfocus/planthealth/plant-pest-and-disease-programs/pests-and-diseases/phytophthora-ramorum

Aylor DE (2003) Spread of plant disease on a continental scale: role of aerial dispersal of pathogen. Ecology 84:1989–1997

Cook D, Proctor W (2007) Assessing the threat of exotic plant pests. Ecol Econ 63:594–604

De la Perriere RAB, Amir H, Bounaga N (1995) Prospects for integrated control of 'bayoud'(Fusarium wilt of the date palm) in Algerian Plantations. Crop Prot 14:227–235

Desprez-Loustau ML, Robin C, Buee M, Courtecuisse R, Garbaye J, Suffert F, Sache I, Rizzo DM (2007) The fungal dimension of biological invasions. Trends Ecol Evol 22:472–480

EPPO (2014) Decision-support scheme for prioritizing action during outbreaks. OEPP/EPPO Bull 44(3):443–456
EPPO (2016) EPPO activities on plant quarantine. European and Mediterranean plant protection organization. https://www.eppo.int/QUARANTINE/quarantine.htm
Foxell JW (2001) Current trends in agroterrorism (antilivestock, anticrop, and antisoil bioagricultural terrorism) and their potential impact on food security. Stud Conflict Terrorism 24:107–129
Frischknecht F (2003) The history of biological warfare. EMBO Rep 4:s47–s52
Gamliel A (2008) High consequence plant pathogens. In: Gullino ML, Fletcher J, Gamliel A, Stack JP (eds) Crop biosecurity: assuring our global food supply. Springer, Dortdrecht, pp 25–36
Gamliel A, Fletcher J (2008) Crop biosecurity: containment and eradication of invasive pathogens. In: Gullino ML, Fletcher J, Gamliel A, Stack JP (eds) Crop biosecurity: assuring our global food supply. Springer, Dortdrecht, pp 73–92
Gamliel A, Gullino ML, Stack JP (2008) Crop biosecurity: local, national, regional and global perspective. In: Gullino ML, Fletcher J, Gamliel A, Stack JP (eds) Crop biosecurity: assuring our global food supply. Springer, Dortdrecht, pp 37–61
Gottwald TR, Graham JH, Schubert TS (2002) Citrus canker: the pathogen and its impact. Online. Plant Health Prog. doi:10.1094/PHP-2002-0812-01-RV
Hopkins DL, Purcell AH (2002) Xylella fastidiosa: cause of Pierces's disease of grapevine and other emergent diseases. Plant Dis 82:1056–1066
Hulme PE (2009) Trade, transport and trouble: managing invasive species pathway in the era of globalization. J Appl Ecol 46:10–18
IPPC (2015a) Glossary of phytosanitary terms. International Standards for Phytosanitary measures. ISPM 5, International Plant Protection Convention. FAO, Rome, 34 pages
IPPC (2015b) Guidelines for surveillance. ISPM 6, International Plant Protection Convention. FAO, Rome, 9 pages
IPPC (2015c) Determination of pest status in an area. ISPM 8, International Plant Protection Convention. FAO, Rome, 14 pages
IPPC (2015d) Guidelines for pest eradication programmes. ISPM 9, International Plant Protection Convention. FAO, Rome, 11 pages
Kaplan S, Garrick BJ (1981) On the quantitative definition of risk. Risk Anal 1:11–27
Koseoglu AT, Tokmak S, Mamol MT (1996) Relationships between the incidence of fire blight and nutritional status of pear trees. J Plant Nutr 19:51–61
Liebhold AM, Brockerhoff EG, Garrett LJ, Parke JL, Britton KO (2012) Live plant imports: the major pathway for forest insect and pathogen invasion of the US. Front Ecol Environ 10(3):135–143
Main CE, Keever T, Holmes G, Davis J (2001) Forecasting long-range transport of downy mildew spores and plant disease epidemics. APSnet Feature, April–May 2001. Online publication. http://apsnet.org/online/feature/forecast/. Accessed 7 Jan 2008
Marshall D, Work TT, Cavey JF (2003) Invasion pathways of Karnal bunt of wheat into the United States. Plant Dis 87:999–1003
Mumford JD (2013) Biosecurity management practices: determining and delivering a response. In: Dobson A, Barker K, Taylor S (eds) Biosecurity: the socio-politics of invasive species and infectious diseases. Earthscan, London, pp 105–119. ISBN 978-0-415-53477-2
Mumford JD, Holt JD, Leach AW, Quinlan MM (2013) National risk registers for plant health: lists, priorities and performance. Acts Phytopathologica Sin 43(Supplement):81–82
Myerson LA, Reaser JK (2002) Biosecurity: moving toward a comprehensive approach. Bioscience 52:593–600
Pujadas MJ (2001) Land-use and socio-economic correlates of plant invasions in European and North African countries. Biol Conserv 100:397–401

Roberts MJ, Schimmelpfennig D, Ashley E, Livingston M (2006) The value of plant disease early-warning systems: a case study of USDA's soybean rust coordinated framework. Online. Economic Research Report No. 18. USDA-ERS, Washington, DC

Schaad NW, Shaw JJ, Vidaver A, Leach J, Erlick BJ (1999) Crop biosecurity. APSnet feature, September–October 1999. Online publication. http://apsnet.org/online/feature/BioSecurity/. Accessed 7 Jan 2008

Seebens H, Essl F, Dawson W, Fuentes N, Moser D, Pergl J, Pyšek P, van Kleunen M, Weber E, Winter M, Blasius B (2015) Global trade will accelerate plant invasions in emerging economies under climate change. Glob Change Biol 21:4128–4140. doi:10.1111/gcb.13021

Stack R (2000) Return of an old problem: fusarium head blight of small grains. APS online. http://www.apsnet.org/online/feature/Fusarium head blight/top.html

Stack JP, Cardwell K, Hammerschmidt R, Byrne J, Loria R, Snover-Clift K, Baldwin W, Wisler G, Beck H, Bostock R, Thomas C, Luke E (2006) The national plant diagnostic network. Plant Dis 90:128–136

Strange RN, Scott PR (2005) Plant disease: a threat to global food Security. Annu Rev Phytopathol 43:83–116

Stukenbrock EH, McDonald BA (2008) The origins of plant pathogens in agro-ecosystems. Annu Rev Phytopathol 46:75–100

Tkacz BM (2002) Pest risks associated with importing wood to the United States. Can J Plant Pathol 24:111–116

Tsror (Lahkim) L, Aharon M, Erlich O (1999) A survey of bacterial and fungal seedborne diseases in imported and domestic potato seed tubers. Phytoparasitica 27:215–226

USDA (2016) Agricultural select agent services. United States Department of Agriculture, Animal and Plant Health Inspection Service. https://www.aphis.usda.gov/aphis/ourfocus/animalhealth/animal-and-animal-product-import-information/sa_ag_select_agent

Venette RC, Kriticos DJ, Magarey RD, Koch FH, Baker RHA, Worner SP, Gomez NN, McKenney RD, Dobesberger EJ, Yemshanov D, De Barr DJ, Hutchison OD, Fowler G, Kalaris TM, Pedlar J (2010) Pest risk maps for invasive alien species: a roadmap for improvement. Bioscience 60:349–362

Westphal MI, Browne M, MacKinnon K, Noble I (2008) The link between international trade and the global distribution of invasive alien species. Biol Invasions 10:391–398

Wheelis M, Casagrande R, Madden LV (2002) Biological attack on agriculture: low tech high impact bioterrorism. Bioscience 52:569–576

Whitby SM (2001) The potential use of plant pathogens against crops. Microbes Infect 3:73–80

Wolken WAM, Tramper J, Van der Werf M (2003) What spores can do for us. Trends Biotechnol 21:338–344

Yang XB, Dowler WM, Royer MH (1991) Assessing the risk and potential impact for an exotic plant disease. Plant Dis 75:976–982

Yemshanov D, Koch FA, Ducey MJ, Siltanen M, Wilson K, Koehler K (2013) Exploring critical uncertainties in pathway assessments of human-assisted introductions of alien forest species in Canada. J Environ Manag 129:173–182

Chapter 6
Integrating Crop Bioterrorism Hazards into Pest Risk Assessment Tools

John D. Mumford, Adrian W. Leach, Johnson Holt, Frédéric Suffert, Ivan Sache, Benedicte Moignot, and R. Alexander Hamilton

Abstract Risks from intentional releases of organisms to agriculture, the food chain or the environment must be assessed to ensure proportionate planning, just as accidental releases from trade or natural spread must be predicted so that management can be organised. Pest risk assessment methods are well established for trade related introductions and it is efficient to build on these and adapt available risk assessment components from agricultural and environmental assessment tools. Some additional risk considerations, particularly related to the motivation, capacity and intended impact of a perpetrator should be included, and some key elements of trade related assessments, such as the volume of trade, may be irrelevant for intentional targeted releases. Risk levels from the various causes and impacts should be comparable to allow authorities to direct responses appropriately. Preventative actions, for both intentional and unintentional introductions, are particularly impor-

J.D. Mumford (✉) • A.W. Leach • J. Holt
Centre for Environmental Policy, Imperial College London,
Silwood Park Campus, Buckhurst Road, Ascot, Berkshire SL5 7PY, UK
e-mail: j.mumford@imperial.ac.uk; a.w.leach@imperial.ac.uk; j.holt@imperial.ac.uk

F. Suffert
UMR BIOGER, INRA, AgroParisTech, Université Paris-Saclay,
78850 Thiverval-Grignon, France

Cropbiosecurity and Agroterrorism Watch, Paris, France
e-mail: frederic.suffert@inra.fr; cropsafexpertise@yahoo.fr

I. Sache
INRA, UMR1290 BIOGER, Campus INRA-AgroParisTech,
F-78850 Thiverval-Grignon, France
e-mail: ivan.sache@versailles.inra.fr

B. Moignot
Anses, Laboratoire de la santé des végétaux, Unité Evaluation des Risques Biologiques,
7 rue Jean Dixméras, 49045 Angers Cédex, France
e-mail: benedictemoignot@gmail.com

R.A. Hamilton
United Nations Interregional Crime and Justice Research Institute (UNICRI),
Viale Maestri del Lavoro, 10, 10127 Turin, Italy
e-mail: hamilton@unicri.it

M.L. Gullino et al. (eds.), *Practical Tools for Plant and Food Biosecurity*, Plant Pathology in the 21st Century 8, DOI 10.1007/978-3-319-46897-6_6

tant. For intentional release this puts emphasis on motivation, capacity and sources. A scenario based approach to assessing intentional release risks is taken to develop a pest risk assessment tool that can cover the range of levels of potential activity. A risk assessment framework is illustrated and a range of example scenarios is described.

Keywords Risk assessment • Intentional release • Plant health • Agroterrorism • Bioterrorism • Biocrime • Biowarfare • Pest risk analysis • Risk model

6.1 Introduction

Traditionally, crop biosecurity efforts have focused on preventing and responding to the natural or unintentional introduction, establishment and spread of pests or pathogens. Government agencies and industries take steps to limit these accidental introductions through quality standards in trade, official rules on risk mitigation measures, public and private surveillance for new organisms, and control planning and capacity. This approach to biosecurity has driven standardised approaches to Pest Risk Analysis (PRA), which have been developed to enable risk managers to identify, assess, manage and communicate risks of this kind (IPPC 2004, 2007; EPPO 2011). Some European countries, such as the United Kingdom and France, have developed extensive national catalogues of conventional plant health risk (Baker et al. 2014; Defra 2015) and methodology for prioritizing plant pests (Moignot and Reynaud 2013), in order to comply with the requirements of the new EU plant health regulation (now agreed for implementation in 2019). However, in recent years there has been growing concern about the possible deliberate misuse of biological agents against agriculture and the food supply with a view to causing economic losses, generating fear and/or undermining social stability (European Commission 2007).

Referred to broadly as agroterrorism,[1] this possibility has, in turn, motivated calls at the national and international level to ensure that public and private responses to possible threats from harmful organisms include both unintended and intentional releases. This requires a reassessment and revision of standard approaches to PRA to explicitly account for the motives and capabilities of potential attackers who might choose to deliberately misuse biological agents in pursuit of particular political or ideological goals. As a consequence, security becomes yet another consideration, in addition to biosecurity, that should be incorporated within a Pest Risk

[1] The term 'agroterrorism' is commonly used to refer to the 'deliberate misuse of biological agents against agriculture by non-state actors' (that is, a subset of 'bioterrorism'). However, in this chapter, we also include 'biocrime' and 'biowarfare' under this general definition. Our rationale is that each of these 'agro-risks' represents a mode of 'deliberate misuse', distinct from traditional views of risk in agriculture, which focus on natural or unintentional outbreaks. Each of these risks possess some specific characteristics, so 'bioterrorism', 'biocrime', and 'biowarfare' are defined in legal terms (see Chap. 7) and we consider a range of deliberate misuse scenarios in our analysis.

Analysis framework. Agroterrorism and other ecologically based security risks should be assessed and managed within a common framework along with more conventional pest risks to ensure that proportionate responses are taken across the full range of risks.

This common framework should be consistent with conventional assessments of unintentional risks from pests and pathogens and should reflect technical values for factors that have been identified as components of risk, as well as an indication of the uncertainty for each component. The outcome of such risk assessments is a distribution of the likelihood and consequences of the threat.

6.2 Agent-Pathway-Receptor Framework

Assessing and managing risks posed by agroterrorism requires thinking differently about potential threats to agriculture and the food supply. Traditional approaches to crop biosecurity have focused on the natural or unintentional introduction of pests or pathogens. Agroterrorism, although it poses similar ecological threats from potential harmful species (European Commission 2007), introduces the further dimension of a rational actor who chooses (to the extent possible) the conditions whereby risks of this kind are generated (Mair and Mair 2003). For example, the choice of biological agent, the specific crops that are targeted, and the location, scale and timing of the potential outbreak are all outcomes that can be determined by the perpetrator of an attack. In this light, biosecurity efforts must not only consider the biological characteristics of harmful invasive species, but also the motives and capabilities of individuals, non-governmental groups or state-sponsored organizations that may attempt to exploit vulnerabilities in agricultural, environmental and social systems with a view to achieving particular political, economic or personal goals. The 'biological' and 'human' dimensions are characteristic of such an 'hydrid threat' (Barbier 2008), is also discussed in Chap. 2.

With this context in mind, in this section we consider how the threat of agroterrorism complicates existing approaches to crop biosecurity. While the threat of agroterrorism shares aspects in common with naturally occurring and unintentional pest and disease outbreaks, in as much as agroterrorism can be viewed as an alternative source of harmful invasive species, the rational actor in the risk calculation presents new challenges to risk assessment and risk management. In particular, risk analysis should consider the motive and capabilities of perpetrators and the vulnerability of receptor targets in more detail than would occur in conventional PRA. In Fig. 6.1, we illustrate the components that make up this expanded range of biosecurity risks. Therefore, although existing PRA standards can be used as a basis for analysing the biological and environmental aspects of agroterrorism risks, they must nonetheless be adapted to reflect the novel aspects associated with the threats from intentional misuse of biological agents against agriculture and the food supply. Closer attention to the novel aspects of agroterrorism can provide a clearer understanding of the risk assessment and risk management strategies needed to account for the possibility of deliberate misuse, and can help to overcome vulnerabilities in existing biosecurity control measures.

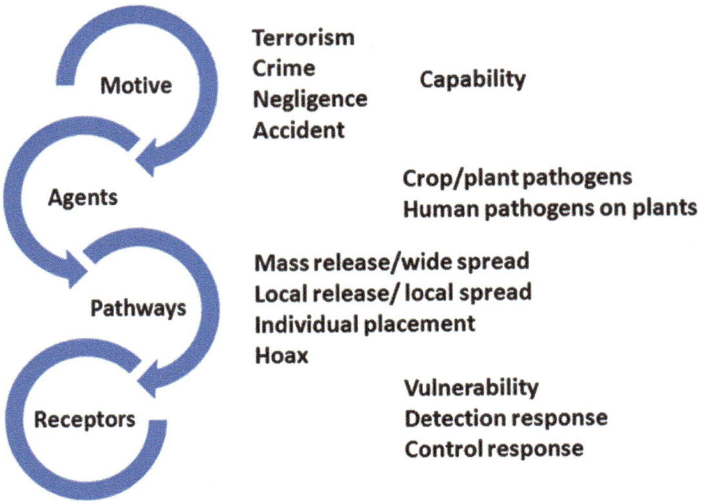

Fig. 6.1 A Motive-Agent-Pathway-Receptor approach to biosecurity risks

6.3 Risk Paradigms for Crop Biosecurity

Pest Risk Analysis attempts to assess the probability of introduction, establishment and spread of a particular pest or pathogen in a specified PRA area and the resulting consequences, usually measured in terms of direct and indirect economic costs, of a potential outbreak (IPPC 2007). In this context, biosecurity risks are represented as the outcome of a unidirectional process that links potential agents (insects, fungi, bacteria or viruses) with vulnerable receptors (crops, forests or broader social and economic systems). Illustrated in Fig. 6.2, the 'Agent-Pathway-Receptor' (A-P-R) model is central to PRA outlined in the international standard by the International Plant Protection Convention (IPPC 2004). Stakeholders may have different priorities in relation to the implementation of biosecurity controls. For example, the IPPC is concerned primarily with facilitating safe trade, recommending export restrictions as a last resort, whereas the Convention on Biological Diversity (CBD) (1992) is concerned primarily with protecting vulnerable ecosystems, habitats and species, advocating a precautionary approach to biosecurity. However, the overriding view of biosecurity risks remains much the same. Biosecurity risks are perceived to arise when vulnerable receptors are brought into contact with harmful biological agents by way of natural or unintentional processes.

This risk paradigm, however, does not fully account for the case of agroterrorism. Agroterrorism, a deliberate act of political, economic or personally motivated violence, is determined by the choices of rational actors (Mair and Mair 2003) or 'intelligent adversaries' (National Research Council 2008). These perpetrators attempt to identify and bypass biosecurity controls that would limit the outcome of the A-P-R process. The probabilities that apply to inspections or quality standards in a conventional pest risk assessment would no longer be valid in an agroterrorism

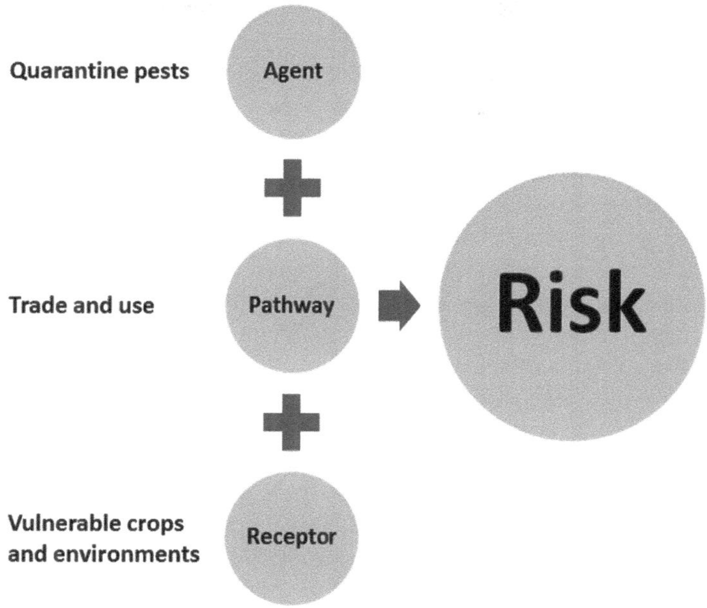

Fig. 6.2 The conventional Agent-Pathway-Receptor model of biosecurity

risk assessment. Perpetrators may also be opportunistic, choosing targets based on perceived vulnerabilities, selecting the agents, pathways and receptors that are best suited to achieving their particular goals, within the limits of their capabilities (Mair and Mair 2003; National Research Council 2008; Radosavljevic and Belojevic 2009). Therefore, while a biological agent, once it has been introduced to a particular PRA area, will behave in the same manner regardless of whether it has been unintentionally or deliberately released, the steps that lead up to the point of introduction depend upon several considerations that are beyond the scope of the standard risk paradigm associated with natural or unintentional outbreaks.

Recent work by agricultural scientists and security experts has helped demonstrate how traditional approaches to PRA can be revised to account for the novel aspects of agroterrorism (European Commission 2007; Latxague et al. 2007; Radosavljevic and Belojevic 2009; Suffert et al. 2009; Ancona et al. 2010). For example, according to Latxague et al. (2007), standard PRA schemes should be amended to account for at least five further variables or criteria, including: (1) the ease of use of the pathogen, (2) the epidemic potential of the pathogen, (3) the importance of the target crop, (4) potential obstacles to swift and effective response, and (5) potential regional or global consequences of a planned attack. These considerations illustrate that in adapting a PRA to the problem of deliberate misuse, risk managers must take into account not only the biological characteristics of a harmful invasive species, but also the *motive* and *capabilities* of perpetrators (Fig. 6.1).

In terms of motive, for example, if the primary objective of a terrorist group is to destabilise a national economy, the group will likely target a key agricultural commodity with a view to generating trade restrictions on valuable export markets. Alternatively, if the primary objective is to generate casualties, the group will attempt to release a pathogen that is capable of generating illness in humans.

In relation to capability, a terrorist group would also take into account the feasibility of an attack given their perceived scientific and technical skills and facilities. Although it may be desirable from the point of view of a terrorist group to undermine a national economy by causing damage to a key agricultural commodity, possibly including large areas of cultivated land, it may not be within the scope of their capabilities to produce and disseminate a biological weapon that is capable of generating large-scale losses of this kind. In practice, there exist significant barriers to acquiring a sui biological agent, culturing sufficient quantities of infectious material, and developing an effective delivery system that is capable of disseminating this material over a large area (Office of Technology Assessment 1993). Based on the principle that the non-proliferation of chemical biological, radiological and nuclear (CBRN) weapons aims to protect the human population worldwide, an international list of biological agents was established for export control, considering that they could be misused as bioweapon against crops (see Chap. 2). The impact of such export controls in the dissemination of these problematic species should not be dismissed, but it cannot completely prevent a malicious, competent perpetrator to procure them. During this multi-step process, referred to as a proliferation pathway, there exist multiple avenues for failure, each of which would undermine the realisation of a particular attack. Consequently, in the absence of significant financial and technical resources, terrorist groups are most likely to pursue relatively simple proliferation pathways, producing a more localised impact on agriculture, the food supply or the environment.

A further dimension of agroterrorism that complicates standard approaches to PRA is the potential for perpetrators to identify and exploit *vulnerabilities* in agricultural, environmental, and social systems. The concept of vulnerability is used to express that some receptors (for example, a particular crop) are more prone to ecological and/or economic damage than others (Mumford et al. 2011). While this observation also applies to natural or unintentional outbreaks, it is especially relevant in relation to agroterrorism. This is because perpetrators can choose (within the limits of their capabilities and motives) specific agents, pathways and receptors, and are thus able to adjust their tactics in relation to perceived vulnerabilities. This means, for example, that a terrorist group might target a crop that is relatively isolated, lacking adequate biosecurity controls, or target a species of tree that is especially valuable or of symbolic significance to a country, or introduce a pathogen for which vaccines are unavailable. Although one cannot account for an infinite range of contingencies, identifying vulnerabilities of this kind is an essential component of PRA adapted to the case of agroterrorism. It also illustrates a significant difference in security and biosecurity risks. The focal points of trade risks are the probabilistic relationship between the quality of the imported material and the efficiency of the controls and inspections on the delivery pathway. For security risks, there is much more emphasis on the possible relationships between perpetrator motives and receptor vulnerabilities, while the delivery aspects are relatively certain, since they are in the hands of the perpetrators unless they are thwarted.

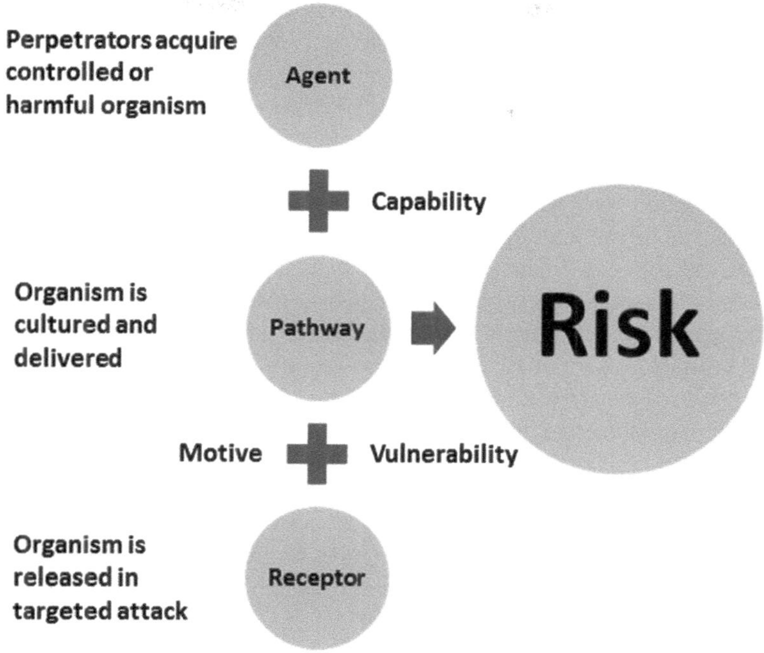

Fig. 6.3 The Agent-Pathway-Receptor model adapted to the case of deliberate threat

In light of these considerations, it is apparent that standard approaches to PRA cannot be directly applied to the case of deliberate misuse. Although these approaches can be used to understand the mechanisms of disease outbreaks, informing the choice of phytosanitary measures, they do not account for the role played by rational actors in influencing the outcome of the A-P-R process. Therefore, an alternative risk paradigm is needed, one that accounts for the motives and capabilities of potential attackers, as well as for their capacity to identify and to exploit vulnerabilities. Conceptually, this new paradigm can once again be represented as a unidirectional process linking potential agents with vulnerable receptors. However, in this instance, as illustrated in Fig. 6.3, this process is driven by rational choices and the motives and capabilities of potential attackers who pursue particular goals in light of perceived vulnerabilities in agricultural, environmental, and social systems.

6.4 Real and Perceived Risk: Understanding the 'Impact' of Agroterrorism

A distinctive feature of agroterrorism is that even an 'unsuccessful' attack or a 'successful' hoax (if made public) could generate fear, reduce consumer confidence, and possibly undermine social stability (Chalk 2001; Turvey et al. 2003; Cupp et al.

2004; Byrne 2006; Eggers et al. 2011). Fischhoff (2011) has observed that terrorist attacks of this kind not only have the potential to "inflict direct damage to the people they injure, to the economies they disrupt, and to the leaders they discredit", but also to "inflict indirect damage by instilling fear over who will be next, by undermining investors' confidence in future economic activity, and by eroding faith in governments that cannot protect their people". Therefore, in addition to the direct and indirect economic costs associated with diminished crop yields, compensation to farmers, export restrictions, and so on, an agroterrorism attack has the potential to generate costs to society that are disproportionate to the actual magnitude of a disease outbreak. In other words, the anticipated 'impact' of agroterrorism needs to take into account the manner in which individuals and communities perceive and respond to attacks on agriculture and the food supply.

Work on risk perception also illustrates how the public might respond to an (anticipated) agroterrorism attack. Slovic (1987) suggests that hazards tend to be assessed on the basis of intuitive judgments that do not necessarily correlate with expert assessments of expected annual fatalities, economic costs, and so on. In particular, hazards that are perceived to be uncontrollable, catastrophic, new or highly uncertain (for example, nuclear accidents) tend to generate considerable concern, even though the hazard itself may be infrequent and of limited magnitude. Consequently, Slovic suggests, risk assessments must take into consideration not only anticipated losses to life, to property, and so on, but also the particular characteristics of an event, which, for many, serve as indicators or signals for the event itself.

Like the threat of nuclear accidents, agroterrorism attacks (both real and foreseen) have the potential to elicit considerable societal alarm. Indeed, research has shown that threats to the food supply, especially when linked with terrorism, strongly resonate with perceptions of catastrophic harm, evoking heightened anxiety comparable to perceptions of natural disasters (Eggers et al. 2011). Moreover, it is necessary to differentiate between different types of attack. For example, agroterrorism attacks employing zoonotic disease agents are likely to be of greater concern than attacks that affect only plant health, due to fears about human contagion and disease (Chalk 2001). An attack (or the threat of an attack) not only represents a source of societal concern, it would also likely have significant repercussions for the agriculture sector and for the food industry, changing consumer buying behaviour, as well as diminishing public confidence in government's capacity to protect the food supply (Turvey et al. 2003).

In this light, if PRA is to be successfully adapted to the case of agroterrorism, expert assessments must take seriously the manner in which individuals and communities understand threats to agriculture and the food supply, and how perceived risks influence individual and collective behaviour. Whereas PRA has traditionally focused on calculating the economic costs (resulting from diminished crop yields, export restrictions, and so on) of natural or unintentional disease outbreaks based on scientific knowledge and statistics (Dahlstrom et al. 2011), PRA adapted to the case of agroterrorism must take into account both the scientific facts about pest or disease transmission and the social ramifications of potential attacks on agriculture and the food supply. Although these social values may be more challenging to assess by strict adherence to probabilistic models, they must nonetheless be accounted for in the risk calculation.

Finally, a more nuanced understanding of how individuals and communities perceive and respond to threats to agriculture and the food supply is also essential for developing communication strategies that limit public anxiety and concern during an agroterrorism event (including a publicised hoax). Terrorist groups may be well aware of the psychological impact of their actions and before, during and after an attack may seek to generate fear, create confusion, and foment dissent in an effort to "advance their cause even when their operations fail" (Fischhoff 2011). In turn, the success of communication strategies will depend upon the capacity of government authorities to provide timely information on the nature of agroterrorist threats, helping to maintain (or restore) public trust and enable rapid recovery.

6.5 Agroteterrorism Scenarios and Risk Analysis

Having highlighted how agroterrorism challenges the standard approaches to PRA and traditional understandings of biosecurity in agriculture, in this section we describe the development of a risk analysis model that can explicitly account for the problem of deliberate misuse or intentional introductions in agriculture, the food chain or the environment, while retaining features of conventional trade related pest risk assessment.

6.5.1 The Role of Scenarios and Why They Were Developed

Agroterrorism is considered in the broad sense to include anti-crop bioterrorism and the use of bioweapons against crops through the "intentional use (as well as the threat or simulation of use) of plant pathogens (fungi, bacteria, viruses) by any human individual or group in order to cause direct damage to crops or forests, or to indirectly affect the agricultural sector" (Latxague et al. 2007). The only modification of this definition is that the term 'plant pathogens' may be extended to 'plant pests' (fungi, viruses, bacteria, nematodes, insects, and so on) in the framework described in the PLANTFOODSEC project.

In the earlier CROPBIOTERROR project, a three-step methodology was developed for risk assessment, involving: (1) building a list of 50 candidate pathogens, (2) a scenario-based investigation of potential deliberate misuse events, and (3) the design of a risk evaluation scheme (RES), derived from a standard Pest Risk Analysis (PRA) scheme, originally used to decide whether an organism should be placed on plant health quarantine lists. In the PLANTFOODSEC project, we developed a risk assessment tool consisting of a foresight exercise (assigning 51 key pests and pathogens to different motive-receptor scenarios and comparing these based on

Table 6.1 Nodes in the R-bNM for assessing scenario risks

Node number	Node description
01	Ease of sourcing the pest/pathogen
02	Ease of pest/pathogen culture
03	Ease of release and inoculation (of the pest/pathogen)
04	Pest/pathogen transfer/infectivity in the environment
05	Persistence (of pest/pathogen) in the environment
06	Rapidity or extent of spread (of pest/pathogen) in the environment
07	Host (plant) importance
08	Pest/pathogen severity to host plant
09	Introduction would damage trade
10	Pathogen toxicity to consumers
11	Frequency of similar malevolent acts
12	Likelihood of intent
13	Likelihood that normal security effort would not prevent the release
14	Negative public reaction
15	Negative reaction of primary stakeholders
16	Assessor rating of overall risk and uncertainty (subjective expert opinion provided independently of the model result)

specific criteria) and an analytical assessment (application of an adapted RES to the key agent-pathway options and qualitative analysis resulting in scenario specific risk profiles). Building these scenarios was an important part of the methodology developed to assess agroterrorism risks in Europe, because we cannot experiment with dangerous pests and because there are very limited historical precedents (beyond the activities of former biological weapons programmes) for agroterrorism, which could otherwise have been analysed. Suffert et al. (2009) give a list of biological weapons threats to agriculture in the twentieth century, most relating to the (alleged) activities of former biological weapons programmes. The historical review of anti-crop bioweapons given in Chap. 2 is the starting point of the characterization and context of the threat in Europe (Table 6.1).

Nine different, non-overlapping scenarios of agroterrorism attacks (Latxague et al. 2007) were developed. These general scenarios were demonstrated by selecting a pest or pathogen taken from the list of candidate organisms, together with a motive-pathway related to the nature of the acts and the agent-receptor relationship that results in their potential consequences. For each scenario, INRA scientists experienced in conventional PRA took the role of an imagined perpetrator: they defined a tangible target and selected the most appropriate pest or pathogen from the list of candidate organisms. Based on this, the risk analysts wrote a brief scenario describing the hypothetical agroterrorist attack and its expected consequences.

The three sections of the scenarios: 'Synopsis' (mode of operation and expected consequences), 'Justification' (geopolitical context and perpetrator motivations), and 'Feasibility' (perpetrator capability to succeed and technical constraints) were substantiated with information extracted from relevant documents or materials that cannot be found easily through conventional systems of publication, bibliographic review, or subscription.

The nine key pathogens used in the demonstration scenarios were *Tilletia indica* (the wheat Karnal bunt fungus), *Phytophthora infestans* (the potato late blight Oomycete) and *Pleospora papaveracea* (a potential opium poppy mycoherbicidal fungus) as biowarfare agents, *Fusarium graminearum* (a grain-infecting, toxinogenic fungus), *Mycosphaerella populorum* (the poplar stem canker fungus) and *Ceratocystis fagacearum* (the oak wilt fungus) as bioterrorism agents, and *Xylella fastidiosa* (the grapevine Pierce's disease bacterium), *Puccinia triticina* (the wheat brown rust fungus) and *Phakopsora pachyrhizi* (the soybean rust fungus) as biocrime agents.

The nine scenarios involving these pathogens were then ranked for specific features and salient components consistently highlighted in the literature on agroterrorism: diversity of impacts (on production, trade, society), motive for a perpetrator to claim responsibility for the attack, availability of technical capabilities (origin of the scientific information and inoculum), possible delivery by air (considered as a classical mode of operation), and potential countermeasures (early detection, and availability of control measures). This was an initial step in choosing appropriate criteria for a risk assessment tool.

6.5.2 The Use of Scenarios in PLANTFOODSEC

In the PLANTFOODSEC project, we used the categorization of scenarios (category Biowarfare BW, Bioterrorism BT, and Biocrime BC, each with three subcases) to describe a sufficient set of cases to cover the wide range of potential threats (described in Table 2 in Chap. 2). For each of the nine subcategories we identified at least three scenarios (only one was considered in CROPBIOTERROR), including organisms encompassing all main pest groups (fungi, viruses, bacteria, nematodes, insects) to ensure the scenarios covered a broad range of threats. Only pathogens (fungi, viruses, bacteria) were considered in the earlier CROPBIOTERROR project. These scenarios (27 in total) were briefly described (with 3–4 lines) rather than the more detailed 'Synopsis', 'Justification' and 'Feasibility' descriptions used in the scenario reports developed in CROPBIOTERROR. These scenarios were used in the subsequent development of a model to serve as a consistent risk assessment framework.

6.5.3 Building and Validating a Novel Risk Comparison Tool Using Scenarios

Building the risk assessment tool involved three sets of experience. UNICRI and INRA experience was used to identify a range of intentional release motives. Imperial College London partners developed a Rule-based Network Model (R-bNM) using GENIE software[2] to assess the relative risk posed by a given scenario (not by a given pest, an important distinction, as it included more than just the agent organism itself). INRA scientists contributed to the development and testing of the model. The tool is described in detail in the next section.

The model has subsequently been tested on 98 Motive-Agent-Pathway-Receptor scenarios and an evaluation has been conducted based on comparison between the assessment of a scenario by an expert (rating + uncertainty) and the output of the model. The expertise required to parameterise the model for the scenarios was already present in the French institutions, INRA and ANSES.

Some example illustrative scenarios include:

- **BW1a** – *Ralstonia solanacearum* in potato (*Solanum tuberosum*) for localised release of *R. solanacearum* in water (contamination of river water used for irrigation) in Northwest France in seed potatoes (plant for planting).
- **BW2a** – *Puccinia graminis* strain Ug99 in wheat (*Triticum aestivum*) with mass production of inoculum and release of spores in Italy.
- **BT2b** – *Microcyclus ulei* in rubber (*Hevea brasiliensis*) for contamination of Indonesian rubber plantations.
- **BT2c** – *Bursaphelenchus xylophilus* in maritime pine (*Pinus pinaster*) for mass release of pine nematode (artificially multiplied in contaminated wood samples collected in Portugal) by air in pine plantations of Southwest France.
- **BT3a** – *Ceratocystis fagacearum* in oak (*Quercus robur*) in simultaneous terrorist attacks against English oaks in a forest and an urban park.
- **BC2c** – *Leptosphaeria maculans* in oilseed rape (*Brassica napus*) for deliberate release of a hyper-resistant and aggressive strain in experimental fields performed by a plant pathologist in response to their recent dismissal.
- **BC3a** – *Phakopsora pachyrhizi* in soybean (*Glycine max*) through introduction of inoculum in Southwest France by an agrochemical company, to make farmers dependent on both tolerant soybean cultivars and a fungicide.
- **BC3c** – *Puccinia striiformis* in wheat (*Triticum aestivum*) in simultaneous deliberate introductions of a new virulent *P. striiformis* strain in several European countries by a breeder company to modify the European panel of cultivated cultivars and promote the sale of one of its resistant cultivars.

Based on a preliminary characterization and contextualization of the threat resulting from plant pathogens misuse as anti-crop bioweapons in Europe (Chap. 2), we determined that the most problematic agroterrorism scenarios for Europe are BW1 (state-sponsored threat to export trade), BW2 (state-sponsored threat to domestic production), BT1 (terrorism threat to domestic production and health), and BC1 (attack by activists or other groups against local production).

[2] See http://genie.sis.pitt.edu/

6.6 A Rule-Based Network Model to Integrate the Components of Risk Associated with Pathogens or Arthropods Used as Crop Bioterrorism Agents

In this section we provide a description of the process outlined above, in which the development of a model and its parameterisation is explained for a set of selected scenarios. A method for integrating PRA components (Holt et al. 2012), developed in the PRATIQUE project and adopted by the European and Mediterranean Plant Protection Organization (EPPO), has been adapted for the paradigms of biowarfare, bioterrorism and biocrime in PLANTFOODSEC. The system uses a combination of logical rules in a Bayesian–type network to simulate how experts express and integrate risks (Holt et al. 2013).

Aspects of pest or pathogen biology, local climate and ecology, ease of introduction and culture are combined with the pest/pathogen potential impact on crops, human health and public alarm to give an overall rating of impact. The use of a Bayesian approach allows expert uncertainty of various inputs to be expressed and explicitly incorporated in the overall summary. It also allows for a variety of rules to be used in combining component values leading to intermediate and final conclusions. Some examples of these combinations are given later in this section. A template for collecting expert ratings was also developed in which ratings of 15 input questions (and associated uncertainty) have been collected for a wide range of scenarios.

6.6.1 *Components of Risk*

There are 15 components of risk distinguished in the model, and an overall subjective assessor rating by a pest risk analyst can be used as a comparator in calibrating the system (Table 6.1).

These components are rated nodes in which the user is required to select or rate the most appropriate choice from a list that describes that component (from very low to very high likelihood or impact, or very difficult to very easy). An expression of uncertainty is also incorporated such that the assessors should first give a rating that is judged most appropriate and then describe their uncertainty (low, moderate or high) that this chosen rating is in fact the correct one. Inputs are elicited from experts using a specifically designed template (Table 6.2). A frequency distribution (Holt et al. 2013) is then generated which describes that component, low uncertainty implies a narrow distribution and high uncertainty, a wider one. For both the scoring for the rating and the uncertainty the expert is expected to provide documentation or comments that allow independent review of the scores.

Table 6.2 Example template of data elicited from experts about a specific scenario

BT23	Scenario :	*Bursaphelenchus xylophilus/Pinus pinaster*/Landes (South West France)			
	Release procedure:	Release of *Monochamus galloprovincialis*, vector of the pine nematode, directly into the forest			
	Assessor:	XXXX			
	Date:	XXXX			
	Is the scenario relevant?	YES			
	Scenario assumption:				
		Expert judgment		Comment	Comment
#	CRITERIA	Rating	Uncertain : Rating		Uncertainty
1	Ease of sourcing pest or pathogen (P)	Moderate	Medium	The easiest way to source *Bursaphelenchus xylophilus* would be to collect it directly from areas within its current distribution. It is present in North America, Asia and Europe (Portugal and Spain).	Medium uncertainty on their scientific knowledge
2	Ease of pathogen culture (P)	Difficult	Medium	Culture of the nematodes from infected wood is possible. This requires females of the vector to oviposit directly into infested wood. Emerging adults would be effective vectors of the nematode and be suitable for releasing in unaffected areas. Adults transmit the nematode in the wood when feeding.	Medium uncertainty on their scientific knowledge
3	Ease of release and inoculation (P)	Easy	Medium	Release of adults of the vector *Monochamus galloprovinciallis* infected by the nematode in the forests of Landes is easy. The access to the forests is easy as there is little access control. The limiting factor is to release the insect at the most efficient time for contamination of the trees, that is in the adult fligh period from May to October. As adults of *Monochamus* are more attracted by weakened or dead trees of most confiderous species, it should more efficient to release the insects close to such trees.	Medium uncertainty on their scientific knowledge

4	Pathogen infectivity in the environment (A)	High	Low	*Bursaphelenchus xylophilus* is transmitted to pines by insects of the genus *Monochamus*. Species of this genus are present in Landes and population density are sufficient for transmission of the nematode to pines.	
5	Persistence in the environment (A)	Very high	Medium	Once *Bursaphelenchus xylophilus* is transmitted in a tree by *Monochamus*, the nematode multiply in the vessels of the plants.	
6	Ability to cover area required for crop damage expected in the scenario	Low	High	In Portugal where the nematode is present it is estimated that it causes 5% wilting of pines. Crop damage caused by the nematode accumulates over time. In Japan in 1960 pine represented 30% of the wood harvest. In 2010 pines represented 4% of the wood harvest. The decrease of pine harvest is due to the nematodes. The ability of nematode to cover the Landes and induce crop damage should be low.	The assessment is based on the phytosanitary situation in Portugal. Impact measure is not very well known in Portugal. So uncertainty is high at present.
7	Host importance (R)	Very high	Low	Pine stands are mainly cultivated for timber in Landes. Pine forestry activities employs many people in the Landes region. Moreover, pinces stands were initially planted to stop sand progression in the lands and have an important environmental role.	
8	Pathogen severity to host plant (A)	High	Medium	Inoculation of pine with *B. xylophilus* leads to the development of embolisms in the xylem within hours. Within a few days (to weeks), the foliage begin to show signs of discolorations, following disturbance to xylem flow.	Medium uncertainty on the real impact of *B. xylophilus* on the pines in the conditions of the Landes

(continued)

Table 6.2 (continued)

9	Introduction would damage trade e.g. quarantine pest (A)	Very high	Low	*Bursaphelenchus xylophilus* has been a quarantine pest since its first report in Europe in the 1980s. Transport of logs is forbidden within a 20 km radius around an infested tree. If *B. xylophilus* is detected in the Landes region the economic consequences will be very important.	
10	Pathogen toxicity to host plant consumers (A)	Very low	Low		
11	Frequency of malevolent acts in similar context (M)	Very low	Medium	No reports in Suffert, 2009 of agroterrorism act based on a nematode	
12	Likelihood of intent of a malevolent act against this receptor (M)	Moderate	High	In the Landes there is only one species : *Pinus pinaster*. As a consequence, biodiversity of the species in the Landes islow.	High uncertainty of eco-warriors groups that claim pines eradication in Landes
13	Likelihood that security conditions allow malevolentact (M)	High	Medium	Pine forests in Landes are easy to access.	
14	Negative public reaction (R)	High	Medium	Forests in the Landes have a high amenity value especially for tourism. As the scenario involves the destruction of whole or part of the pine forest in the Landes, public reaction should be highly negative.	Medium uncertainty depending on the media communication
15	Negative reaction of stake holders of the agricultural sector (R)	High	Low	The detection of *Bursaphelenchus xylophilus* in Landes and the associated loss of timber yield is expected to be very worrying to stakeholders	
16	Assessor's perception of overall risk				

Criteria relate to Motive (M), Agent (A), Pathway (P) and Receptors (R) in the M-A-P-R framework. Abbreviated comments are shown in the example

6.6.2 Combining Risk Components

The components are integrated two at a time in a sequence of nodes describing more aggregated concepts. They are combined two at a time to make it easier to describe the logic of the way the components should come together to affect the final result. Figure 6.4 shows the topology of the model as a screenshot of the working tool, with some demonstration results with parameters estimated for pine wood nematode in maritime pine.

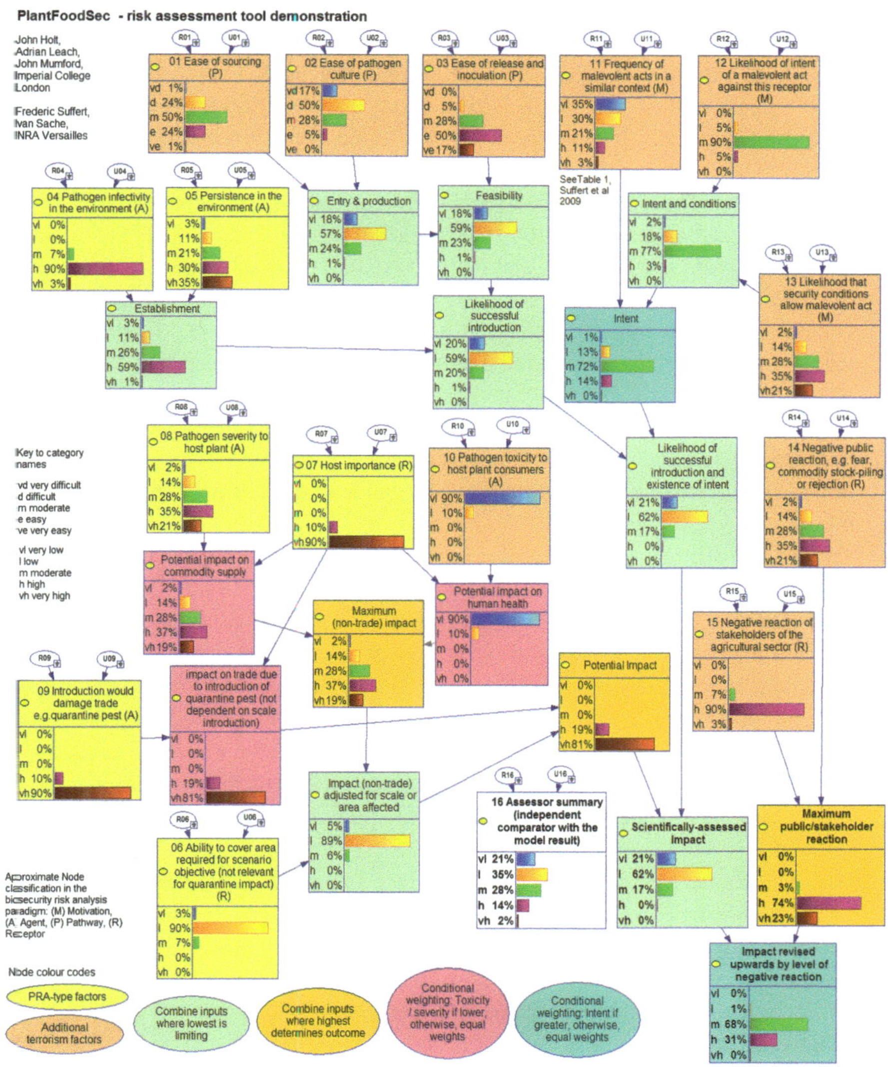

Fig. 6.4 An example screenshot of the PLANTFOODSEC R-bNM tool, integrating risk components with approximate parameterisation of the rated nodes (01) to (15) for pine wood nematode in maritime pine and an example distribution for an independent subjective assessment (16)

The combination logic is as follows:

- 'Entry and production' takes the lower of its two inputs ('01 Ease of importation', '02 Ease of pathogen culture'), the logic being that both are required for the pathogen to pose a risk, so the lower of the two is the limiting factor. The same logic applies when combining the result with '03 Ease of release and inoculation'; the result 'Feasibility' is constrained by the minimum of the three components: 01, 02 and 03.
- 'Establishment' takes the larger of its two inputs ('04 Pathogen infectivity in the environment', '05 Persistence of pathogen in the environment') so whichever is the higher of these two properties is taken as the best indicator of likelihood of establishment.
- Successful introduction requires both 'Feasibility' and 'Establishment', so the node 'Likelihood of successful introduction' takes the lower of its two inputs. 'Extent of infection' also requires spread, so again this node takes the lower of its two inputs.
- The other main branch of the network concerns impact and two aspects are considered, impact on trade and non-trade values. For non-trade '10 Pathogen toxicity to host plant consumers' (due to mycotoxins) and '08 Pathogen severity to host plant' (impacting on food supply) are considered, while it is necessary to consider '07 Host importance' in both cases. For trade values '07 Host importance' and'09 Introduction would damage trade' are also considered. The impact on health or on the commodity is determined by the toxicity/severity when the crop importance is greater and not regarded as limiting. Otherwise the impact is reduced by lower crop importance with toxicity or severity and commodity importance having equal weights and the nodes take the average of their two inputs.
- The potential impact combines the impacts on health and on the commodity to reflect the greatest in combination.
- Whether this impact actually occurs or not is calculated in the node called 'Scientifically assessed impact' which is a conditional node integrating the 'Potential Impact' and the 'Likelihood of successful introduction'. The Impact is determined by its potential when the extent of infection is greater than the potential impact and so regarded as not limiting. Otherwise the impact is reduced by the likelihood of introduction and, assuming a uniform distribution, the result lies in the interval between the ratings for potential impact and the extent of infection.
- We incorporate the impact of alarm entirely separately from the branch of the network concerned with technical issues. Two interest groups are distinguished: those directly concerned with the problem, '15 Primary stakeholder alarm', and the general public, '14 Public alarm'. These are brought together by a node which indicates whether either group is alarmed.
- Finally, alarm and impact are integrated using a conditional logic. The final impact including public and stakeholder alarm is at least as great as the calculated impact so the final impact is determined by the calculated scientific impact

if its value is greater than that of social alarm. If scientific impact is less than social alarm then the result is the average; in effect, therefore, a lower level of impact is augmented in response to public alarm.

- In addition to the ratings for the individual model components, assessors were also asked to provide an overall rating and uncertainty for the risk posed by the pathogen. This provides a basis for comparison between expert opinion and model logic.

The underlying calculation in each node determines the joint distribution of rating frequencies and according to the node rule or logic, calculates the resulting distribution seen in that node. The logic used in combining the components is both deterministic and constant. The same logic therefore applies to all cases (to aid direct comparisons between scenarios) and uncertainty is expressed in the risk components but not in the logic of their integration. The model is parameterised individually for each pathogen/host scenario and required 15 ratings (one for each rated node) and 15 expressions of uncertainty, one associated with each rating.

The underlying mechanism of the Bayesian model is consistent with expert thinking for many of the scenarios tested and for others it stimulated valuable discussion which prompted changes to the model as well as re-evaluation of inputs by the experts. The system presents a visual description of the mechanisms involved and a rational and consistent basis for the evaluation of preventative or risk mitigation strategies. A framework that is applicable to a broad set of possible scenarios has been developed and provides a benchmark to improve consistency of biosecurity assessments.

6.6.3 *Overall Classification of Generic Scenarios by Integration on a Set of Pest-Crop Pairs*

We undertook a systematic exploration of the nine general classes of scenarios by considering each case with a range of agent-receptor pairs to test the applicability of the criteria in the risk assessment and to evaluate outcomes against subjective assessments by expert assessors across all the scenarios. The goal was to assess the threat posed by each of these scenarios, irrespective of the pest and crop that is chosen. We selected 15 crop-pest pairs, from a list of 51 harmful organisms and a comprehensive list of European crops and environments of high value (Chap. 2 gives lists of pests and crops/trees). For each of these nine generic scenarios we considered plausible examples that could include each of the 15 agent and receptor combinations, although not all were relevant. There were 98 cases where some of the 15 pest-crop pairs fit into one of the nine general scenarios. Criteria included those most feasible for a perpetrator to use, and that would result in the highest impact; the 15 selected crop-pest pairs could be different for each scenario. The input data used by the PLANTFOODSEC Rule-based network model was elicited from experts for each of the 15 crop-pest pairs, in the context of each of the 98 scenarios.

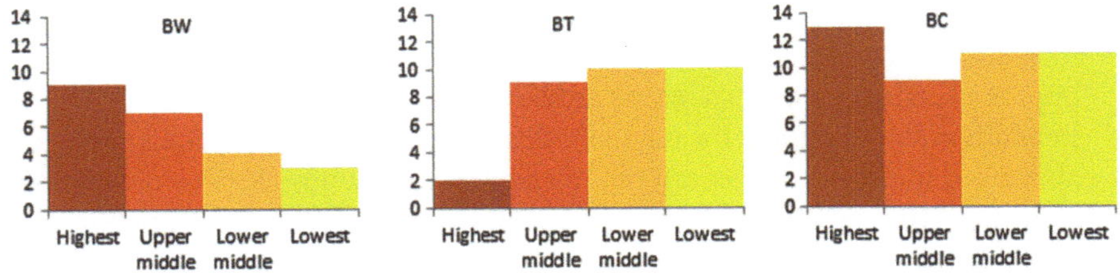

Fig. 6.5 Distribution of risks in the 98 scenarios assessed (BW is Bio-warfare; BT is Bio-terrorism; BC is Bio-crime)

It was necessary to identify relevant panels of experts for each crop-pest pairing and to ensure linguistic uncertainty is minimized through a formal elicitation process. Of the 98 cases, 23 were Biowarfare, 31 were Bioterrorism, and 44 were Biocrime (Fig. 6.5). The smaller scale of biocrime scenarios may account for the greater applicability of these cases. While these cases were not selected as being representative, the range of cases allows some general observations. Biowarfare scenarios had the highest median risk score, both in terms of scientific assessment as well as with public/stakeholder impacts included. These scenarios involve high impact with substantial resources and capability. Biocrime scenarios figure prominently in the top 10 risks, probably because their small scale increases their practicality and hence likelihood. Bioterrorism scenarios had the lowest median risk scores. The risk scores take into account the full set of 15 criteria listed above.

6.7 Conclusion

There is a need for an agroterrorism risk assessment method that is consistent with pest risk assessments for introductions of pests through conventional trade or natural spread of pest organisms. This would allow agroterrorism risks to be prioritised in a proportionate manner to similar biosecurity risks. A modified Agent-Pathway-Receptor risk model adds Motive as a consideration. Agroterrorism risks are characterised by the Motive-Receptor relationship, which is best described in a risk scenario, rather than as either an agent or pathway based analysis, which is common in conventional pest risk analysis. Various agents and pathways may fit within a scenario. A broad series of hypothetical agroterrorism scenarios has been described, built on past experience from the CROPBIOTERROR project and further elaborated in PLANTFOODSEC, in which specific 'Motive-Agent-Pathway-Receptor' sets can be analysed as examples. A rule-based network model was developed to process relevant scoring and confidence values to rate and give a preliminary demonstration of how agroterrorism scenarios could be assessed using skills, experience, methods and terms common to plant health risk assessors. This has shown that agroterrorism risks can be assessed within a comprehensive risk analysis framework that is broadly compatible with other plant health risks.

The creation of a broad list of scenarios across the spectrum of potential threat types and calculation of risk scores from elicitations of common component values from expert assessors establishes a base of risk scores. The PLANTFOODSEC project built up a base collection of 98 such scenarios relevant across participating countries. New scenarios can be added individually, and relatively quickly, as threat scenarios arise and their relative positions can be set against the background of the scenarios already established. Scenarios can be compared across the full set of cases, or within narrower threat types, as a guide to the severity of the risk and proportionality of possible responses. The performance of management actions can be estimated against relevant components in the model to evaluate the likely return on mitigation efforts, and the addition of cost functions could provide an economic evaluation of management effort. Any of these analyses require a fundamental core that uses a standardised risk assessment framework.

References

Ancona V, Appel D, Figueiredo P (2010) *Xylella fastidiosa*: a model for analyzing agricultural biosecurity. Biosecurity and Bioterrorism: Biodefense strategy, practice, and science. Mary Ann Leibert, Inc. http://online.liebertpub.com/doi/pdf/10.1089/bsp.2009.0021

Baker RHA, Anderson H, Bishop S, MacLeod A, Parkinson N, Tuffen MG (2014) The UK plant health risk register: a tool for prioritizing actions. EPPO Bulletin 44:187–194

Barbier M (2008) Expertise in the making about agroterrorism: issues and paradox of rationalizing hybrid threats. Presented at the Society for Social Studies of Science (4S) & European Association for the Study of Science and Technology (EASST), Rotterdam

Byrne R (2006) Horizon scanning rural crime – agroterrorism an emerging threat? Paper presented at the University of New England, Armidale, NSW, Australia, 30 November 2006

CBD (1992) Convention on biological diversity. United Nations Environment Program, New York, USA. 1760 UNTS 79; 31 ILM 818 (1992)

Chalk P (2001) Terrorism, infrastructure protection, and the U.S. Food and Agricultural Sector. RAND Testimony before the Subcommittee on Oversight of Government Management, Restructuring and the District of Columbia, US Senate, Washington, DC, USA

Cupp OS, Walker DEII, Hillison J (2004) Agroterrorism in the U.S.: key security challenge for the 21st century. Biosecur Bioterror: Biodefense Strateg Pract Sci 2(2):97–105

Dahlstrom A, Hewitt CL, Campbell ML (2011) A review of international, regional and national biosecurity risk assessment frameworks. Mar Policy 35:208–217

Department for Environment, Food and Rural Affairs (Defra) (2015) UK plant health risk register. Defra, London, UK. https://secure.fera.defra.gov.uk/phiw/riskRegister/

Eggers S, Verrill L, Bryant CM, Thorne SL (2011) Developing consumer-focused risk communication strategies related to food terrorism. Int J Food Saf Nutr Public Health 4:45–62

European and Mediterranean Plant Protection Organisation (EPPO) (2011) Decision-support scheme for quarantine pests. PM 5/3(5). EPPO, Paris, France. http://www.eppo.int/QUARANTINE/Pest_Risk_Analysis/PRA_intro.htm

European Commission (EC) (2007) Green paper on biopreparedness, 11 July 2007, Brussels. Report available at: http://ec.europa.eu/food/resources/gp_bio_preparedness_en.pdf

Fischhoff B (2011) Communicating about the risks of terrorism (or anything else). Am Psychol 66:520–531

Holt J, Leach AW, Knight JD, Griessinger D, MacLeod A, van der Gaag DJ, Schrader G, Mumford JD (2012) Tools for visualising and integrating pest risk assessment ratings and uncertainties. EPPO Bull 42:35–41

Holt J, Leach AW, Schrader G, Petter F, MacLeod A, van der Gaag DJ, Baker RHA, Mumford JD (2013) Eliciting and combining decision criteria using a limited palette of utility functions and uncertainty distributions: illustrated by application to Pest Risk Analysis. Risk Anal 34:4–16. doi:10.1111/risa.12089

International Plant Protection Convention (IPPC) (2004) Pest risk analysis for quarantine pests, including analysis of environmental risks and living modified organisms. International Standard for Phytosanitary Measures no. 11. FAO, Rome, Italy (later updates can be checked at www.ippc.int)

International Plant Protection Convention (IPPC) (2007) Framework for pest risk analysis. International Standard for Phytosanitary Measures no. 2. FAO, Rome, Italy (later updates can be checked at www.ippc.int)

Latxague E, Sache I, Pinon J, Andrivon D, Barbier M, Suffert F (2007) A methodology for assessing the risk posed by the deliberate and harmful use of plant pathogens in Europe. EPPO Bull 37:427–435

Mair M, Mair JS (2003) A complementary approach to bioterrorism prevention. Nonproliferation Rev Fall-Winter 2003:1–11

Moignot B, Reynaud P (2013) Developing a methodology for the prioritisation of pests in plant health. Euroreference 9:5–9

Mumford J, Baker R, Bremmer J (2011) Enhancement of pest risk analysis techniques: a manual and computerized consistent framework for assessing the vulnerability of receptor environments. PRATIQUE Deliverable number: 2.4. https://secure.fera.defra.gov.uk/pratique/publications.cfm#fldr_D2

National Research Council (NRC) (2008) Department of homeland security bioterrorism risk assessment: a call for change. National Academies Press, Washington, DC

Office of Technology Assessment (OTA) (1993) Technologies underlying weapons of mass destruction, OTA-BP-ISC-115. US Government Printing Office, Washington, DC

Radosavljevic V, Belojevic G (2009) A new model of bioterrorism risk assessment. Biosecur Bioterror: Biodefense Strateg Pract Sci 7(4):443–451

Slovic P (1987) Perception of risk. Science 236(4799):280–285

Suffert F, Latxague É, Sache I (2009) Plant pathogens as agroterrorist weapons: assessment of the threat for European agriculture and forestry. Food Secur 1:221–232

Turvey CG, Mafoua E, Schilling B, Onyango B (2003) Economics, hysteresis and agroterrorism. Principal paper presented at the Canadian Agricultural Economics Society 2003 Annual Meeting Montreal, Quebec, 27–30 July 2003

Chapter 7
Detection of Mycotoxins in Food: Applications of Rapid and Reliable Tools in a Biosecurity Context

Abraham Gamliel, Heinz W. Dehne, Petr Karlovsky, and Jacqueline Fletcher

Abstract Several fungi, including plant pathogens and endophytes, produce secondary metabolites with biological activity traits (e.g. antibiotics, insecticides, and mycotoxins). Mycotoxins can cause severe intoxication of livestock and humans who consume food that is contaminated. Over 350 mycotoxins which may impact food safety, have been recognized, and probably many more exist. Awareness of the significant impacts of mycotoxins on animal and human health has led to the development of analytical methods for their identification and surveillance in food and feed. The wide range of crops, commodities and agricultural systems in which mycotoxins can be found, presents a challenge for effective analyses. The reliability of quantitative analysis depends on careful execution of all component steps from sampling through the extraction and cleanup. Traditional methods, such as chromatography, together with new and improved ones, can meet these needs. Sophisticated UHPLC–MS/MS technologies are currently the cutting-edge methodology for simultaneous multi-mycotoxin analysis in a wide range of matrices. A combination of the cutting-edge technology with effective sample preparation, can provide robust and practical answers for mycotoxin detection. On the other hand, rapid, field deployable methods (such as dipsticks and biosensors) are significantly less expensive while still providing acceptable accuracy. These techniques can be applied and adapted for the specific requirements of biosecurity. The likelihood of discovery of yet-unknown mycotoxins, and the specific context of biosecurity, calls for additional improved technologies for rapid and robust analysis.

A. Gamliel (✉)
Laboratory for Pest Management Research, Institute for Agricultural Engineering, ARO Volcani Center, P.O. Box 6, 50250 Bet Dagan, Israel
e-mail: agamliel@volcani.agri.gov.il

H.W. Dehne
Institute of Plant Pathology, University of Bonn, Bonn, Germany

P. Karlovsky
Department of Molecular Phytopathology and Mycotoxin Research, University of Goettingen, Goettingen, Germany

J. Fletcher
Department of Entomology and Plant Pathology, Oklahoma State University, Stillwater, OK, USA

M.L. Gullino et al. (eds.), *Practical Tools for Plant and Food Biosecurity*, Plant Pathology in the 21st Century 8, DOI 10.1007/978-3-319-46897-6_7

Keywords Food security • Food safety • Toxigenic fungi • Analytical methods • Secondary metabolites • Plant biosecurity

7.1 Introduction

Fungi comprise an important element of biological diversity in various ecological niches. Certain fungi, including many that colonize plants, produce secondary metabolites such as antibiotics and mycotoxins that impact the living communities around them in a variety of ways. Over 350 mycotoxins have been recognized and probably many more than this impact food safety. Mycotoxins can cause severe intoxication of livestock and humans who consume food that is contaminated. Accute lethal poisoning by mycotoxins has rarely been reported in recent decades; chronic non-lethal exposure to mycotoxins, however, poses a serious threat to public health (Hussein and Jeffrey 2001; Wild and Gong 2010). Apart from the accumulation of mycotoxins in tissues of crop plants infected by plant-pathogenic fungi in the field, mycotoxins may be produced by spoilage fungi that colonize harvested plant commodities in storage and during value-added food production processes (Scudamore 2008, Kabak 2009). Because mycotoxins can withstand many food processing conditions, such as cooking at high temperatures and fermentation, there is a risk that of mycotoxin contamination even in processed food such as grains, soft drinks and wines (Molinie et al. 2005).

Increased awareness of mycotoxins, and of the risks they pose to humans and animals, comes from both the growing evidence of mycotoxin toxicity and the development of new analytical tools that enable detection of very small amounts of these chemicals in food and feed (Ndube et al. 2009; Wang et al. 2010; Maragos and Busman 2010; Hickert et al. 2015).

7.2 Common and Abundant Mycotoxins

Aflatoxins The fungi *Aspergillus flavus* and *A. parasiticus* are the most important producers of aflatoxins (AFT). AFTs comprise 20 different compounds, of which B_1 is the most important. They are toxic and carcinogenic to animals and humans and are common contaminants of a wide variety of agricultural products, including corn, peanuts, cottonseed, and tree nuts (Filtenborg et al 1996).

AFT B_1, the most potent natural carcinogen known, is the major AFT produced by most toxigenic strains (Squire 1981). Considering the extremely high carcinogenicity of aflatoxins, most developed nations set permissible levels as low as reasonably achievable. In the European Union (EU) maximum permitted levels are 2 μg/kg for aflatoxin B1 and 4 μg/kg for total aflatoxins (B1, B2, G1, and G2) in various products (EU Commission 2006).

Ochratoxins Ochratoxin A (OTA), the most important mycotoxin in this group, can contaminate meat, fruit juices, cereals, coffee, grapes and dried fruits (Studer-Rohr et al 1995). It is produced by several *Aspergillus* and *Penicillium* species that are natural opportunistic biodeterioration agents. The occurrence of these fungi is widespread since both of these genera thrive in a wide range of substrates, pH and moisture values and temperatures (Ramos et al 1998; Lee and Magan 2000). In the EU, the maximum allowable level for ochratoxins in foodstuffs is 5 μg/kg (EU Commission 2006).

Trichothecenes Trichothecenes, produced mainly by *Fusarium, Trichoderma* and *Stachybotrys* species, are divided into three groups characterized by specific structural features: type A, which lacks a carbonyl group at position C-8, and type B, which has one (Ueno 1983), and macrocyclic trichothecenes type D (for a review see McCormick et al. 2011). T-2 and HT-2 toxins (type A) and desoxynivalenol (DON) and nivalenol (NIV) (type B) have gained the most attention due to their high toxicity and their prevalence. Macrocyclic trichothecenes such as roridin and satratoxin are regarded as substantially more toxic than trichothecenes type A and B. Recent finding of *Stachybotrys chartarum* in culinary herbs (Biermaier et al. 2015) raised toxicological concerns because this species is known to produce these trichothecenes.

Several human diseases have been directly correlated to trichothecene intoxication, such as the outbreaks of alimentary toxic aleukia in Russia in 1913 and 1944. Due to the toxicity and prevalence of these agents, several countries have established legal regulations or recommendations for DON, HT-2 and T-2 toxins. The EC set an advisory level of 750 μg/kg in cereal products intended for human consumption (EU Commission 2006).

Fumonisin Produced mainly by fungi of the genus *Fusarium*, fumonisins comprise 15 different compounds, of which fumonisin B1, produced *by Fusarium proliferatum, F. verticiloides* and some other species, is the most important. They cause severe animal disorders such as equine leukoencephalomalacia, pulmonary oedema in pigs, equine leukoencephalomalacia (ELEM) in horses and oesophageal and hepatic cancer in horses and rats (Zöllner and Mayer-Helm 2006). Although the acute toxicity of fumonisin is lower than that of other mycotoxins, fumonisins occur in higher concentrations in the food, hence the risk from this mycotoxin can be significant. Fumonisin can be found in cereals (mainly maize), onions and dried fruits. In humans fumonisins are suspected to cause cancer of the esophagus and neural tube defects (Wild and Gong 2010). The European Union (EU) has legislated maximum permitted levels of 2000 μg/kg for fumonisins in corn-based foods (EU Commission 2006), and the European Commission suggested a maximum tolerable total fumonisin intake of 2 μg/kg body weight (EC Scientific Committee 2003).

Zearalenone Zearalenone (ZON) is a nonsteroidal estrogenic mycotoxin with a phenolic resorcyclic acid lactone structure. Sometimes found together with low amounts of α-zearalenol, it is frequently produced by *Fusarium* species that colonize grains such as maize, oat, barley, wheat and sorghum under prolonged cool and wet weather conditions in temperate and warm regions (Filtenborg et al 1996). ZON has low toxicity and there is limited evidence for its carcinogenicity based on animal studies. On the other hand, this mycotoxin is agonistic to the estrogen receptor, targeting estrogenic and anabolic properties in several animal species and resulting in severe effects on the reproductive system (EU Health Consumer 2000). The European Union (EU) has legislated maximum permissible levels of 75 μg/kg for zearalenone in food (EU Commission 2006).

Patulin Produced mainly by fungi of the genera *Penicillium* and *Aspergillus*, this compound can be produced at low temperature and during storage. The most commonly contaminated foods are apples and apple products (juice, cider). The maximum permissible levels in the EU are 50 μg/kg for fruit juices (10 μg/kg for infant fruit juices) (Whitworth 2013).

Other Mycotoxins Other known mycotoxins include ergot compounds, moniliformin, enniatins, and beauvericin (Hussein and Jeffrey 2001). In addition, there are likely to be many other mycotoxic compounds yet to be isolated and described.

Because some mycotoxins are harmful at very low concentrations, highly sensitive methods are required for their detection and quantification in food matrices. A possible role for mycotoxins in crop bioterrorism and crime creates further justification for research and development to increase capabilities for detection and mitigation of mycotoxin effects.

7.3 Processes of Mycotoxin Analysis and Quantification

Awareness of the significant impacts of mycotoxins on animal and human health has led to the development of analytical methods for their identification and surveillance in food and feed. The wide range of crops, commodities and agricultural systems in which mycotoxins can be found, presents a challenge for effective analyses. The reliability of quantitative analysis depends on careful execution of all component steps from sampling through the extraction and cleanup (Turner et al 2009). The values which are obtained by chemical analysis reflect only the mycotoxin in an analyte-free matrix spiked with a known amount of the reference analyte. Analytical results for unknown samples are calculated from recorded signal intensities and calibration curves, both of which are affected by random and systematic errors that influence the reliability and utility of results. Quantitative estimates of errors and other performance characteristics of an analytical method are key to effective analysis and credible results. The most important performance parameters are the relative and comparable suitability of the methods for specific applications. Because performance parameters are defined independently of the technical principles underlying the analytical methods, they can be used to compare the performance of multiple methods.

Key performance requirements for quantitative analytical methods are precision and accuracy. When no analyte is detected or the level of an analyte is too low to be quantified, the limit of detection (LOD) and limit of quantification (LOQ) become the most important performance parameters. Unless these parameters are specified, reporting how many samples were found negative or positive for a particular analyte is meaningless. Other important quality criteria are assay repeatability, reproducibility, and ruggedness. Interestingly, linearity is not required; response curves of some methods are inherently nonlinear, and nonlinear calibration can be used successfully with any method. Determination of the range of linearity, however, is useful for methods, including most chromatographic assays, having linear response curves because it reveals analyte concentrations below and above which the signal is strongly affected by factors other than analyte concentration.

Mycotoxins that are produced by fungi within a plant host may either bind chemically to polar groups on certain plant compounds or be structurally altered by enzymatic reactions, either of which can hamper their detection, extraction and quantification. These so-called "masked mycotoxins" remain present in the plant tissue but are currently neither routinely screened for in foods nor regulated by legislation (Berthiller et al. 2013). Toxicological data are scarce, but several studies highlight the potential threat to consumer safety from these substances. In particular, the possible hydrolysis of masked mycotoxins back to their toxic original forms during mammalian digestion raises concerns (Berthiller et al. 2014, Nakagawa et al 2011).

7.3.1 Mycotoxins in the Context of Plant Biosecurity and Food Safety

What is unique about mycotoxin detection and identification in the context of crop biosecurity? In such cases there is a consideration of possible criminal intent, the need for high levels of assay standardization, validation, repeatability, and robustness; it is crucial to employ rapid and accurate multi-toxin screening tests for effective surveillance and analysis in food and feed. In the EU, assays must comply with EU regulations and be acceptable for standard practice. Several reviews discuss methods for mycotoxin analysis in food safety applications (Maragos and Busman 2010, Koppen et al 2010; Vaclavikova et al 2014).

Procedures for effective detection and accurate quantification of mycotoxins in plant and food matrices may be organized in four essential and sequential steps:

- Sampling from the batch of a commodity, grain silo, or food product
- Extraction of the mycotoxin fraction and its cleanup from the sample
- Additional separation of the target mycotoxins from the extract (using analytical instrumentation or other measures)
- Detection and quantification of the extracted mycotoxin

This review focuses on the methods and technologies available for rapid separation and detection of mycotoxins in the view of biosecurity and food safety needs. However, because sampling and extraction procedures also are crucial steps, and are potentially bottlenecks in the detection and analytical steps, these are also reviewed briefly.

7.3.2 Mycotoxin Detection and Quantification – Is Assay Standardization Still Meaningful?

Traditional approaches to mycotoxin analysis have emphasized a need for strict standardization across investigations and testing facilities, facilitating comparative study and equivalent conclusions. Because of its high specificity and suitability for simultaneous analysis of many mycotoxins, HPLC-MS/MS has been regarded as a "gold standard" for mycotoxin detection and quantification. Unfortunately, HPLC-MS/MS is vulnerable to matrix effects, which jeopardize the reliability of results obtained with external calibration. The use of matrix-matched standards alleviates the problem but does not solve it reliably because matrix effects vary among samples. Only the use of isotope-labelled internal standards adequately compensates for matrix effects; due to the high costs of these standards and limited availability of labelled standards for many mycotoxins, internal standards labelled with stable isotopes are not used in commercial analytical services and official mycotoxin monitoring. The recent rise in the number and diversity of alternative technologies available for mycotoxin analysis, while offering greater flexibility, sensitivity and accuracy, also has introduced new complexity for attempts to standardize approaches in a "one size fits all" context. In this paper we present data on existing standards (EU directives and methods approved by AOAC International) formerly the Association of Official Analytical Chemists) and review methods used currently by mycotoxin labs. The paper will demonstrate that strict standardization may not always be meaningful because many validated protocols exist, each with its advantages and drawbacks; mycotoxin analysis technology is developing very quickly, outdating any "standard" protocol; and rigorously validated in-house methods are often as reliable as official methods. Furthermore, restricting mycotoxin analysis in a critical situation to approved/official methods could delay a critical response without a gain in reliability. In such cases, a readily available method that has been rigorously validated in-house should be acceptable.

7.4 Sampling and Sample Preparation

A primary challenge in mycotoxin analysis is obtaining a representative sample of the commodity to be tested. Official and accredited sampling methods have been established for a limited number of mycotoxins (e.g. aflatoxin, ochratoxin A, and fumonisin). Because concentrations may vary widely within any batch of food or feed, large sample quantities may be needed to optimize the chance for toxin detection. Detailed procedures for sampling for mycotoxin analysis can be found in EU Regulation 691/2013.

Sample preparation for mycotoxin analysis involves three steps that may be carried out in various sequences, or even as separate procedures: (1) reducing the size of the sample, (2) reducing the size of the sample particles, and (3) mixing of the sample for uniformity (U.S. Food and Drug Administration 2013). For example, in some cases grinding provides both particle size reduction and mixing. Final method(s) selection depends on whether the samples are moist or relatively high in oil content, yielding a paste when ground (e.g. peanuts, tree nuts, dates), or dry, yielding a powder or dry particles when ground (e.g. corn, small grains).

7.5 Extraction and Cleanup Procedures

7.5.1 Extraction

Sample purity will significantly affect the detection assay sensitivity. Plant and food matrices are complex in structure and content; cleanup procedures, which allow maximal purification from interfering compounds, are essential for accurate and reliable assay results. Because extraction and cleanup steps consume up to two-thirds of the total analysis time, their selection influences the final choice of the detection procedure. Method choices, which depend on the matrix from which the extraction is made, include selection of appropriate solvents and their relative proportions. Various approaches have been developed for mycotoxin extraction (U.S. Food and Drug Administration 2013). One of the most popular extraction approaches, which is perhaps the most appropriate method in the context of biosecurity, is the QuEChERS (or Quechers), which was developed initially for analysis of pesticide residues from biological matrices (Anastassiades et al 2003; Schenck and Hobbs 2004), QuEChERS is an acronym for quick, easy, cheap, rugged and safe and is a highly beneficial analytical approach that simplifies the analysis of multiple pesticide residues in fruit. QuEChERS was quickly modified for extraction of other contaminants in plant tissues and various food matrices, including popcorn (Ferreira et al 2012), cereal syrups (Arroyo-Manzanares et al 2015), and pomegranate fruits and juices (Myresiotis et al 2015). When a sample is mixed with an organic solvent and water the extract moves into the organic layer. Simultaneously, the aqueous and organic phases are separated by the addition of salt. The solid-phase extraction is finished by centrifugation. The various applications of the QuEChERS protocol in mycotoxin analysis were reviewed (Vaclavikova et al 2014; Walorczyk 2014).

7.5.2 *Cleanup and Purification*

Various procedures and protocols, developed and adapted for mycotoxin separation from matrix components, have been reviewed. The most common are described briefly below.

- Liquid–liquid extraction (LLE): Based on the solubility of a mycotoxin in aqueous/organic substrates, it may be captured within that phase. The addition of a modified salting-out assisted liquid–liquid extraction procedure has recently improved this approach (Songa et al 2013).
- Solid phase extraction (SPE): Extracted mycotoxins are loaded into small disposable cartridges, which are then rinsed. The mycotoxin adsorbs to the cartridge surface and most contaminants are removed (Baggiani et al 2007).
- Solid phase microextraction (SPME) and solid bar microextraction (SBME): In this modified SPE, the cartridge, a solid filament support (Gonzalez-Penas et al 2004) or a miniaturized devise resembling a stir bar (Al-Hadithi et al. 2015) is exposed to the sample for a defined period of time until a partitioning equilibrium between the sample matrix and extraction phase is reached.
- Pressurized liquid extraction (PLE): A technique to extract mycotoxins from a solid matrix such as food or plant tissue (Carabias-Martınez et al 2005), PLE involves exposure of the matrix to an appropriate solvent under pressure and elevated temperature. High temperatures reduce solvent viscosity and promote the cleavage of the matrix-mycotoxin bonds. The major advantages of PLE over other extraction/cleanup techniques are reduced performance time and efficiency. PLE is used mainly in automated and on-line extraction and separation techniques.
- Immunoaffinity columns (IACs): IACs rely upon antibodies specific to the analyte of interest, but the method is expensive, requires complex purification systems, is limited to a single or a few mycotoxins and may result in low recovery levels for some mycotoxins. Another problem of IACs is their limited capacity, which leads to underestimation of mycotoxin levels in highly contaminated samples when the operating range of the cartridge is exceeded.

7.6 Analytical Methods for Quantification of Mycotoxins

Several analytical methods (separation and detection) have been developed and adopted for the quantification of mycotoxins in food. The methods are described below and notes on their accuracy and their relevance to the needs of biosecurity are provided. In each method the separation methods and their relevant detectors are described together.

7.6.1 Chromatography Methods

7.6.1.1 Thin Layer Chromatography (TLC)

TLC, a separation technique used widely for aflatoxin analysis (Park et al 1994), is based on a stationary matrix composed of silica, aluminum or cellulose immobilized on glass or plastic. A mobile phase is comprised of solvents; for aflatoxins the matrix is a mixture of methanol, acetonitrile, and water. The distribution of mycotoxin between the mobile and stationary phases depends primarily on differences in solubility of the mycotoxin in the two phases. TLC, useful for the quantification of aflatoxins in food and grains (Gulyas 1985), can detect as little as 1–20 ppb of aflatoxin (Trucksess et al 1984) and can detect multiple mycotoxins in a single test sample. However, it requires a highly skilled technician and extensive sample pretreatment, and it is imprecise because of the likelihood of accumulated errors at multiple points along the process. TLC is a very useful method for detecting known mycotoxins present at high concentrations, especially when state of the state of the art equipment (such as LC/MS) is not available or a large number of samples have to be analyzed in a limited time.

7.6.1.2 High-Performance Thin-Layer Chromatography (HPTLC)

HPTLC, a form of TLC in which the sample application, plate development, and data interpretation are done automatically, is efficient and precise (Ramesh et al 2013). Nevertheless, the requirement for skilled operators and the availability of the equipment limits its use. Moreover, the need for extensive sample pretreatment limits the implementation of HPTLC to the laboratory.

7.6.1.3 Gas Chromatography (GC)

GC separates mycotoxins by the movement of a carrier gas (mobile phase) through a column (stationary phase) that consists of a liquid coated onto inert solid particles (Cunha and Fernandes 2010). The gaseous mycotoxin is separated from the other sample components as it moves across the column. The volatile mycotoxin is then detected using either a flame ionization detector (FID) or an electron capture detector (ECD). Although a mass spectrometer (MS) could also be used for mycotoxin detection, it is a separate, stand-alone piece of equipment and not an integral part of the GC.

Since most mycotoxins are not volatile their separation in GC requires derivatization of the molecule to a detectable, volatile form. Moreover, GC requires a preliminary cleanup step before analysis and it is therefore limited to a few specific mycotoxins. GC combined with MS is highly effective in quantitative analysis of A- and B-trichothecenes but since HPLC-MS became widely available, the latter is a preferred option for MS coupling in mycotoxin analysis.

7.6.1.4 High Performance Liquid Chromatography (HPLC)

High performance liquid chromatography (HPLC) is the main pillar of modern analysis of mycotoxins (Sulyok et al. 2010). Similar to other chromatography methods, HPLC separates mycotoxins within a carrier liquid (mobile phase) through a column (stationary phase, which contains various adsorbents depending on the physical and chemical structure of the mycotoxin). The liquid sample is carried through the column by the carrier solvents, and the mycotoxin is separated from other components during the extraction. Protocols used for HPLC separation differ mainly in the details of column type and the carrier liquids. A few detectors, such as ultraviolet and fluorescent light, are coupled to the HPLC. For example, fluorescent ochratoxin A, aflatoxin and citrinin can be detected directly by fluorescence detectors. Non-fluorescent mycotoxins, such as fumonisin, cannot be detected in this way unless they have been derivatized previously.

Today, most mycotoxins are detected by mass spectrometry (MS). Ultrahigh performance liquid chromatography (UHPLC), coupled with tandem mass spectrometry (MS/MS), has become very popular. The use of HPLC-MS/MS enables the simultaneous detection of mycotoxins belonging to different chemical families and efficient quantitative screening for the most important mycotoxins in food commodities. The main advantage of tandem MS application is the ability to do quantitative analysis without the need for derivatisation (Aguilera-Luiz et al. 2011, Perez-Ortega et al. 2010, Malachová et al 2014; Wen et al 2014).

High resolution (HR) mass spectrometry has also been applied to screening of several mycotoxins in a single analysis (Russell et al. 1997, Zachariasova et al 2010). Although HR-MS can simultaneously detect and confirm multiple mycotoxins it is not yet extensively used for this type of analysis. Among the reasons are a narrow dynamic range of microchannel plates used as detectors in time-of-flight (TOF) mass spectrometers and high costs and high demands on space of Fourier-transform ion cyclotron resonance (FT-ICR) mass spectrometry. The growing number of installed Orbitrap systems which are easy to operate and have small footprints might help overcoming these drawbacks. Current EU legislation requires certain conditions to be fulfilled and validated before data of mycotoxin analysis with HR-MS is accepted.

7.6.2 Methods Based on Mass Spectrometry Without Previous Analyte Separation

Techniques for mass spectrometry, a rapid, no-separation technique, include matrix-assisted laser desorption MS, ambient ionization MS, and ion mobility spectrometry. A few examples of applications of these techniques are provided below:

- Matrix-assisted laser desorption ionization (MALDI-MS) – In this method the mass spectrum of a thin tissue is recorded while it is moved in two dimensions. This method is useful in tissue-based studies as it involves rapid preparation and is non-destructive to the tissue structure (Powers et al. 2014).

- Ambient ionization mass spectrometry – This popular form of ionization takes place outside the mass spectrometer without the need for either sample preparation or separation. Ions can be formed by (1) extraction into charged electrospray droplets, thermal desorption and chemical ionization, or (2) laser desorption or ablation and post-ionization prior to entry into the mass spectrometer. DART (direct analysis in real time), developed in 2005, is now marketed commercially. Because no sample preparation is required, both solid and liquid materials can be analyzed by MS in their native state (Vaclavik et al 2010).
- Ion mobility spectrometry (IMS) – In this gas-phase ion separation technique ion mobility measurement is based on the drift velocities of ions in an electric field at ambient pressure (Kanu et al 2008). A simple multiplier is often used as a detector with IMS, rather than an MS detector. Due to its excellent sensitivity and rapid operation, IMS has gained widespread use in many applications for the detection of contaminants. Its main advantages include low detection limits, rapid response, simplicity, portability, and relatively low cost.

7.6.3 *Infrared Spectroscopy*

Infrared spectroscopy (IR), especially the near IR (NIR) and mid IR (MIR) range, is a rapid, nondestructive analytical technique commonly utilized in monitoring food quality (McMullin et al 2015). IR-based methods require little sample preparation or labor; however, they do require extensive calibration. Mycotoxins currently cannot be detected directly in a complex matrix such as food and plant tissue because of the limited sensitivity of currently available IR-based methods. However IR can be employed as a detection tool following an appropriate separation procedure such as HPLC. IR-based detectors for HPLC are in development but they are not yet available commercially. The limited sensitivity is the main problem of IR detection in mycotoxin analysis.

7.6.4 *Capillary Electrophoresis (CE)*

Sophisticated and highly sensitive, capillary electrophoresis enables the separation of closely related mycotoxins based on their mass and charge, which are reflected in their rates of migration in an electrical field (Arroyo-Manzanares et al 2010). Separations can be further expedited by CE in aqueous buffer solutions, without the need for organic solvents. CE is a useful tool for detection of fumonisins and aflatoxins. Recently, a method using capillary zone electrophoresis with laser-induced fluorescence methods has been developed, allowing detection of aflatoxin, fumonisin and ochratoxin at levels commonly found in naturally contaminated food samples (Bueno et al 2015). The drawback of CE in mycotoxin analysis is that the analytes have to be converted into charged forms in order to be separated in an electric field. Moniliformin is the only ionic compound among mycotoxins but many mycotoxins possess carboxyl or amine groups which dissociate or bind protons, respectively, at suitable pH range.

7.6.5 Immunoassays and Biosensors

Immunoassays are used frequently as rapid tests for mycotoxins, especially for the screening of raw materials, and are available commercially for many mycotoxins. Immunoassays applied in mycotoxin detection include enzyme-linked immunosorbent assay (ELISA), fluorescence polarization immunoassay, surface plasmon resonance (SPR), multiplex flow cytometric immunoassay, magnetoresistive sensors and other biosensors. These technologies are highly sensitive, specific and rapid (Tang et al. 2009; Maragos and Busman 2010), but they require special instruments. Independent component analysis (ICA), in which antigens for aflatoxin, ochratoxin A, and zearolenone were immobilized as test lines, enabled detection of all three mycotoxins within a matrix (Li et al 2013). A recent review delineates currently available immunoassay-based kits that effectively detect low levels of several mycotoxins in food and commodities (Selvaraj et al 2015). Important factors include the method's precision profile and the false negative rates of the tested samples. Lattanzio et al (2015) have successfully applied commercial lateral flow immunoassay for detection of deoxynivalenol in wheat. Their validation design provided information on the containing deoxynivalenol above the legal limit according to guidelines set in EU Regulation 519/2014/EU, which specifies validation criteria for mycotoxin screening methods used for official control purposes.

Other mycotoxin-applicable immunoassays include:

- Surface plasmon resonance. This method measures adsorption of material onto metal surfaces (planar or nanoparticles), and is also the basis for color-based biosensor and chip sensor applications. Advantages include the need for only small sample volumes, the use of reusable metal chips, and portable equipment (Koppen et al 2010). The technique has recently been adopted for detection of aflatoxin M1, DON and for fumonisin B1 (Dunne et al 2005).
- Fluorescence polarization immunoassays (FPI). An analyte labeled with a fluorophore (fluorescein) competes with free analyte for specific antibody-binding sites in the extract solution. During this process the fluorescence polarization of the fluorescein label is measured. FPI has been used for detection of DON, ZON, and OTA in various food samples. The method is sensitive and rapid, and provides acceptable mycotoxin recoveries (Maragos and Busman 2010; Li et al 2015).
- Enzyme linked immunosorbent assay (ELISA). ELISA is based on a competitive assay format that uses either a primary antibody specific for the target molecule or a conjugate of an enzyme and the required target (Dos Santos et al 2011). A complex, formed by the binding of the ELISA construct to the mycotoxin in

the sample, interacts with a chromogenic substrate to yield a measurable product. This assay is popular for mycotoxin detection due to its specificity, relatively low cost and the availability of rapid, portable, user-friendly commercial kits. Limitations of ELISA kits include their "single-use" nature, specificity for a specific mycotoxin, and limited sensitivity. Hence ELISA is used mainly for screening and not for generating analytical results that comply with formal regulations. However, efforts are being made to develop more sensitive ELISA based kits for the screening of mycotoxins in food (Anfossi et al 2015). A recent study proposes a variant of ELISA for the detection and quantification of aflatoxin B1, ochratoxin A, and zearalenone, with detection limits of 0.24, 1.2 and 3 ng/g, respectively (Urusov et al. 2015). Even higher sensitivity of 0.1 ng/g was achieved in a recently developed ELISA for mycophenolic acid (Dietrich and Maertlbauer 2015).
- The use of nanobiosensors following rapid developments in nanotechnology begins to provide sensitive methods for the rapid detection of mycotoxins. The fabrication of nanobiosensors and their application to detect mycotoxins in food and feed involves the use of carbon nanotubes, nanowires, nanoparticles, quantum dots, nanorods and nanofibers (Rai et al 2015). In addition, there are already available nanobiosensors that can be used for specific mycotoxins (Rai et al 2015). An electrochemical sensor for zearalenone based on multi-welled carbon nanotubes can serve an example of the application of nanotechnology in mycotoxin detection (Afzali et al. 2015).
- Aptamers, also known as chemical antibodies, are an emerging field of mycotoxin detection (Toh et al 2015). Aptamers are single strand DNA or RNA that bind to a wide range of molecules with high specificity. One advantage of using aptamers includes production by chemical synthesis, which provides consistency among production batches. Also, aptamers can reform to their original configuration when set back to their optimal conditions. The use of this technology for mycotoxin detection was reported with relevance to the food industry such as in the production of chocolate (Mishra et al 2015) and coffee (Jo et al 2015).

7.6.6 *Metabolomics Methods*

Metabolomics is a holistic study of small compounds in complex biological systems under a given set of conditions. When applied to mycotoxins, it involves comprehensive, qualitative and quantitative analysis of all related metabolites within a fungus or plant. Metabolomics can potentially provide information critical to understanding mycotoxin biosynthesis and regulatory pathways. Metabolomics-based studies rely upon advanced platforms for analysis of numerous chemically diverse metabolites over large concentration ranges. These are naturally linked to NMR spectroscopy, GC–MS, and MS, LC–MS. However, practical applications of metabolomics for mycotoxin detection remain to be developed (Cajka et al 2014; Xu et al 2014).

7.7 Challenges in Mycotoxin Analysis in the Context of Biosecurity

A drawback of most multi-target methods is that, because they require extensive, time- and cost-consuming validation, they are often used only for semi-quantitative screening purposes. Most multi-target methods were developed for use with raw cereals, and data on the performance characteristics in other matrices, such as nuts, are scarce (Varga et al 2013).

Because fungi generally develop in isolated pockets within a plant or other matrix and are not evenly distributed in stored commodities protocols should ensure that samples are representative of the whole consignment. "Grab samples" have been reported generally to give very low estimates of mycotoxin content. In fact, nearly 90 % of the error associated with mycotoxin assays could be attributed to how the original sample was collected (Turner et al 2009).

Improvements in sampling and sample preparation methods used to detect mycotoxins and other quality attributes in food and feed products continues to be a high priority among regulatory agencies, international organizations and commodity industries worldwide (Whitaker 2003). In recent years, more emphasis in studies of mycotoxin distribution is placed on the effects of sample selection methods (Berthiller et al 2013). For instance, the Food and Agriculture Organization (FAO) is developing a tool to assist in the design and performance characterization of sampling plans for mycotoxin detection. Existing mycotoxin contamination data (including specific mycotoxin-commodity combinations, seasonal and regional variations, etc.) will be used to create a database to serve as the basis for the tool. FAO, in collaboration with various research institutes and other international organizations, already has started collecting existing mycotoxin contamination data (CAC 2012).

The choice of extraction solvent for multiple analytes with different physicochemical properties is a challenge that often involves a compromise that allows acceptable extraction recoveries for the majority of the analytes. Extraction of polar metabolites, such as the fumonisins, requires the presence of both water and organic solvents (Shephard 1998), while extraction of hydrophobic toxins such as AFT relies only on the use of organic solvents. These can be direct extractions, or may be partitioned with other solvents, such as n-hexane for partial clean up, to remove excess components of the biological matrix (Holcomb et al 1992). To promote reliable analytical measurements, the European Union Reference Laboratory (EURL) for Mycotoxins, established in 2006, acts together with the National Reference Laboratories (NLRs) to assure the harmonization of mycotoxin measurements in food and feed. The EURL conducts comparative/proficiency-tests (PTs) to benchmark laboratory performance of NLRs and associated reference laboratories, identifying bottle necks and pitfalls in analytical procedures. Future procedure improvements are expected to enhance common standardization of these methods (Stroka 2014).

7.7.1 *Standardization of Analytical Methods*

To assure test results that can answer specific questions, analytical methods are required to meet quality standards for selected performance parameters. To guarantee these quality criteria, standardization procedures for analytical methods have been established at several levels.

7.7.2 *Standard Operating Procedures (SOPs)*

A SOP is a detailed, explicit and unambiguous description of all steps of a method. It will generally include details such as which reaction vessels and tools are used and how solutions are prepared. Each SOP is specific to a particular laboratory, and must be modified if any instrument or other tool is updated. Thus, SOPs are internal management tools of analytical laboratories that facilitate enforcement of standardized protocols, limit operator-dependent variation and protect against protocol modifications after staff replacement.

7.7.3 *Recommended/Approved Methods*

Selected validated methods regarded as reliable and robust have been recommended or enforced by national and international regulatory bodies and food safety authorities. For instance, the United States' Environmental Protection Agency issued a list of approved methods for water quality analysis (EPA 2014). The EPA's methods approval process is exemplary by its efficiency and flexibility; any organization or institution may submit their own analytical method for evaluation. If that method's performance characteristics fall within the EPA's set range for the same contaminant and other requirements are fulfilled, approval is expedited. For example, EPA announced approval of 84 analytical methods in May 2013 and another 21 methods in June 2014, all for drinking water sampling and analysis. New technological advancements can be applied immediately after they have been implemented in methods. Similarly, the U.S. Food and Drug Administration issues guidance for validation of analytical procedures at regular intervals. The most recent draft, published in February 2014, supersedes the previous Analytical Procedures and Methods Validation of 2000

7.7.4 *Minimum Limits of Performance Criteria*

Although the EU has relied, in the past, upon the establishment of a limited number of specific, universally applied "approved methods" for mycotoxin detection, extraction and analysis, this may not be the most effective or feasible approach to mycotoxin risk management. Analytical technologies develop rapidly, outdating established methods. Furthermore, there often are several validated methods for the same analyte. The performances of alternative protocols are often comparable but each method possesses specific advantages and drawbacks and therefore one method may be more suitable than another in a particular setting. New European regulations on mycotoxin analysis, such as the European Commission's Regulation (EC) No 401/2006, instead defines performance criteria that validated methods must fulfill. When no suitable validated method is available, a "fitness-for-purpose" approach may be used. This more reasonable approach allows for timely implementation of the newest technologies and minimizes bureaucratic burdens without compromising the reliability of the results.

The not-for-profit organization AOAC International (Rockville, MD, U.S.A.) holds a special position in the standardization of analytical methods. For over a century this international association of analytical chemists has developed and validated methods, publishing them as AOAC Official Methods. Government agencies and civil organizations often require that laboratories use official AOAC methods.

7.8 Concluding Remarks

Surveillance and detection of mycotoxins in agricultural systems, commodities, food and feed are major criteria for assuring food and feed quality and safety. Considering the increasing number of known mycotoxins. the likelihood of discovery of yet-unknown mycotoxins, and the specific context of biosecurity, there is a critical need for rapid and robust analytical strategies. Traditional methods, such as chromatography, together with new and improved ones, can meet these needs. Sophisticated UHPLC–MS/MS technologies are currently the cutting-edge methodology for simultaneous multi-mycotoxin analysis in a wide range of matrices. A combination of the cutting-edge technology with effective sample preparation (such as QuEChERS) can provide robust and practical answers for mycotoxin detection. On the other hand, rapid, field deployable methods (such as dipsticks and biosensors) are significantly less expensive while still providing acceptable accuracy. These techniques can be applied and adapted for the specific requirements of biosecurity.

References

Afzali D, Padash M, Mostafavi A (2015) Determination of trace amounts of zearalenone in beverage samples with an electrochemical sensor. Mycotoxin Res 31:203–208

Aguilera-Luiz P, Plaza-Bolanos R, Romero-Gonzalez JL, Martınez-Vidal A (2011) Comparison of the efficiency of different extraction methods for the simultaneous determination of mycotoxins and pesticides in milk samples by ultra high-performance liquid chromatography-tandem mass spectrometry. Anal Bioanal Chem 399:2863–2875

Al-Hadithi N, Kössler P, Karlovsky P (2015) Determination of ochratoxin A in wheat and maize by solid bar microextraction with liquid chromatography and fluorescence detection. Toxins 7:3000–3011

Anastassiades M, Lehotay SJ, Stajnbaher D, Schenck FJ (2003) Fast and easy multiresidue method employing acetonitrile extraction/partitioning and "dispersive solid phase extraction" for the determination of pesticide residues in produce. J AOAC Int 86:412–431

Anfossi L, Di Nardo F, Giovannoli C, Passini C, Baggiani C (2015) Enzyme immunoassay for monitoring aftatoxin in eggs Food Control 57:115–121

Arroyo-Manzanares N, Gamiz-Gracia L, Garcia-Campana AM, Soto-Chinchilla JJ, Garcia-Ayuso LE (2010) On-line preconcentration for the determination of aflatoxins in rice samples by micellar electrokinetic capillary chromatography with laser-induced fluorescence detection. Electrophoresis 31:2180–2185

Arroyo-Manzanares N, Huertas-Perez JF, Gamiz-Garcia L, Garcia-Campana AM (2015) Simple and efficient methodology to determine mycotoxins in cereal syrups. Food Chem 177:274–279

Baggiani C, Anfossi L, Giovannoli C (2007) Solid phase extraction of food contaminants using molecular imprinted polymers. Anal Chim Acta 591:29–39

Berthiller F, Crews C, Dall'Asta C, De Saeger S, Haesaert G, Karlovsky P, Oswald IP, Seefelder W, Speijers G, Stroka J (2013) Masked mycotoxins: a review. Mol Nutr Food Res 57:165–186

Berthiller F, Burdaspal PA, Crews C, Iha MH, Krska R, Lattanzio VM, MacDonald S, Malone RJ, Maragos C, Solfrizzo M, Stroka J, Whitaker TB (2014) Developments in mycotoxin analysis: an update for 2012–2013. World Mycotoxin J 7:3–33

Bueno D, Istamboulie G, Munoz R, Marty JL (2015) Determination of mycotoxins in food: a review of bioanalytical to analytical methods. Appl Spectrosc Rev 50:728–774

CAC (2012) Report of the sixth session of the Codex Committee on Contaminants in Foods. Food and Agriculture Organization of the United Nations, Rome

Cajka T, Vaclavikova M, Dzuman Z, Vaclavik L, Ovesna J, Hajsolva J (2014) Rapid LC–MS-based metabolomics method to study the *Fusarium* infection of barley. J Sep Sci 37:912–919

Carabias-Martınez R, Rodrıguez-Gonzalo E, Revilla-Ruiz P, Hernandez-Mendez J (2005) Pressurized liquid extraction in the analysis of food and biological samples. J Chromatogr A 1089:1–17

Commission Regulation (EU) No 519/2014 of 16 May 2014 amending Regulation (EC) No 401/2006 as regards methods of sampling of large lots, spices, and food supplements, performance criteria for T-2 and HT-2 toxin and citrinin and screening methods of analysis. L147/29

Cunha SC, Fernandes JO (2010) Development and validation of a method based on a QuEChERS procedure and heart-cutting GC-MS for determination of five mycotoxins in cereal products. J Sep Sci 33:600–609

Dietrich R, Maertlbauer E (2015) Development and application of monoclonal antibodies against the mycotoxin mycophenolic acid. Mycotoxin Res 31:185–190

Dos Santos SJ, Takabayashi CR, Ono EYS (2011) Immunoassay based on monoclonal antibodies versus LC-MS: deoxynivalenol in wheat and flour in Southern Brazil. Food Addit Contam A 28:1083–1090

Dunne L, Daly S, Baxter A, Haughey S, O'Kennedy R (2005) Plasmon resonance-based immunoassay for the detection of aflatoxin B_1 using single-chain antibody fragments. Spectrosc Lett 38:230–245

EPA (2014) Analytical methods approved for drinking water compliance monitoring of organic contaminants. http://water.epa.gov/scitech/drinkingwater/labcert/upload/815b14003.pdf

European Commission (2000) Health consumer protection directorate-general. Opinion of the scientific committee on food and Fusarium toxins. Part 2, Zearalenone

European Commission (2003) Scientific Committee on Food. Updated opinion of the Scientific Committee, Fumonisin B1, B2 and B3. Off J Eur Union 325:9–30

European Commission (2006) Regulation (EC) No 1881/2006 of 19 December 2006 setting maximum levels for certain contaminants in foodstuffs. Off J Eur Union 364:5–24

Ferreira I, Fernandes JO, Cunha SC (2012) Optimization and validation of a method based in a QuEChERS procedure and gas chromatography–mass spectrometry for the determination of multi-mycotoxins in popcorn. Food Control 27:188–193

Filtenborg O, Frisvad JC, Thrane U (1996) Moulds in food spoilage. Int J Food Microbiol 33:85–102

Gonzalez-Penas E, Leache C, Viscarret M, Perez de Obanos A, Araguas V, Lopez de Cerain A (2004) Determination of ochratoxin A in wine using liquid-phase microextraction combined with liquid chromatography with fluorescence detection. J Chromatogr 1025:163–168

Gulyas H (1985) Determination of aflatoxins b1, b2, g1, g2 and m1 by high pressure thin layer chromatography. J Chromatogr 319:105–111

Hickert S, Gerding J, Ncube E, Huebner F, Flett B, Cramer B, Humpf H-U (2015) A new approach using micro HPLC-MS/MS for multi-mycotoxin analysis in maize samples. Mycotoxin Res 31:109–115

Holcomb M, Wilson DM, Trucksess MW, Thompson HC (1992) Determination of aflatoxins in food products by chromatography. J Chromatogr 624:341–352

Hussein SH, Jeffrey MB (2001) Toxicity, metabolism, and impact of mycotoxins on humans. Toxicology 167:101–134

Jo EJ, Mun H, Kim SJ, Shim WB, Kim MG (2015) Detection of ochratoxin A (OTA) in coffee using chemiluminescence resonance energy transfer (CRET) aptasensor. Food Chem 194:1102–1107

Kabak B (2009) The fate of mycotoxins during thermal food processing. J Sci Food Agric 89:549–554

Kanu AB, Dwivedi P, Tam M, Matz L, Hill HH (2008) Ion mobility-mass spectrometry. J Mass Spectrom 43(1):1–22

Koppen R, Koch M, Siegel D, Merkel S, Maul R, Nehls I (2010) Determination of mycotoxins in foods: current state of analytical methods and limitations. Appl Microbiol Biotechnol 86:1595–1612

Lattanzio VMT, Ciasca B, Powers S, Von Holst C (2015) Validation of screening methods according to regulation 519/2014/eu. determination of deoxynivalenol in wheat by lateral flow immunoassay: a case study. Trends Anal Chem, http://dx.doi.org/doi: 10.1016/j.trac.2015.10.009

Lee HB, Magan N (2000) Impact of environment and interspecific interactions between spoilage fungi and Aspergillus ochraceus on growth and ochratoxin production in maize grain. Int J Food Microbiol 61:11–16

Li X, Li P, Zhang Q, Li R, Zhang W, Zhang Z, Ding X, Tang X (2013) Multi-component immunochromatographic assay for simultaneous detection of aflatoxin B1, ochratoxin A and zearalenone in agro-food. Biosens Bioelectron 49:426–432

Li C, Mi T, Oliveri Conti G, Yu Q, Wen K, Shen J, Ferrante M, Wang Z (2015) Development of a screening fluorescent polarization immunoassay for simulataneous detection of fumonisins B1 and B2 in maize. J Agric Food Chem 63:4090–4096

Malachová A, Sulyok M, Beltrán E, Berthiller F, Krska R (2014) Optimization and validation of a quantitative liquidchromatography–tandem mass spectrometric method covering 295 bacterial and fungal metabolites including all regulated mycotoxins in four model food matrices. J Chromatogr 1362:145–156

Maragos CM, Busman M (2010) Rapid and advanced tools for mycotoxin analysis: a review. Food Addit Contamin 27:688–700

McCormick SP, Stanley AM, Stover NA, Alexander NJ (2011) Trichothecenes: from simple to complex mycotoxins. Toxins 3:802–814

McMullin D, Mizaikoff B, Krska R (2015) Advancements in IR spectroscopic approaches for the determination of fungal derived contaminations in food crops. Anal Bioanal Chem 407:653–660

Mishra RK, Hayat A, Catanate G, Ocana C, Marty JL (2015) A label free aptasensor for ochratoxin A detection in Cocoa beans: an application to chocolate industry. Anal Chim Acta 889:106–112

Molinie A, Faucet V, Castegnaro M, Pfohl-Leszkowicz A (2005) Analysis of some breakfast cereals on the French market for their contents of ochratoxin A, citrinin and fumonisin B1: development of a method for simultaneous extraction of ochratoxin A and citrinin. Food Chem 92:391–400

Myresiotis CK, Testempasis S, Vryzas Z, Karaoglanidis GS, Papadopoulou-Mourkidou E (2015) Determination of mycotoxins in pomegranate fruits and juices using a QuEChERS-based method. Food Chem 182:81–88

Nakagawa H, Ohmichi K, Sakamoto S, Sago Y, Kushiro M, Nagashima H, Yoshida M, Nakajima T (2011) Detection of a new Fusarium masked mycotoxin in wheat grain by high-resolution LC–Orbitrap™ MS. Food Addit Contamin 28:1447–1456

Ndube N, van der Westhuizen L, Shephard GS (2009) Determination of fumonisins in maize by HPLC with ultraviolet detection of *o*-phthaldialdehyde derivatives. Mycotoxin Res 5:225–228

Park DL, Trucksess MW, Nesheim S, Stack M, Newell RF (1994) Solvent-efficient thin-layer chromatographic method for the determination of aflatoxins B1, B2, G1, and G2 in corn and peanut products: collaborative study. J AOAC Int 77:637–646

Perez-Ortega P, Gilbert-Lopez B, Garcıa-Reyes JF, Molina-Dıaz A (2010) Generic sample treatment method for simultaneous determination of multiclass pesticides and mycotoxins in wines by liquid chromatography-mass spectrometry. J Chromatogr A 1249:32–40

Powers TW, Neely BA, Shao Y, Tang H, Troyer DA, Mehta AS, Haab BB, Drake RR (2014) MALDI imaging mass spectrometry profiling of N-glycans in formalin-fixed paraffin embedded clinical tissue blocks and tissue microarrays. PLoS One 9:106255

Rai M, Jogee PS, Ingle AP (2015) Emerging nanotechnology for detection of mycotoxins in food and feed. Int J Food Sci Nutr 66:363–370

Ramesh J, Sarathchandra G, Sureshkumar V (2013) Analysis of feed samples for aflatoxin b1 contamination by hptlc-a validated method. Int J Curr Microbiol Appl Sci 2:373–377

Ramos AJ, Labernia N, Marin S, Sanchis V, Magan N (1998) Effect of water activity and temperature on growth and ochratoxin production by three strains of Aspergillus ochraceus on a barley extract medium and on barley grains. Int J Food Microbiol 44:133–140

Russell D, Edmondson H, High RD (1997) High resolution mass spectrometry and accurate mass measurements with emphasis on the characterization of peptides and proteins by matrix-assisted laser desorption/ionization time-of-flight mass spectrometry. J Mass Spectrom 32:263–276

Schenck FJ, Hobbs JE (2004) Evaluation of the Quick, Easy, Cheap, Effective, Rugged, and Safe (QuEChERS) approach to pesticide residue analysis. Bull Environ Contam Toxicol 73:24–30

Scudamore KA (2008) Fate of Fusarium mycotoxins in the cereal industry: recent UK studies. World Mycotoxin J 1:315–323

Selvaraj JN, Zhou I, Wang Y, Zhao Y, Xing F, Dai X, Liu Y (2015) Mycotoxin detection- recent trends at global level. J Integr Agric 14:2265–2281

Shephard G (1998) Chromatographic determination of the fumonisin mycotoxins. J Chromatogr A 815:31–39

Songa S, Njumbe Ediagea E, Wub A, De Saeger S (2013) Development and application of salting-out assisted liquid/liquid extraction for multi-mycotoxin biomarkers analysis in pig urine with high performance liquid chromatography/tandem mass spectrometry. J Chromatogr A 1292:111–120

Squire R (1981) Ranking animal carcinogens: a proposed regulatory approach. Science 214:877–880

Stroka J (2014) The approach of the European Union Reference Laboratory to promote reliable measurements. In 36th Mycotoxin Workshop, Göttingen

Studer-Rohr I, Dietrich DR, Schlatter J, Schlatter C (1995) The occurrence of ochratoxin A in coffee. Food Chem Toxicol 33:341–355

Sulyok M, Krska R, Schuhmacher R (2010) Application of an LC–MS/MS based multi-mycotoxin method. Food Chem 119:408–416

Tang D, Sauceda JC, Lin Z, Ott S, Basova E, Goryacheva I, Biselli S, Lin J, Niessner R, Knopp D (2009) Magnetic nanogold microspheres-based lateral-flow immunodipstick for rapid detection of aflatoxin B2 in food. Biosens Bioelectron 25:514–518

Toh SY, Citartan M, Gopinath SCB, Tang TH (2015) Aptamers as a replacement for antibodies in enzyme-linked immunosorbent assay. Biosens Bioelectron 64:392–403

Trucksess M, Brumley W, Nesheim S (1984) Rapid quantitation and confirmation of aflatoxins in corn and peanut butter, using a disposable silica gel column, thin layer chromatography, and gas chromatography/mass spectrometry. J AOAC Int 67:973–975

Turner NW, Subrahmanyam S, Piletsky SA (2009) Analytical methods for determination of mycotoxins: a review. Anal Chim Acta 632:168–180

U.S. Food and Drug Administration (2013) ORA laboratory manual: mycotoxin analysis. FDA Office of Regulatory Affairs. Section 7. Document IV-07, Version 1.6. http://www.fda.gov/ScienceResearch/FieldScience/default.htm

Ueno Y (1983) Trichothecenes: chemical, biological and toxicological. Elsevier Press, Amsterdam/New York

Urusov AE, Zherdev AV, Petrakova AV, Sadykhov EG, Koroleva OV, Dzantiev BB (2015) Rapid multiple immunoenzyme assay of mycotoxins. Toxins 7:238–254

Vaclavik L, Zachariasova M, Hrbek V, Hajslova J (2010) Analysis of multiple mycotoxins in cereals under ambient conditions using direct analysis in real time (DART) ionization coupled to high resolution mass spectrometry. Talanta 82:1950–1957

Vaclavikova M, Vaclavik L, Cajka T (2014) High-throughput analysis of mycotoxins. In: Wang PG, Vitha MF, Kay JF (eds) High-throughput analysis for food safety. Wiley, Hoboken, pp 231–266

Varga E, Glauner T, Berthiller F, Krska R, Schuhmacher R, Sulyok M (2013) Development and validation of a (semi-)quantitative UHPLC-MS/MS method for the determination of 191 mycotoxins and other fungal metabolites in almonds, hazelnuts, peanuts and pistachios. Anal Bioanal Chem 405:5087–5104

Walorczyk S (2014) Validation and use of a QuEChERS-based gas chromatographic–tandem mass spectrometric method for multiresidue pesticide analysis in blackcurrants including studies of matrix effects and estimation of measurement uncertainty. Talanta 120:106–113

Wang J, Wang X, Zhou Y, Duc L, Wanga Q (2010) Fumonisin detection and analysis of potential fumonisin-producing *Fusarium* spp. in asparagus (*Asparagus officinalis* L.) in Zhejiang Province of China. J Sci Food Agric 90:836–842

Wen J, Kong W, Hu Y, Wang J, Yang M (2014) Multi-mycotoxins analysis in ginger and related products by UHPLC-FLR detection and LC-MS/MS confirmation. Food Control 43:82–87

Whitaker TB (2003) Standardisation of mycotoxin sampling procedures: an urgen necessity. Food Control 14:233–237

Whitworth J (2013) More than half of apple juice samples above legal patulin limit, says study. Food Quality News, 14 Jun 2013

Wild CP, Gong YY (2010) Mycotoxins and human disease: a largely ignored global health issue. Carcinogenesis 31:71–82

Xu YJ, Wang C, Ho WE, Ong CN (2014) Recent developments and applications of metabolomics in microbiological investigations. Trends Anal Chem 56:37–48

Zachariasova M, Lacina O, Malachova A (2010) Novel approaches in analysis of Fusarium mycotoxins in cereals employing ultra performance liquid chromatography coupled with high resolution mass spectrometry. Anal Chim Acta 662:51–61

Zöllner P, Mayer-Helm B (2006) Trace mycotoxin analysis in complex biological and food matrices by liquid chromatography–atmospheric pressure ionisation mass spectrometry. J Chromatogr A1136:123–169

Chapter 8
Containment of Mycotoxins in the Food Chain by Using Decontamination and Detoxification Techniques

Davide Spadaro and Angelo Garibaldi

Abstract Mycotoxin contamination of agricultural products is a global problem but is most severe in tropical and subtropical regions. The Food and Agriculture Organization estimated that up to 25 % of the world food crops are significantly contaminated with mycotoxins. The most effective tools against mycotoxins are essentially based on the prevention of mould growth in each stage of the food chain. A strategy to reduce the risk of mycotoxin contamination should include prevention practice in the field and during the postharvest phase, and control measures. However, when contamination is not prevented during the preharvest and postharvest periods, several approaches can be employed to help remove mycotoxins from the subsequently contaminated commodities, including physical, chemical, and biological techniques. The main decontamination and detoxification strategies, and food processing, used to reduce the mycotoxin in contaminated food or feed are considered in this chapter. Techniques for food decontamination are based on the collection and removal of the contaminated parts from a mass of product. Detoxification processes should destroy or inactivate mycotoxins, generate no toxic products, guarantee the nutritional value of the food, and induce no modification to the technological properties of the product. Detoxification processes effective *in vitro* do not necessarily retain their efficacy when tested *in vivo*. Although certain treatments are effective in reducing specific mycotoxins in foods and feeds, no single method is equally effective against the wide variety of mycotoxins occurring in different commodities. More work is needed to study the fate of mycotoxins during decontamination, detoxification, and food processing. Future studies should focus on the reduction of toxicological risk associated with processed commodities

D. Spadaro (✉)
Centro di competenza per l'Innovazione in campo agro-ambientale (AGROINNOVA), Università degli Studi di Torino, Largo P. Braccini 2, 10095 Grugliasco (TO), Italy
e-mail: davide.spadaro@unito.it

A. Garibaldi
DISAFA – Dipartimento di Scienze Agrarie, Forestali ed Alimentari, Università degli Studi di Torino, Largo P. Braccini 2, 10095 Grugliasco (TO), Italy
e-mail: angelo.garibaldi@unito.it

M.L. Gullino et al. (eds.), *Practical Tools for Plant and Food Biosecurity*, Plant Pathology in the 21st Century 8, DOI 10.1007/978-3-319-46897-6_8

contaminated with mycotoxins and on the prevention of recontamination during storage.

Keywords Adsorption • Aflatoxins • Aluminosilicates • Ammoniation • *Aspergillus* • Bacteria • Biological treatments • Chemical treatments • Containment • Control • Cooking • Decontamination • Detoxification • Extrusion • Feed • Fermentation • *Flavobacterium aurantiacum* • Flotation • Food • Food and Agriculture Organization • Food processing • Frying • Fumonisins • Fungi • *Fusarium* • Grinding • Heating • Irradiation • *Metschnikowia pulcherrima* • Microwave • Milling • Mould • Mycotoxin • Nixtamalization • Ochratoxins • Ozone • Patulin • *Penicillium* • Physical treatments • Plasma • Postharvest • Prevention • Roasting • Solvent extraction • Sorting • Transgenic plants • Trichothecenes • Ultrasound • Washing • Yeast

8.1 Introduction

Most food products are prone to fungal contaminations during the production phases, from farm to fork. The development of moulds can lead to a reduction of food quantity and quality, but also to the formation of fungal metabolites with toxic activity, such as mycotoxins. Mycotoxins are secondary metabolites produced by fungi belonging predominantly to the genera *Alternaria, Aspergillus, Claviceps, Fusarium,* and *Penicillium*. Although *Aspergillus* and *Penicillium* species are generally found as contaminants in food during drying and storage, *Alternaria, Claviceps* and *Fusarium* species can produce mycotoxins before or after harvesting (Sweeney and Dobson 1999). The mycotoxins of most significance from both a public health and agronomic perspective include aflatoxins, ochratoxin A, trichothecenes, zearalenone, fumonisins, and patulin. However, there are over 400 metabolites with toxigenic potential produced by fungi, some of them emerging as potential human risks, such as alternariatoxins (Prelle et al. 2013a) or phomopsins (Battilani et al. 2011). Exposure of humans and animals to mycotoxins occurs primarily through food and can cause acute and/or chronic intoxication, known as mycotoxicosis, which depend on various factors including intake levels, duration of exposure, toxin type, mechanisms of action, metabolism, and defence mechanisms (Hussein and Brasel 2001). Mycotoxins exhibit four basic kinds of toxicity: acute, chronic, mutagenic, and teratogenic. Mycotoxin contamination of agricultural products is a global problem but is most severe in tropical and subtropical regions. The Food and Agriculture Organization (FAO 1996) has estimated that up to 25 % of the world food crops are significantly contaminated with mycotoxins.

Due to the high resistance of some mycotoxins to the physical, chemical, or biological practices used for food processing, storage and sanitization, the most effective tools against mycotoxins are essentially based on the prevention of mould growth in each stage of the food chain. One approach to reduce the risk of mycotoxin contamination could be the development of an integrated system like HACCP

(Hazard Analysis and Critical Control Point), which should use the general principles of Good Agricultural Practice (GAP), Good Hygienic Practice and Good Manufacturing Practice (GMP).

Due to heterogeneity of food matrices contaminated, and chemical structure of mycotoxins, there are several control methods to ensure the reduction of mycotoxins. A strategy to reduce the risk of mycotoxin contamination should include prevention practice in the field and during the postharvest phase, and control measures. Prevention of mycotoxin formation in the field is based on the use of appropriate agricultural practices, such as the use of cultivars resistant to attack by mycotoxigenic fungi, appropriate techniques of irrigation and fertilization of crops, pesticide use and crop rotation (Kabak et al. 2006).

Postharvest strategies to maintain the health of grains and nuts and to prevent the contamination from mycotoxigenic fungi, include the drying of foodstuffs, appropriate use of synthetic or natural antimicrobial agents, storage in controlled atmosphere, low temperature and low relative humidity. To avoid the development of mycotoxigenic fungi on fresh produce, such as fruit and vegetables, use of fungicides, biocontrol agents, thermal treatments, essential oils or combinations of these treatments should be used (Lopez-Reyes et al. 2010; Spadaro and Droby 2016). Appropriate control measures should be developed at every level and they should be universally accepted able to protect public health and promote national and international trade.

Preventive measures aimed at the inhibition of mycotoxin formation in agricultural products are the most effective approach for avoiding consumer exposure. However, when contamination is not prevented during the preharvest and postharvest periods, several approaches can be employed to help remove mycotoxins from the subsequently contaminated commodities, including physical, chemical, and biological techniques. Different methods are used to decontaminate or detoxify food and feed from mycotoxins before ingestion. Decontamination aims at the removal of contaminated parts, while detoxification involves the destruction or inactivation of mycotoxins *in situ*. Presently, regulations do not permit the decontamination of food that exceeds the concentration threshold limits (Bullerman and Bianchini 2007). Both removal and detoxification of mycotoxins have been studied using physical, chemical or biological methods.

8.2 Food Decontamination Techniques

Techniques for food decontamination are based on the collection and removal of the contaminated parts from a mass of product. This practice is particularly effective for products that have discrete parts, such as cereals, legumes, nuts, fruit or vegetables. Decontamination can be easily applied to mycotoxins because, in contaminated food, they tend to concentrate on a relatively small number of seeds or kernels.

8.3 Physical Decontamination

Cleaning and mechanical separation of the contaminated parts are non-invasive valid systems for partial, but not complete, decontamination of the grain. Cleaning of grain can remove significant amounts of mycotoxins, particularly sclerotia of *Claviceps* spp. present in wheat and rye, and fumonisin (FUM) from maize.

Sorting techniques are the most used physical decontamination methods used by industrial production/processing of peanuts, oils and grains. The principle is to identify and remove visibly mouldy or presumably contaminated products, which could be broken, small, discoloured, malformed or lighter. The removal of contaminated portions is realized by manual, mechanical or electronic sorting techniques.

Sorting of kernels and seeds has been widely used for decontamination of aflatoxins (AF) from almonds, walnuts, and peanuts, while it is more difficult to apply to cereals (Bata and Lasztity 1999). Segregation by fluorescent light is another technique widely used to decontaminate peanuts, maize, cotton seeds and Brazil nuts from AF, when the portions appear fluorescent after exposure to ultraviolet light at 365 nm (Hocking 1997). This method has the disadvantage of producing false positives. Sorting of maize, to remove lighter, broken or wrinkled kernels, was also effective for FUM decontamination. Removal by sieving of the husk from barley, wheat, and rye contaminated with deoxynivalenol (DON) and zearalenon (ZEA) was effective in reducing the 40–100 % contamination levels of both toxins in the finished product. Sieving of flour of barley, wheat and maize, is able to significantly reduce the content of DON and ZEA, since both toxins tend to concentrate in fractions with smaller particle size.

The removal of damaged seeds of maize and peanut, presumably contaminated by AFs, can also be realized by flotation and density segregation (Huff and Hagler 1985), with the disadvantage of increasing the moisture content of the seeds and of requiring an additional cost due to drying of the treated products.

Since AFs are often associated with small particles in suspension, appropriate filtration systems can lead to effective AF decontamination during the processing of some types of peanut oils.

Grinding can be useful for contaminated maize, because mycotoxins are distributed in different milling fractions (Bullerman and Bianchini 2014). Wet grinding, mainly used in processing of maize contaminated with aflatoxin B1 (AFB1) leads to a distribution of the toxin in: steep liquor (42 % of the initial content), bran (38 %), gluten, germ and starch. Even wet milling of maize contaminated by DON, leads to an accumulation of toxin in the waters of maceration, although not negligible quantities are detectable in the starch. In the case of maize contaminated by ZEA, the toxin is concentrated mainly in the gluten, while the starch is almost free from contamination. Similarly, the starch obtained by wet milling of maize contaminated by FUMs contains negligible amounts of fumonisin B1 (FUMB1), while bran, gluten and germ have a concentration of toxin equal to 10–40 % of the initial concentration. Finally, wet milling of maize contaminated with T-2 toxin and ochratoxin A

(OTA) leads to a concentration of both toxins in the water of maceration, respectively up to 67 % and 43 %.

Dry grinding does not significantly reduce the levels of contamination by ZEA and DON, while it is effective for FUM and AF. In particular, after dry milling of maize, the highest concentration of FUMs are recorded in the bran, those of AFs in the germ and chaff.

At industrial level, an accurate combination of grain cleaning, grinding and sieving constitutes a valid decontamination system, able to reduce the risks of exposure to mycotoxins. However, for this strategy to be effective decontamination, it is important to know the distribution of the various mycotoxins in the milling fractions and to remove those where mycotoxins are concentrated.

Monitoring the fruit quality may be considered the primary measure of control of patulin adoptable by the fruit juice industry, not only for apple (Spadaro et al. 2007), but also for pear, peach and apricot juices and nectars (Spadaro et al. 2008). The use of batches with low incidence of rotten fruit is of extreme importance to avoid contamination of healthy fruit. According to FAO (2002), batches of apples of inferior quality (with high percentage of fruits damaged or rotten) should not be accepted for processing, since it would be very difficult to manually select apples on a lot with over 10 % rotten fruit and to obtain an acceptable patulin level in the final product.

During the early stages of fruit processing, washing fruits or removal of rotten areas do not necessarily imply the elimination of all the patulin present, as a percentage, albeit minimal, could also be found in the inner apparently healthy parts of the fruit. Removal of the rotten part is able to remove about 99 % of patulin present, but in apples, also at 1 or 2 cm from the affected area, significant amounts of patulin can be detected (Rychlik and Schieberle 2001).

8.4 Solvent Extraction

Chemical decontamination is based on the removal of mycotoxins by extraction with organic solvents, and has a very limited application for both undesirable organoleptic changes of the products and doubts about safety of treated products. The chemical decontamination has found wide application only for the sanitation of cooking oils, which can be decontaminated during alkaline refining process. In particular, AFs can be extracted from the contaminated oils with mixtures containing acetone-hexane-water, hexane-methanol, or with an aqueous solution of isopropanol (Rustom 1997). This method, while removing almost completely AFs, has the disadvantage of depleting important nutritional components, such as lipids and proteins, and to leave solvent residues.

Washing with distilled water removes 65–69 % DON and 2–61 % ZEA from contaminated barley and maize, while washing with bicarbonate solutions extracts up to 74 % DON and 87 % ZEA. However, washing with aqueous solutions of cereals may be considered a convenient decontamination system only used prior to wet milling or alcoholic fermentation. Otherwise, the costs for drying decontaminated grains make chemical decontamination economically disadvantageous on industrial scale.

8.5 Detoxification Processes

Detoxification processes should destroy or inactivate mycotoxins, generate no toxic products, guarantee the nutritional value of the food and induce no modification to the technological properties of the product. Removal of mycotoxins from contaminated commodities is rather difficult to achieve at industrial level, but the exploitation of practices of inactivation or destruction of mycotoxins by physical, chemical and/or biological agents may be useful.

Due to the high stability of mycotoxins, food should be treated with drastic detoxifying agents, with resulting profound modifications to achieve an acceptable degree of detoxification. Detoxification processes are often partial and, for the frequent presence of residues, reserve considerable doubts about the health and safety of the treated products. Several strategies are available for the detoxification of mycotoxins. These can be classified as physical, chemical, physicochemical and biological approaches (Karlovsky 1999; Park and Liang 1993; Ramos et al. 1996; Scott 1998; Varga and Toth 2004).

8.6 Heating Treatments

Physical treatments are generally included in the food processing chain and include cooking, boiling, roasting, microwave heating, extrusion, and irradiation. Mycotoxins are relatively heat-stable and tolerant to heat treatments (Bullerman and Bianchini 2007). Mycotoxin detoxification by thermal process depends on temperature, moisture content, and length. In some cases, a change in pH, pressure and humidity can help to facilitate the process of detoxification.

AFs are very resistant to most of the transformation processes that involve the use of heat, such as cooking in water or in autoclave. Partial removal of these toxins can be obtained by oil roasting or dry roasting. Interesting results were obtained by roasting hazelnuts contaminated by AFs in fixed air oven or in infrared oven: the second type of roasting was able to eliminate the four types of AFs on hazelnuts (Spadaro et al 2014a).

Cooking at high pressure (about 15 psi) allows a significant reduction of AF content (72 % in rice) and the preservation of the nutritional value of the finished product. After cooking pasta contaminated by AFs for 10 min in boiling water, 66 % of the toxins can be found in the finished product.

DON is the most heat resistant mycotoxin and, therefore, the least affected by food processing, including baking (Wolf and Bullerman 1998). Also FUMs are rather heat stable (Castelo et al. 1998). FUMB1 was not affected during polenta preparation after cooking for 20–30 min in boiling water, while a reduction of 16–28 % FUMB1 was recorded after cooking maize for 20 min at 175 °C or 200 °C. On the contrary, frying polenta reduced 70–80 % FUMB1 in the finished product. Nixtamalization (alkaline cooking in water and lime), which is used during the production of maize tortillas, tortilla chips and maize chips, is a process which can greatly reduce the AF levels maize (Torres et al. 2001). It also lowers considerably the content of FUMB1 in the finished products (Voss et al. 2012).

About ZEA, cooking naturally contaminated maize at 150 °C for 44 h led to a 28 % reduction of ZEA (Ryu et al. 2003). Studies on the thermal inactivation of OTA in samples of white flour indicated a considerable reduction of the toxin (76 %) after baking at 250 °C for 40 min.

Finally, the ergoline alkaloids are not very stable to heat and almost all the toxins present in contaminated cereals can be inactivated after oven baking.

8.7 Other Physical Detoxification Processes

The process of extrusion of maize, widely used by food manufacturers for the preparation of cornflakes and snacks, is able to reduce the levels of FUM contamination. In particular, in a study conducted to evaluate the stability of FUMs during the preparation of cornflakes, only 30–40 % of FUMs are recovered in the finished product, after extrusion and roasting (Voss et al. 2008).

Gamma rays can reduce AFs in maize and soybeans (Rustom 1997) and reducing the levels of T-2 toxin, DON and ZEA (in wheat, maize and soya beans) by 16 %, 25 % and 33 %, respectively, and for DON and FUM (in maize), respectively, 13 % and 20 % (Raso and Barbosa-Canovas 2003). Microwaves can reduce aflatoxin content in peanuts (Farag et al. 1996) and trichotecenes in corn (Scott 1998).

A detoxification of 70–90 % trichothecenes was observed after treatment with ultrasound of contaminated maize, without alteration in appearance or taste.

8.8 Mycotoxin Adsorption

Among the physical methods of detoxification for feed intended for animal consumption, the addition of adsorbents (sodium and calcium aluminosilicates, zeolites, active carbon, bentonite, clays or specialty polymers) represents the most innovative and effective to reduce the risk of poisoning by mycotoxins and to avoid, the transfer through carry-over of toxins in the products of animal origin (Jard et al. 2011). Adsorbents are inert materials able for their structure and physical-chemical properties to stably bind mycotoxins and reduce absorption in the gastrointestinal tract. Many of these materials are available on the market and widely used among farmers for prevention and treatment of mycotoxicosis. However, the effectiveness of most of these compounds and the real benefits remain uncertain because of the lack of knowledge about their ability to sequester mycotoxins *in vivo* and their potential adverse effects, including depletion of essential nutritional factors.

Among the most promising absorbents to prevent the risk of aflatoxicosis, there are aluminosilicates, derived from natural zeolites (Phillips et al. 2008). Aluminosilicates can also reduce the residues of aflatoxin M1 in cow and goat milk. The main limitations of aluminosilicates are to absorb nutrients and to have a rather narrow spectrum of activity, being unable to protect the animals from other important mycotoxins. Activated carbons, obtained by the pyrolysis of organic materials, are able to sequester *in vitro* most of the mycotoxins (Avantaggiato et al. 2004). Bentonite, commonly used as binding and lubricant agent in the preparation of pelleted feeds (Kurtbay et al. 2008), is capable of reducing the gastrointestinal absorption of T-2 toxin in rats, alleviating the symptoms of intoxication. Other clays (kaolin, sepiolite and montmorillonite) added to animal feed contaminated with AFB1 are able to reduce the symptoms of aflatoxicosis, by sequestering the mycotoxin.

Among the synthetic anion exchange resins, synthetic zeolites are capable of reducing in rats the intestinal absorption of ZEA (Tangni et al. 2006). Polymers of styrene-divinylbenzene are able to reduce *in vivo* the toxic effects produced in rats by ingestion of T-2 toxin and ZEA. The mixture polyvinylpolypyrrolidone/bentonite was able to normalize blood parameters of hens altered after ingestion of feed contaminated with AFB1.

Despite the increasing number of products on the market, only a few adsorbents were carefully evaluated in experiments *in vivo* with animals.

8.9 Chemical Detoxification

A large variety of products, including acids, bases (ammonia, sodium hydroxide), oxidizing agents (hydrogen peroxide, ozone), reducing agents (bisulphite), chlorinating agents (chlorine), salts, formaldehyde, were assayed to verify their ability to degrade mycotoxins in foodstuffs.

Acid treatment is based on the use of aqueous solutions of strong acids and it is able to produce a significant reduction of AFB1 and aflatoxin G1 (AFG1). This method, however, has the drawback of being ineffective for the detoxification of aflatoxins B2 (AFB2) and G2 (AFG2), and it is difficult to apply on a large scale. In contrast, hydrogen peroxide is commonly used in India on a commercial scale, for the detoxification of peanuts contaminated by AF. Sodium bisulphite, widely used in the food industry for its preservative and antioxidant properties, is able to react with the main AFs forming water-soluble products that can be easily removed from the matrix. Moreover, the association of sodium bisulphite with hydrogen peroxide and heat is able to enhance its effectiveness in detoxifying AFB1 in dried figs. Satisfactory results were obtained at industrial level after detoxification of AFs in oilseeds with aqueous solutions of calcium hydroxide and monomethylamine.

Monomethylamine is also able to produce a reduction of 50–99 % of T-2 toxin and diacetoxiscirpenol in contaminated feed. In a recent study, lactic acid has been shown to efficiently degrade aflatoxin B1 into aflatoxin B2, with aflatoxin B2 as the major degradation product under heat treatment (Aiko et al. 2015).

Reductions up to 95 % in the level of DON were recorded after maize treatment in autoclave for 1 h at 121 °C in the presence of bisulphite solution. The sodium bisulphite transformation of DON to DON-sulfonate, which is less toxic than DON, was reported as an effective tool to overcome the depressive effects of DON on feed-intake in piglets (Dänicke et al. 2005). This detoxification treatment does not work for flour in that, while being able to inactivate the toxin without releasing toxic secondary metabolites, it has the drawback of altering the rheological properties.

Ozone treatment can reduce up to 90 % of DON in maize and AFs in hazelnuts (Spadaro et al. 2014b), while it has little effect on the content of DON in wheat, and it is ineffective for the detoxification of FUMB1.

During the last years, plasma technology was used for degradation of fungal spores, but it could also be used for the degradation of mycotoxins. The operating parameters of cold atmospheric pressure plasma were optimized to reduce the presence of aflatoxins on hazelnuts. First, the effect of different gases, power, and exposure time were optimized. On hazelnuts, with plasma treatment at 1000 W for 12 min, a reduction in the concentration of total aflatoxins and AFB_1 of over 70 % was obtained (Siciliano et al. 2016). Cold atmospheric pressure plasma could be a promising method for degradation of aflatoxins in food.

The heating of aqueous solutions of FUMB1 with reducing sugars, such as fructose or D-glucose, originates N-(carboxymethyl)-derivatives which are much less cytotoxic. Further studies are required to verify the effectiveness of this treatment for the detoxification of FUMs.

Several alkaline treatments, including hydrogen peroxide/sodium bicarbonate with or without calcium hydroxide, are able to inactivate almost completely FUMB1 in maize. Treatment with formaldehyde, in aqueous solution or vapour state, is able to reduce the levels of ZEA in maize. An elimination of 100 % of ZEA in maize grits was observed after treatment for 16 h at 50 °C with a solution at 3.7 % formaldehyde.

Finally, exposure for 30 min to 30 % chlorine gas was able to completely pull down DON in maize.

8.10 Ammoniation

Among the chemical methods of detoxification of AF, ammoniation is effective and it has a practical use on industrial scale for feed detoxification (Huwig et al. 2001). Ammoniation provides the use of ammonia in the liquid or gaseous state and at a concentration lower than 7 %. This treatment may be performed at ambient temperature and pressure for several days (14–42 days), or at high temperatures (70–120 °C) and pressure (around 35–50 psi) for few hours. Ammoniation at room temperature and ambient pressure is preferred in treatments to be performed directly on farms as it is cheaper. Ammoniation at high pressure and temperature requires, however, the use of adequate facilities and staff, and it is performed mainly in food processing plants. This process is legally permitted in certain countries for detoxification of cotton seed, maize and peanut flour. The detoxification process by ammonia is irreversible and can lead, if properly applied, to the total inactivation of AFs. *In vivo* studies of toxicity did not show any toxic effect due to ammoniation process.

The effect of ammoniation on FUM detoxification is less clear. In some studies, ammoniation at high pressure and room temperature or at low pressure and high temperature was able to reduce by 79 % FUMB1 in maize. In other studies, however, a similar treatment of maize reduced the mycotoxin only by 24–45 % FUMB1.

The exposure of maize contaminated by DON to ammonia for 18 h resulted in a reduction of the toxin by 85 %. Also the exposure for 16 h to a solution of ammonium hydroxide to 3 % has, however, reduced by 64 % the content of ZEA in maize naturally contaminated.

8.11 Biological Detoxification

Besides physical and chemical methods, also microbes or their enzymes could be applied for mycotoxin detoxification, due their ability to degrade or enzymatically transform the mycotoxin, by eliminating or reducing its toxicity (Schatzmayr et al. 2006). Biological detoxification requires the use of specific biotic agents (bacteria, moulds, yeasts, plants or their metabolites), which should be selected for their ability to inactivate one or more mycotoxins.

Among bacteria, *Flavobacterium aurantiacum* (also known as *Nocardia corynebacteriodes*).seems to be the most effective species to detoxify AFs in milk or oil. Studies performed using radioactive AFB1 have shown that the bacterium is able to metabolize the toxin and to form non-toxic derivatives. The red-orange pigmentation associated with the growth of the microorganism is a limitation to its use. Biological detoxification of AFs can also be done with some fungi as *Trichoderma viride* and *Aspergillus niger*, however, these microorganisms can produce secondary metabolites of unknown toxicity. Teniola et al. (2005) isolated the extracellular enzymes from *Rhodococcus erythropolis*, responsible for the biotransformation of AFB1.

Some bacteria isolated from soil were able to degrade, in *in vitro* tests, DON to 3-keto-deoxynivalenol, a less immunotoxic derivative. *Gliocladium roseum* resulted able to irreversibly detoxify ZEA, by causing the breakdown of the lactone with subsequent decarboxylation.

Several microbes and their enzymes are capable of detoxifying OTA, including bacteria, yeasts (Angioni et al. 2007; Molnar et al. 2004; Schatzmayr et al. 2006), filamentous ascomycetes (Varga et al. 2000, 2005) and basidiomycetes (Engelhardt 2002). Three yeast, *Metschnikowia pulcherrima, Pichia guilliemonodii* and

Rhodococcus erythropolis were selected for their ability to degrade OTA: 30 °C was the optimum temperature for yeast growth and OTA degradation. In particular, *M. pulcherrima* effectively degraded OTA (>80 %) at 30 °C after 15 days incubation and few amounts of OTA adsorption was observed in the yeast cell wall (Patharajan et al. 2011). LC-MS studies revealed that no by-products like α-ochratoxin or phenylalanine were found during the degradation process.

M. pulcherrima was also able to reduce the development of blue mould of apple (Spadaro et al. 2013) and to completely (100 %) degrade patulin within 48 h. Patulin was not absorbed in yeast cell walls, it was completely degraded, and the mycotoxin did not affect yeast cell concentration during growth (Reddy et al. 2011). Therefore, these yeast strains could potentially be used for the reduction of patulin in naturally contaminated fruit juices.

Two strains of yeast (including the *Exophiala spinifera*) and a bacterial strain belonging to the genus *Caulobacter*, isolated from maize, have the ability to grow using FUMB1 as the sole source of carbon and degrade it completely. The enzymes responsible for the degradation of FUMB1 were isolated and expressed in a variety of transgenic maize which, in cultivation trials, were able to accumulate a minor amount of toxin compared to traditional varieties.

Other varieties of transgenic maize (Bt maize), genetically modified to express proteins from *Bacillus thuringiensis* toxic to certain insects, presented a higher resistance to the attacks of the European corn borer and consequently to the colonization of the ears by toxigenic strains of *Fusarium verticillioides*, with lower content of FUM in the maize (Wu 2006).

8.12 Fermentation Processes

Several studies have been performed to investigate the possibility of producing ethanol or fermented food products from food matrices highly contaminated with mycotoxins. However, it is important to verify that the presence of mycotoxins in the matrix does not affect yeast viability or alter the properties of the finished product.

AFB1 is partially removed during the process of beer production, and a quantity of toxin equal to 18–27 % is recoverable in the finished product. Moreover, AFs tend to accumulate in solid exhausted residues, which are often destined to animal feed.

In alcoholic fermentation with maize contaminated by ZEA, the presence of the toxin had no effect on the final alcohol yield and the alcohol produced was free from ZEA. On the contrary, using maize contaminated by ZEA for the beer production, an accumulation of toxin was observed in the residual solid, and a concentration equal to 51 % was registered in the product finished. In presence of *Saccharomyces cerevisiae*, 69 % of ZEA present in the wort was converted into beta-zearalenol, a metabolite with lower toxicity.

OTA can be recovered both on beers (Prelle et al. 2013b) and wines (Spadaro et al. 2010), after the fermentation processes. In the beer production, the fermentation process seems to have no effect on the stability of OTA: a concentration of toxin equal to 96 % of the initial one is recovered in the finished product. The fermentation process does not affect also FUMB1 and 85 % of the toxin can be recovered in the final products.

The alcoholic fermentation of apple juice, patulin is completely detoxified; similarly, wine produced from mouldy grapes were free from patulin contamination.

8.13 Conclusions

A wide range of decontamination and detoxification strategies are available to reduce the mycotoxin in contaminated food or feed. Detoxification processes effective *in vitro* do not necessarily retain their efficacy when tested *in vivo*. Although certain treatments are effective in reducing specific mycotoxins in foods and feeds, no single method is equally effective against the wide variety of mycotoxins occurring in different commodities. More work is needed to study the fate of mycotoxins during decontamination, detoxification, and food processing. Future studies should focus on the reduction of toxicological risk associated with processed commodities contaminated with mycotoxins and on the prevention of recontamination during storage.

References

Aiko V, Edamana P, Mehta A (2015) Decomposition and detoxification of aflatoxin B1 by lactic acid. J Sci Food Agric. doi:10.1002/jsfa.7304

Angioni A, Caboni P, Garau A, Farris A, Orro D, Budroni M (2007) In vitro interaction between ochratoxin A and different strains of *Saccharomyces cerevisiae* and *Kloeckera apiculata*. J Agric Food Chem 55:2043–2048

Avantaggiato G, Havenaar R, Visconti A (2004) Evaluation of the intestinal absorption of deoxynivalenol and nivalenol by an *in vitro* gastrointestinal model, and the binding efficacy of activated carbon and other adsorbent materials. Food Chem Toxicol 42:817–824

Bata A, Lasztity R (1999) Detoxification of mycotoxin contaminated food and feed by microorganisms. Trends Food Sci Technol 10:223–228

Battilani P, Gualla A, Dall'Asta C, Pellacani C, Galaverna G, Giorni P, Caglieri A, Tagliaferri S, Spadaro D, Pietri A, Dossena A, Marchelli R, Gullino ML, Costa LG (2011) Phomopsins: an overview of phytopathological and chemical aspects, toxicity, analysis and occurrence. World Mycotoxin J 4:345–359

Bullerman LB, Bianchini A (2007) Stability of mycotoxins during food processing. Int J Food Microbiol 119:140–146

Bullerman LB, Bianchini A (2014) Good food processing techniques: stability of mycotoxins in processed maize-based foods. In: Leslie JF, Logrieco AF (eds) Mycotoxins in food chains. Wiley Blackwell, Ames, pp 89–100

Castelo MM, Sumner SS, Bullerman LB (1998) Stability of fumonisins in thermally processed corn products. J Food Prot 61:1030–1033

Dänicke S, Valenta H, Gareis M, Lucht HW, von Reichenbach H (2005) On the effects of a hydrothermal treatment of deoxynivalenol (DON)-contaminated wheat in the presence of sodium metabisulphite (Na2S2O5) on DON reduction and on piglet performance. Anim Feed Sci Technol 118:93–108

Engelhardt G (2002) Degradation of ochratoxin A and B by the white rot fungus *Pleurotus ostreatus*. Mycotoxin Res 18:37–43

FAO (Food and Agriculture Organization) (1996) Basic facts on the world cereal situation, Food outlook, 5/6. Food and Agriculture Organization, Rome

FAO/WHO (Food and Agriculture Organisation of United Nations/World Health Organisation) (2002) Proposed draft code of practice for the prevention of Patulin contamination in apple juice and apple juice ingredients in other beverages. In: Summary Meeting reports, Codex Committee on Food Additives and Contaminants, 34th Session Rotterdam, Netherlands, 11–15 March 2002

Farag RS, Rashed MM, Abo Hagger AA (1996) Aflatoxin destruction by microwave heating. Int J Food Sci Nutr 47:197–208

Hocking AD (1997) Toxigenic *Aspergillus* species. In: Doyle MP, Beuchat LR, Montville TJ (eds) Food microbiology, fundamentals and frontiers. ASM press, Washington, DC, pp 393–405

Huff WE, Hagler WMJ (1985) Density segregation of corn and wheat naturally contaminated with aflatoxin, deoxynivalenol and zearalenone. J Food Prot 48:416–420

Hussein HS, Brasel JM (2001) Toxicity, metabolism, and impact of mycotoxins on humans and animals. Toxicology 167:101–134

Huwig A, Freimund S, Käppeli O, Dutler H (2001) Mycotoxin detoxification of animal feed by different adsorbents. Toxicol Lett 122:179–188

Jard G, Liboz T, Mathieu F, Guyonvarch A, Lebrihi A (2011) Review of mycotoxin reduction in food and feed: from prevention in the field to detoxification by adsorption or transformation. Food Addit Contam A 28:1590–1609

Kabak B, Dobson ADW, Var I (2006) Strategies to prevent mycotoxin contamination of food and animal feed: a review. Crit Rev Food Sci Nutr 46:593–619

Karlovsky P (1999) Biological detoxification of fungal toxins and its use in plant breeding, feed and food protection. Nat Toxins 7:1–23

Kurtbay HM, Bekci Z, Merdivan M, Yurdakoc K (2008) Reduction of ochratoxin a levels in red wine by bentonite, modified bentonites, and chitosan. J Agric Food Chem 56:2541–2545

Lopez-Reyes JG, Spadaro D, Garibaldi A, Gullino ML (2010) Efficacy of plant essential oils on postharvest control of rot caused by fungi on four cultivars of apples *in vivo*. Flavour Fragrance J 25:171–177

Molnar O, Scahtzmayr G, Fuchs E, Prillinger H (2004) Trichosporon mycotoxivorans sp. nov., a new yeast species useful in biological detoxification of various mycotoxins. Syst Appl Microbiol 27:661–671

Park DL, Liang B (1993) Perspectives on aflatoxin control for human food and animal feed. Rev Trends Food Sci Technol 4:334–342

Patharajan S, Reddy KRN, Karthikeyan V, Spadaro D, Lorè A, Gullino ML, Garibaldi A (2011) Potential of yeasts antagonists on *in vitro* biodegradation of ochratoxin A. Food Control 22:290–296

Phillips TD, Afriyie-Gyawu E, Williams J, Huebner H, Ankrah NA, Ofori-Adjei D, Jolly P (2008) Reducing human exposure to aflatoxin through the use of clay: a review. Food Addit Contam A 25:134–145

Prelle A, Spadaro D, Garibaldi A, Gullino ML (2013a) A new method for detection of five alternaria toxins in food matrices based on LC-APCI-MS. Food Chem 140:161–167

Prelle A, Spadaro D, Denca A, Garibaldi A, Gullino ML (2013b) Comparison of clean-up methods for ochratoxin A on wine, beer, roasted coffee and chili commercialized in Italy. Toxins 5:1827–1844

Ramos AJ, Fink-Gremmels J, Hernandez E (1996) Prevention of toxic effects of mycotoxins by means of non-nutritive adsorbent compounds. J Food Prot 59:631–641

Raso J, Barbosa-Canovas GV (2003) Nonthermal preservation of foods using combined processing techniques. Crit Rev Food Sci Nutr 43:265–285

Reddy KRN, Spadaro D, Gullino ML, Garibaldi A (2011) Potential of two *Metschnikowia pulcherrima* (yeast) strains for *in vitro* biodegradation of patulin. J Food Prot 74:154–156

Rustom IYS (1997) Aflatoxin in food and feed: occurrence, legislation and inactivation by physical methods. Food Chem 59:57–67

Rychlik M, Schieberle P (2001) Model studies on the diffusion behavior of the mycotoxin patulin in apples, tomatoes, and wheat bread. Eur Food Res Technol 212:274–278

Ryu D, Hanna MA, Eskridge KM, Bullerman LB (2003) Heat stability of zearalenone in an aqueous buffered model system. J Agric Food Chem 51:1746–1748

Schatzmayr G, Zehner F, Taubel M, Schatzmayr D, Klimitsch A, Loibner AP (2006) Microbiologicals for deactivating mycotoxins. Mol Nutr Food Res 50:543–551

Scott PM (1998) Industrial and farm detoxification processes for mycotoxins. Rev Med Vet 149:543–548

Siciliano I, Spadaro D, Prelle A, Vallauri D, Cavallero MC, Garibaldi A, Gullino ML (2016) Use of cold atmospheric plasma to detoxify hazelnuts from aflatoxins. Toxins 8:125

Spadaro D, Droby S (2016) Development of biocontrol products for postharvest diseases of fruit: the importance of elucidating the mechanisms of action of yeast antagonists. Trends Food Sci Technol 47:39–49

Spadaro D, Ciavorella A, Frati S, Garibaldi A, Gullino ML (2007) Incidence and level of patulin contamination in pure and mixed apple juices marketed in Italy. Food Control 18:1098–1102

Spadaro D, Garibaldi A, Gullino ML (2008) Occurrence of patulin and its dietary intake through pear, peach and apricot juices in Italy. Food Addit Contam B 1:134–139

Spadaro D, Lorè A, Garibaldi A, Gullino ML (2010) Occurrence of ochratoxin A before bottling in DOC and DOCG wines produced in Piedmont (northern Italy). Food Control 21:1294–1297

Spadaro D, Lorè A, Garibaldi A, Gullino ML (2013) A new strain of *Metschnikowia fructicola* for postharvest control of *Penicillium expansum* and patulin accumulation on four cultivars of apple. Postharvest Biol Technol 75:1–8

Spadaro D, Prelle A, Gullino ML, Garibaldi A (2014a) Studio di parametri di tostatura con forni ad infrarossi ed ad aria per la detossificazione di nocciole da aflatossine. Protezione delle Colture 7(2):84

Spadaro D, Prelle A, Cavallero MC, Vallauri D, Gullino ML, Garibaldi A (2014b) Impiego di plasma a freddo e ozono per la decontaminazione di nocciole da aflatossine e di grano da deossinivalenolo. Protezione delle Colture 7(2):83

Sweeney MJ, Dobson ADW (1999) Molecular biology of mycotoxin biosynthesis. FEMS Microbiol Lett 175:149–163

Tangni EK, Simonis J, Larondelle Y, De Meeus D, Argenteuil L (2006) Biological method for detoxification of a liquid food medium. US patent 2006/0263410 A1. Clark & Brody, Washington, DC.

Teniola OD, Addo PA, Brost IM, Färber P, Jany KD, Alberts JF, van Zyl WH, Steyn PS, Holzapfel WH (2005) Degradation of aflatoxin B-1 by cell-free extracts of *Rhodococcus erythropolis* and *Mycobacterium fluoranthenivorans* sp. nov. DSM44556(T). Int J Food Microbiol 105:111–117

Torres P, Guzman-Ortiz M, Ramirez-Wong B (2001) Revising the role of pH and thermal treatments in aflatoxin content reduction during the tortilla and deep frying processes. J Agric Food Chem 49:2825–2829

Varga J, Toth B (2004) Strategies to control mycotoxins in feeds. Acta Vet Hung 53:189–203

Varga J, Rigo K, Teren J (2000) Degradation of ochratoxin A by *Aspergillus* species. Int J Food Microbiol 59:1–7

Varga J, Peteri Z, Tabori K, Teren J, Vagvolgyi C (2005) Degradation of ochratoxin A and other mycotoxins by *Rhizopus* isolates. Int J Food Microbiol 99:321–328

Voss KA, Bullerman LB, Bianchini A, Hanna MA, Ryu D (2008) Reduced toxicity of fumonisin B1 in corn grits by single-screw extrusion. J Food Prot 71:2036–2041
Voss KA, Riley RT, Moore ND, Burns TD (2012) Alkaline cooking (nixtamalization) and the reduction in the in vivo toxicity of fumonisin-contaminated corn in a rat feeding bioassay. Food Addit Contam A 30:1415–1421
Wolf CE, Bullerman LB (1998) Heat and pH the concentration of deoxynivalenol in an aqueous environment. J Food Prot 61 365–367
Wu F (2006) Mycotoxin reduction in Bt corn: potential economic, health, and regulatory impacts. Transgenic Res 15:277–289

Chapter 9
Decision Tool for Assessing the Likelihood of an Intentional Foodborne Illness Outbreak

Li Maria Ma, Jacqueline Fletcher, John D. Mumford, Johnson Holt, and Adrian W. Leach

Abstract Outbreaks of foodborne illnesses are frequent, but intentional cases are rare. Authorities dealing with suspected intentional foodborne illness outbreaks need a decision support tool to help distinguish accidental or natural outbreaks from intentional cases. Two broad discrimination models are available that cover biological warfare agents but these are not fully relevant to the scale and nature of intentional foodborne illness outbreaks. Two new models are proposed, one involving a scoring system, evaluated on total points, and another based on a Bayesian Network. The Bayesian Network model is more complex, but deals with uncertainty explicitly. The two proposed models are demonstrated by assessing four known outbreaks, two intentional and two accidental.

Keywords Foodborne outbreaks • Decision tool • Intentional • Scoring system • Bayesian network • Assessment

9.1 Introduction

A foodborne illness outbreak (FBIO) is defined as the occurrence of two or more cases of a similar illness resulting from the ingestion of a common food (Gould et al. 2013). Each year, large numbers of foodborne illness outbreaks are reported worldwide. According to the European Food Safety Authority and the European Centre for Disease Prevention and Control, a total of 5196 foodborne illness outbreaks were reported in the European Union in 2013 (EFSA 2015). In the United States, there were 831 foodborne illness outbreaks in 2012 (CDC 2014) and 10,409

L.M. Ma (✉) • J. Fletcher
National Institute for Microbial Forensics & Agricultural Biosecurity, Department of Entomology and Plant Pathology, Oklahoma State University, Stillwater, OK, USA
e-mail: li.ma@okstate.edu

J.D. Mumford • J. Holt • A.W. Leach
Centre for Environmental Policy, Imperial College London, London, UK

M.L. Gullino et al. (eds.), *Practical Tools for Plant and Food Biosecurity*, Plant Pathology in the 21st Century 8, DOI 10.1007/978-3-319-46897-6_9

foodborne illness outbreaks reported to CDC between 2002 and 2011 (CSPI 2014). Most of these were considered to be "natural" or "accidental" events, in which the presence of the causal agents in the food vehicle was believed to be the result of improper handling. To date, there have been only two confirmed and well-documented foodborne illness outbreaks attributable to intentional contamination in the world (Carus 2001). These are the salmonellosis outbreak at The Dalles, Oregon, in 1984 (Torok et al. 1997) and the shigellosis outbreak in Dallas, Texas, in 1996 (Kolavic et al. 1997). The infrequency of known intentional events may indicate that foodborne illness outbreaks due to intentional contamination are rare, or that there is difficulty in identifying such events. Insufficient data has been the main obstacle in identifying a biological attack (bioterror or biocrime) in general (Rediel 2004) and this factor may be more pronounced with foodborne illness outbreaks. Unlike most biological warfare agents, the biological agents causing foodborne illnesses, such as *Salmonella* and Shiga toxin-producing *E. coli*, are routinely carried by animals and are easily and frequently encountered in nature. Nevertheless, the ability to differentiate between "natural" and "deliberate" foodborne illness outbreaks, especially in the early stages of an outbreak investigation, is very critical. Identifying hoaxes or other false claims is also important to reduce panic or unnecessary disruption to normal food trade.

In this chapter, we have adapted and demonstrated a tool that could be used by investigators to assess the likelihood that a foodborne illness outbreak may have been intentionally caused. Relevant data and information were collected on the two intentional foodborne illness outbreaks that have been confirmed and this was used to assess the applicability of two existing models (Grunow and Finke 2002; Rodosavljevic and Belojevic 2012) developed for differentiating between natural and deliberate outbreaks of a wider set of biological warfare agents. A set of assessment criteria was developed to fit the nature and scope of foodborne illness outbreak events and subsequently developed into two new prototype decision tool models, one based on a scoring system and the other on a Bayesian network.

9.2 Collection of Data on Confirmed Intentional Foodborne Illness Outbreaks

Relevant data and information were obtained by a literature review on the salmonellosis outbreak at The Dalles, Oregon, 1984 and the shigellosis outbreak at Dallas, Texas, 1996. The data are described below.

9.2.1 Salmonellosis-The Dalles, Oregon, 1984

From 9–18 September, and 19 September to 10 October, 1984, two waves of salmonellosis cases occurred in The Dalles, Oregon, U.S.A. Local health authorities received illness reports first on 17 September, 1984. Case interviews by health officials revealed a connection between dining at two local restaurants and the illness, associated primarily with eating food items from salad bars. *Salmonella* Typhimurium isolates were obtained from clinical specimens. As gastroenteritis cases occurred in increasing numbers, health authorities closed all salad bars in The Dalles on 25 September, 1984 while they conducted an increasingly extensive investigation. *S.* Typhimurium was isolated from 388 patients in this outbreak. There were only eight isolates of *S.* Typhimurium that had been collected by the local health department during the 4 years before the outbreak. Eventually 751 salmonellosis cases were identified, with patients ranging from newborns to 87 years old. Most cases were associated with dining in one of ten local restaurants.

Epidemiological analysis revealed that multiple food items in the salad bars were involved. The local health department and the US Department of Agriculture (USDA) investigated the distribution chain and suppliers of foods used in these restaurants but no suppliers common to all ten were identified. Sanitarians inspected the restaurants and collected and analyzed tap water, but found no Salmonella. The state public health laboratory serotyped and performed antibiotic-susceptibility testing on the human clinical isolates.

One year later, in October 1985, the US Federal Bureau of Investigation (FBI) investigated a nearby religious cult (Rajneeshees) for unrelated criminal violations. In the course of the FBI search of the premises a vial containing a culture of *S.* Typhimurium was discovered in the cult's health clinic laboratory (Carus 2001). Records were found documenting its purchase prior to the outbreak. One commune member then admitted placing bacterial cultures in salad dressing (Carus 2001).

9.2.2 Shigellosis – Dallas, Texas, 1996

Between 29 October and 1 November, 1996, twelve clinical laboratory workers at the St Paul Medical Center in Dallas developed a severe acute diarrheal illness. *Shigella dysenteriae* type 2 was cultured from stools collected from these workers. This was an uncommon strain of *Shigella* that had not been associated with outbreaks since 1983. Interviews with the patients revealed an association between eating pastries (muffins and doughnuts) left in their break room on 29 October, 1996 and the illness. Another individual became ill with *S. dysenteriae* type 2 after eating the same pastries, brought home by a laboratory worker. In total, five patients were treated in hospital emergency departments and released; four were admitted for hospital care.

Interviews of laboratory employees by the investigating epidemiologists revealed that an anonymous email had been sent from a supervisor's computer inviting workers to eat pastries in the laboratory break room while the supervisor was away from the office. The pastries were prepared by a commercial bakery, but no other cases were reported in the community. Everyone who ate a muffin or doughnut became ill (100 % attack rate, percentage of people sick over the total number of people who have consumed the implicated food). No links were found between illness and consumption of other food or beverages from the break room refrigerator or cafeteria. *S. dysenteriae* type 2 was also isolated from a leftover muffin specimen.

Examination of the hospital laboratory storage freezer revealed tampering with reference cultures of *S. dysenteriae* type 2. DNA fingerprinting by pulsed-field gel electrophoresis revealed that patterns among the reference culture and the isolates from patients and the muffin specimen were indistinguishable.

After the epidemiological report was published, it was hypothesized that someone had removed a laboratory reference culture, possessed the laboratory skills to culture the pathogen and to inoculate the pastries, and also had access to the locked break room (Kolavic et al. 1997). A police investigation was also launched during the epidemiological study and attention was focused on a laboratory technician (Carus 2001). Nearly a year later, on 28 August, 1997, investigators indicted the laboratory technician who had access to the laboratory culture stocks on three charges of tampering with a food product and infecting 12 co-workers with *S. dysenteriae* type 2. She was eventually sentenced to 20 years in prison (Carus 2001).

9.3 Assessment of Confirmed Intentional Outbreaks of Foodborne Illnesses by Existing Models

There are two existing models for differentiating between natural and deliberate outbreaks of biological warfare agents (Grunow and Finke 2002; Radosavljevic and Belojevic 2012). The Grunow and Finke (2002) model consists of two categories of criteria, 11 non-conclusive (shown in Table 9.1) and two conclusive (shown in Table 9.2).

In the Grunow and Finke model, the likelihood that an epidemic has an intentional cause is determined by first calculating the total number of points and then comparing the total with an arbitrary scale of likelihood as shown in Table 9.3. The logic of the scale is based on the lowest third of possible scores being unlikely, the middle third doubtful, and the upper third likely or high likely.

The second assessment model, published by Radosavljevic and Belojevic (2012) consists of 10 indicators describing an epidemic as shown in Table 9.4. Each indicator is scored with 0 or 1, for a low or high probability of an unnatural outbreak, respectively. Evaluation is made in three categories, based arbitrarily on approximate thirds of the total possible score.

In this work, we examined whether these two general decision tool models (Grunow and Finke 2002; Rodosavljevic and Belojevic 2012) could be useful when applied to investigations of illness outbreaks caused by foodborne human pathogens. To do this, we assessed available data from the two known incidents of intentional contamination of food items. Our assessment of the 1984 Salmonellosis outbreak in The Dalles, Oregon (The Dalles, 1984), using the Grunow and Finke model is summarized in Table 9.5.

Table 9.1 Non-conclusive criteria for assessing an epidemic in order to rule out the use of a biological warfare agent

No.	Criterion	Assessment (possible points)[a]	Weighting factor	Maximum number of points[b]
1.	Existence of a biological risk	0–3	2	6
2.	Existence of a biological threat	0–3	3	9
3.	Special aspects of the biological agent	0–3	3	9
4.	Peculiarities of the geographic distribution of the biological agent	0–3	1	3
5.	High concentration of the biological agent in the environment	0–3	2	6
6.	Peculiarities of the transmission mode of the biological agent	0–3	1	3
7.	Peculiarities of the intensity and dynamics of the epidemic	0–3	2	6
8.	Peculiarities of the time of the epidemic	0–3	1	3
9.	Unusually rapid spread of the epidemic	0–3	1	3
10.	Limitation of the epidemic to a specific population	0–3	1	3
11.	Peculiarities of the clinical manifestation	0–3	1	3
	Total			54

Reproduced from Grunow and Finke (2002)
[a]Assessment of a criterion
0 = Criterion ruled out or no data available (giving "no data" cases a zero is conservative, requiring "proven" evidence to contribute to the result)
1 = Existence of peculiarities or suspicions, but uncertain and indistinct
2 = Existence of obvious peculiarities or indications, causes yet to be clarified for certain
3 = Existence of considerable peculiarities or deviations from expected norm, clear indication or proof of an intentional release
[b]Assessment score × weighting factor

Table 9.2 Conclusive criteria for assessing an epidemic in order to rule out the use of a biological warfare agent[a]

No.	Criterion	Assessment (possible point)	Weighting factor	Maximum number of points
1.	Identification of the agent as a biological warfare agent	0–3	N/A	N/A
2.	Proof of the release of the agent as a biological weapon	0–3	N/A	N/A

Reproduced from Grunow and Finke (2002)
[a]Conclusive criteria are not weighted, since they are, by definition, conclusive proof. For this reason, they are taken into special consideration in the final result

Table 9.3 Assessing the likelihood of the use of a biological warfare agent based on non-conclusive criteria from Table 9.1

Number of points	Assumption of an intentional biological attack
51–54	Highly likely
36–50	Likely
18–35	Doubtful
0–17	Unlikely

Table 9.4 Assessment for differentiation between natural, accidental and deliberate outbreaks

No	Indicators	Score[a]
1.	Unusual/atypical disease/manifestation or unexpected fulminant course of disease in humans and/or animals	0, 1
2.	Failure of patient to respond to usual therapy or illness in a population despite immunizations	0, 1
3.	Several unusual/unexplained syndromes coexisting in the same case without any other explanation	0, 1
4.	Sudden unexplainable increase in the number of cases or deaths in human populations	0, 1
5.	Higher than expected morbidity and/or mortality rates	0, 1
6.	Clustering of patients with fever only or with fever and other symptoms	0, 1
7.	Disease identified in the region for the first time, again after a long period of time	0, 1
8.	A disease with an unusual/atypical seasonal distribution	0, 1
9.	Simultaneous occurrence of epidemics and/or epizootics	0, 1
10.	Explosive epidemics/outbreaks with indicators of a point-source origin	0, 1
11.	Disease with an unusual geographic distribution	0, 1
12.	Occurrence of a non-endemic or previously eradicated disease	0, 1
13.	Epidemiological data suggesting a common exposure	0, 1
14.	Simultaneous epidemics and/or epizootics occur at different locations	0, 1
	Total score[b]	0–14

Reproduced from Radosavljevic and Belojevic (2012)
[a]1 = High probability of a deliberate outbreak; 0 = Low probability of a deliberate outbreak
[b]Evaluation of scores: 1–4, probably natural outbreak; 5–9, possibly deliberate outbreak; 10–14, probably deliberate outbreak

In summary, our analysis of this case yielded a total of 33 points using the non-conclusive criteria of the Grunow and Finke model. Using the intent "likelihood ranges" (Table 9.3) the likelihood that this Salmonellosis outbreak was intentionally incited was "Doubtful". There were three criteria in the model that were not applicable to foodborne illness outbreaks. If we gave N/A a mean value of the possible score for that criteria instead of zero (to have a neutral effect on the overall evaluation categories), it would add 6 points overall. This would make 39 points in total and shift it into the "likely" category.

Table 9.5 Scoring of the Salmonellosis outbreak according to the model of Grunow and Finke (2002) by non-conclusive criteria

No.	Criterion	Assessment (possible point)[a]	Weighting factor	Maximum number of points[b]
1.	Existence of a biological risk	3	2	6
2.	Existence of a biological threat	3	3	9
3.	Special aspects of the biological agent	3	3	9
4.	Peculiarities of the geographic distribution of the biological agent	N/A	N/A	N/A
5.	High concentration of the biological agent in the environment	N/A	N/A	N/A
6.	Peculiarities of the transmission mode of the biological agent	0	1	0
7.	Peculiarities of the intensity and dynamics of the epidemic	3	2	6
8.	Peculiarities of the time of the epidemic	N/A	N/A	N/A
9.	Unusually rapid spread of the epidemic	3	1	3
10.	Limitation of the epidemic to a specific population	0	1	0
11.	Peculiarities of the clinical manifestation	0	1	0
	Total			33

[a]Assessment of a criterion
0 = Criterion ruled out or no data available
1 = Existence of peculiarities or suspicions, but uncertain and indistinct
2 = Existence of obvious peculiarities or indications, causes yet to be clarified for certain
3 = Existence of considerable peculiarities or deviations from expected norm, clear indication or proof of an intentional release
[b]Assessment score × weighting factor

Rationale for Table 9.5 Scoring

No.1. "A biological risk is considered to be the presence of a political or terrorist environment from which a biological attack could originate. Such biological risk arises if states, groups, or individual persons have access to biological warfare agents, the necessary means of distributing them and are willing to use them" (Grunow and Finke 2002). The political environment at The Dalles, Oregon during that time was tense (Carus 2001); the religious cult had access to both the agent (documented purchase from a commercial supplier and the discovery of a culture vial in its clinical laboratory) (Carus 2001), and the food vehicles (salad bars, which were accessible to all customers at all

(continued)

the restaurants involved). Although these facts were not immediately apparent during the outbreak investigation, it should and is fully scored retrospectively (score 6).

No. 2. "A biological threat is to be assumed if, in the environment of a biological risk, states, groups, or individual persons openly threaten to use biological warfare agents or if a specific interest in their use can be assumed"(Grunow and Finke 2002). In this case, specific interest in using the pathogen was confirmed by the confession of a cult member a year later after the outbreak (Carus 2001), therefore, similar to No. 1, a full score is given retrospectively (score 9).

No. 3. Special aspects of the biological agent might include cases in which "a potential aggressor has genetically manipulated certain characteristic of pathogens or toxins prior to their use as biological warfare agents" (Grunow and Finke 2002, or the agent was previously eradicated or does not naturally occur in the outbreak area. Although foodborne pathogens generally are considered "endemic", the *S.* Typhimurim strain used in this outbreak was not a "natural" or "environmental" isolate as would be expected in a typical natural outbreak, but rather was a strain used commonly as a control by licensed clinical laboratories to meet quality assurance requirements (i.e., a laboratory strain) (Carus 2001). For this reason, a full score is given in this criterion (score 9).

No. 4. Peculiarities of the geographic distribution of the disease are primarily defined as "if the disease, and thus the pathogen species or strain, is identified in region concerned for the first time ever or again after a long period of time" (Grunow and Finke 2002). Given the "ubiquitous" nature of foodborne pathogens, this criterion could not be applied in this case.

No. 5. The concentration of the biological agent in the environment, in this model, is related to situation in which if a biological agent is released intentionally, it would be expected to occur in unusually high concentrations in the air, soil, and/or drinking or surface water over a large area. This criterion is not applicable to foodborne pathogens, as the target sites for their release are specific food items rather than the environment.

No. 6. Peculiarities of the transmission mode of the biological agent are referred to any deviation from the feature paths of transmission that are typical for the pathogen and its hosts in natural epidemics. Food items involved in this outbreak could have been contaminated naturally by *Salmonella*; therefore, a score of 0 was given.

No. 7. Peculiarities of the intensity and dynamics of the epidemic are referred to deviations from the typical epidemiological curve of a disease outbreak (characterized by the number of cases of a disease per unit of time or the total number of cases)(Grunow and Finke 2002). For example, it would be

(continued)

peculiar for an outbreak that typically produces explosive epidemic curve (sudden and sharp increase in cases number over a short time) to have a slow start. The epi curve for this outbreak can be considered as explosive with high intensity (Fig. 9.1) comparing to the typical Salmonella outbreaks, therefore was scored as 6.

No. 8. Certain disease outbreaks tend to occur during certain seasons, such as influenza in winter. Peculiarities of the time of the epidemic are refereed to any deviations from such seasonality of a specific disease outbreak. Since foodborne illness outbreaks could occur at any time, and there are no recognized temporal patterns, this criterion does not apply.

No. 9. The rate of spread of an epidemic is usually determined by the virulence, resistance, and concentration of the pathogen, and the contagiousness of the disease. This outbreak was explosive, with a large number of cases appearing in a short time, therefore, a full score (3) was given.

No. 10. When an epidemic is limited to a specific demographic population the possibility exists that a group was targeted In this outbreak, however, no particular population appeared to be targeted (score 0).

No. 11. The clinical manifestation in this outbreak is typical of Salmonellosis, therefore, the criterion was scored as 0.

However, proof of the intentional release of the agent was provided by a cult member a year after the outbreak. Therefore, using Grunow and Finke's second decision tool model, based on "conclusive criteria" (Table 9.2), the outcome is much different and suggests that this outbreak was likely due to a biological attack. The neutral scoring for N/A in non-conclusive criteria is appropriate in this case.

Next, we assessed the same Salmonellosis outbreak (The Dalles, 1984) using the second model (Radosavljevic and Belojevic 2012); our scores related to these criteria are shown in Table 9.6.

In summary, a total of 6 points was generated using this model, resulting in a conclusion that the outbreak was "Possibly deliberately caused". However, many of the indicators in this model do not apply to foodborne illness and foodborne pathogens. An evaluation neutral approach, giving N/A a score of 0.5, would add 2.5 to the total score. The overall value of 8.5 is nearly the top of the "Possibly deliberately caused" category, a stronger indication of deliberate cause.

Table 9.6 Scoring of the Salmonellosis outbreak according to the model of Radosavljevic and Belojevic (2012)

No	Indicators	Score[a]
1.	Unusual/atypical disease/manifestation or unexpected fulminant course of disease in humans and/or animals	0
2.	Failure of patient to respond to usual therapy or illness in a population despite immunizations	0
3.	Several unusual/unexplained syndromes coexisting in the same case without any other explanation	0
4.	Sudden unexplainable increase in the number of cases or deaths in human populations	1
5.	Higher than expected morbidity and/or mortality rates	1
6.	Clustering of patients with fever only or with fever and other symptoms	N/A
7.	Disease identified in the region for the first time, again after a long period of time	N/A
8.	A disease with an unusual/atypical seasonal distribution	N/A
9.	Simultaneous occurrence of epidemics and/or epizootics	1
10.	Explosive epidemics/outbreaks with indicators of a point-source origin	1
11.	Disease with an unusual geographic distribution	N/A
12.	Occurrence of a non-endemic or previously eradicated disease	N/A
13.	Epidemiological data suggesting a common exposure	1
14.	Simultaneous epidemics and/or epizootics occur at different locations	1
	Total score[b]	6

[a]1 = High probability of a deliberate outbreak; 0 = Low probability of a deliberate outbreak
[b]Evaluation of scores: 1–4, probably natural outbreak; 5–9, possibly deliberate outbreak; 10–14, probably deliberate outbreak

Rationale for Table 9.6 Scoring

Indicator 1. See No. 11 in Grunow and Finke model.

Indicators 2 and 3. Neither of these criteria was reported for this outbreak.

Indicator 4. This outbreak is considered as sudden increase in the number of cases in human populations, but the association with eating at local restaurants was made quickly.

Indicator 5. There were no direct reports on typical morbidity and/or mortality rates for Salmonellosis patients at The Dalles during the period of the outbreak. However, the morbidity rate of this outbreak could be considered as higher than usual due to the large case number (751) involved.

Indicator 6. The common symptoms of foodborne illness are mild to severe diarrhea (sometimes bloody), abdominal cramping and pain, nausea, vomiting (sometimes), and fever (sometimes). Therefore, this indicator doesn't apply to foodborne illness.

Indicator 7. Since foodborne pathogens can be found anywhere this indicator doesn't apply to foodborne illness.

Indicator 8. Doesn't apply, see No. 8 in the Grunow and Finke model

Indicator 9. There were two waves of cases of Salmonellosis during this outbreak, so it was scored as 1.

(continued)

Indicator 10. This outbreak was explosive, with indicators on point-source origins, so 1

Indicator 11. Doesn't apply; see No. 4 in in Grunow and Finke model.

Indicator 12. Doesn't apply, see Indicator 7.

Indicator 13. Common exposure was evidenced during the investigation. However, this indicator overlaps with indicator 10.

Indicator 14. Multiple restaurants were implicated.

9.4 Assessment of a Confirmed Intentional Outbreak of Shigellosis by Existing Models

Our assessment of the Shigellosis outbreak in Dallas, Texas, 1996, using the Grunow and Finke (2002) model, is shown in Table 9.7.

Table 9.7 Scoring of the 1996 Dallas, TX, Shigellosis outbreak according to the model of Grunow and Finke (2002) using non-conclusive criteria

No.	Criterion	Assessment (possible point)[a]	Weighting factor	Maximum number of points[b]
1.	Existence of a biological risk	3	2	6
2.	Existence of a biological threat	3	3	9
3.	Special aspects of the biological agent	3	3	9
4.	Peculiarities of the geographic distribution of the biological agent	N/A	N/A	N/A
5.	High concentration of the biological agent in the environment	N/A	N/A	N/A
6.	Peculiarities of the transmission mode of the biological agent	0	1	0
7.	Peculiarities of the intensity and dynamics of the epidemic	3	2	6
8.	Peculiarities of the time of the epidemic	N/A	N/A	N/A
9.	Unusually rapid spread of the epidemic	3	1	3
10.	Limitation of the epidemic to a specific population	3	1	3
11.	Peculiarities of the clinical manifestation	0	1	0
	Total			36

[a]Assessment of a criterion
0 = Criterion ruled out or no data available
1 = Existence of peculiarities or suspicions, but uncertain and indistinct
2 = Existence of obvious peculiarities or indications, causes yet to be clarified for certain
3 = Existence of considerable peculiarities or deviations from expected norm, clear indication or proof of an intentional release
[b]Assessment score × weighting factor

Rationale for Table 9.7 Scoring

No.1 and No. 2. Full scores are given based on the fact that the individual had access to the agent, the necessary means of distributing them, and was willing to, and did, use them.

No. 3. Full score is given due to the laboratory origin of the specific pathogen strain.

No. 4 and No. 5. Given the "ubiquitous" nature of foodborne pathogens and their release target sites are specific food items rather than the environment, these criteria are not applicable to foodborne illness outbreaks.

No. 6. Although unlikely, the pastries involved in this outbreak could have been inadvertently contaminated by *Shigella* by a sick employee due to improper handling, therefore a score 0 was given.

No. 7. The high attack rate observed in this outbreak is unusual; therefore, a full score was given.

No. 8. Since foodborne illness outbreaks could occur at any time, and there are no recognized temporal patterns, this criterion does not apply.

No. 9. The speed at which an epidemic spreads is usually determined by the virulence, resistance, and concentration of the pathogen, and the contagiousness of the disease. This outbreak had attack rate of 100 %, which suggests either high virulence or high concentration of the pathogen, or both; therefore, a full score was given.

No. 10. Limitation of the epidemic to a specific population seems to apply to this case, in which the attack appeared to be directed toward the small group of people who received the email invitation; therefore, a full score is given.

No. 11. In this case the clinical manifestation was typical of shigellosis; therefore a 0 score was given.

In summary, a total of 36 points was generated using the non-conclusive criteria of the Grunow and Finke model, resulting in a conclusion that a biological attack was at the lowest end of "Likely". Three criteria in this model do not apply to foodborne illness and foodborne pathogens. Adding the Evaluation neutral scores for the three lines rated N/A would add 6 points overall, giving a final score of 42, much more firmly in the "Likely" category.

We also assessed the same Shigellosis outbreak in Dallas, Texas, using the Radosavljevic and Belojevic (2012) model, as shown in Table 9.8.

Table 9.8 Scoring of the 1996 Shigellosis outbreak according to the model of Radosavljevic and Belojevic (2012)

No	Indicators	Score[a]
1.	Unusual/atypical disease/manifestation or unexpected fulminant course of disease in humans and/or animals	0
2.	Failure of patient to respond to usual therapy or illness in a population despite immunizations	0
3.	Several unusual/unexplained syndromes coexisting in the same case without any other explanation	0
4.	Sudden unexplainable increase in the number of cases or deaths in human populations	1
5.	Higher than expected morbidity and/or mortality rates	1
6.	Clustering of patients with fever only or with fever and other symptoms	N/A
7.	Disease identified in the region for the first time, again after a long period of time	N/A
8.	A disease with an unusual/atypical seasonal distribution	N/A
9.	Simultaneous occurrence of epidemics and/or epizootics	0
10.	Explosive epidemics/outbreaks with indicators of a point-source origin	1
11.	Disease with an unusual geographic distribution	N/A
12.	Occurrence of a non-endemic or previously eradicated disease	N/A
13.	Epidemiological data suggesting a common exposure	1
14.	Simultaneous epidemics and/or epizootics occur at different locations	0
	Total score[b]	4

[a]1 = High probability of a deliberate outbreak; 0 = Low probability of a deliberate outbreak
[b]Evaluation of scores: 1–4, probably natural outbreak; 5–9, possibly deliberate outbreak; 10–14, probably deliberate outbreak

Rationale for Table 9.8 Scoring

See explanation for the assessment of this outbreak using the Grunow and Finke (2002) model for indicator scoring. There were no simultaneous epidemics in this outbreak therefore no points were given for indicators 9 and 14.

In summary, a total of 4 points were generated by this model, resulting in a conclusion that the incident was "probably a natural outbreak". However, since many of the indicators of the Radosavljevic and Belojevic do not apply to foodborne illness and foodborne pathogens the assessment is less robust. An Evaluation neutral scoring for the N/A indicators would add 2.5 points, giving a total of 6.5, well into the "possibly deliberate outbreak" category.

Overall, our assessments of two confirmed intentional foodborne illness outbreaks, performed using two existing decision tool models created for use in investigation with outbreaks caused by biological warfare agents, were of limited value. The models do not consider uncertainty and do not allow for "bonus" points for extreme values in particular lines, which might add additional weight or certainty to a conclusion. Due to the nature and scope of foodborne illness, a number of the criteria or indicators in these models were inappropriate for the evaluation of a foodborne illness outbreak. Unlike most other agents used in biological warfare, foodborne pathogens are naturally "ubiquitous" and can be found worldwide. Although certain subtypes of a particular foodborne pathogen may be more prevalent in one location or continent than others, increasing globalization of the food supply chain is undermining such separation. Foodborne illness outbreaks are not rare events and could happen at any time.

9.5 Selection of Criteria for the Development of a New Decision Tool Model Appropriate for Use with Foodborne Illness Outbreaks

Based on the two intentional events described above, it can be speculated that features unique to foodborne illness outbreaks due to intentional contamination occur in two areas: (1) epidemiological and (2) potential perpetrator(s). Many control measures have been put into place to safeguard food, throughout its production chain, from contamination by foodborne pathogens. Consequently, concentration of the pathogen(s) in a "naturally" or "accidentally" contaminated food would be expected to be generally low, resulting in low number of cases and a low food-specific attack rate (the ratio between the number of people who ate a specific food and became sick and the total number of people who ate that food). On the other hand, a relatively high concentration of the pathogen in the implicated food is expected in an intentional event, and typically would result in large number of cases and a high food specific attack rate. It is also possible, in an intentional event, to have multiple epidemics in a short time at different locations, with single or multiple food vehicles involved without common suppliers.

Interestingly, both pathogen strains used in the two intentional events were from laboratory cultures. With current whole genomic sequencing approaches in epidemiological investigation of foodborne illness outbreaks, such laboratory origin of an outbreak strain would now become apparent early on during an investigation (Sjodin et al. 2013). The origin of an outbreak strain should serve as an indicator for a possible intentional outbreak, as accidental release of laboratory strain to the environment and subsequent contamination of a food source is a very rare event. Other features of the outbreak strain should be taken into consideration as well, such as unusual virulence characteristics and the possibility of genetic manipulation influencing certain characteristics of the pathogen or associated toxins.

Certain features of the food vehicle(s) implicated in a foodborne illness outbreak can also serve as indicators for intentional contamination. The food vehicles involved in the two intentional outbreaks described herein were both contaminated at the point-before-consumption, resulting in point-source outbreak epidemiological curves, which are characterized as having a rapid rise, followed by a brief plateau, then a quick drop in case numbers. Multiple food items at the salad bar were implicated in the Salmonellosis outbreak in The Dalles, Oregon, but no common connection could be identified, such as possible cross contamination by employees or common suppliers. The vulnerability of the implicated food item to tampering could also be taken into account (although not included in the criteria in this work). An additional feature could be any "unusual" association of a particular pathogen/toxin with a particular food vehicle. For example, the presence of botulinum toxin in fresh produce would be an unusual association and could signal the possible 'intentional" act. Botulinum toxin could only be produced under strict anaerobic conditions by spore-former *Clostridium botulinum* or few other *Clostridium* species. Such conditions could rarely exist during fresh produce production and minimal processing before consumption.

Organizations or individuals that could have access to the food items at the point of contamination should be evaluated for potential motivation in inflicting an intentional contamination event, the degree to which the pathogen is accessible to them, their ability/skill to culture the pathogen, and the accessibility of the food vehicle(s) to them.

At minimum, necessary components for determining that an outbreak of foodborne illness was due to intentional tampering should include investigation of the epidemiological and clinical characteristics of the outbreak, particularly any indications of unusually large numbers of cases; high attack rate; point-source epi curve and/or multiple epidemics; unusual clinical manifestation of illness; and/or unusual pathogen strain (either origin or genetic or virulent characteristics). Information regarding the potential perpetrator(s) should include the motivation, their accessibility to the outbreak strain and implicated food vehicle(s).

9.6 Development of New, Foodborne Illness Outbreak-Specific Decision Tools

The assessments described above indicate that current models for evaluating the likelihood that an intentional outbreak of foodborne illness are insufficient, and that the development of new decision tools, designed specifically for that application, is needed. Therefore, we propose two new models based on criteria of importance in the evaluation of disease caused by foodborne pathogens outlined above. One of the models is based on criterion scoring (Table 9.9) and the other on a Bayesian network analysis (Table 9.10).

9.6.1 New Model Based on Scoring System

The new scoring system is developed based on nine criteria (listed in Table 9.9) that could be derived from epidemiological investigation and possible perpetrator(s) analysis of an outbreak of foodborne illness. There are three assessment points for

Table 9.9 Proposed criteria for a new assessment model for outbreaks of foodborne illness, based on a criterion scoring system[c]

No.	Criterion	Assessment (possible point)[a]	Weighting factor	Maximum number of points[b]
	Epidemiological			
1.	Unusually large number of cases	0, 1, 2	1	2
2.	Point-source epi curve	0, 1, 2	3	6
3.	Multiple epidemics	0, 1, 2	2	4
4.	High attack rate	0, 1, 2	3	6
5.	Unusual clinical manifestation of illness	0, 1, 2	3	6
6.	Unusual strain	0, 1, 2	3	6
	Perpetrator			
7.	Motivation (risk)	0, 1, 2	3	6
8.	Accessibility to the strain	0, 1, 2	3	6
9.	Accessibility to the food vehicle	0, 1, 2	2	4
	Total[c]			46

[a]Assessment of a criterion
0 = Criterion ruled out
1 = Existence of peculiarities or suspicions, but uncertain and indistinct
2 = Existence of obvious peculiarities or indications, causes yet to be clarified for certain
[b]Assessment score × weighting factor
[c]Evaluation of scores: 0–14, Low probability of an intentional outbreak; 15–29, medium probability of an intentional outbreak; 30–46, high probability of an intentional outbreak

Table 9.10 Comparative scoring of demonstration outbreaks using the new scoring model

		Salmonellosis	Shigellosis	*Salmonella*	*E. coli* O104:H4 infections
No.	Criterion	The Dalles, OR	Dallas, TX	Saintpaul infections, US	Germany
1.	Unusually large number of cases	2	0	2	2
2.	Point-source epi curve	6	6	0	0
3.	Multiple epidemics	4	0	4	4
4.	High attack rate	3	6	3	3
5.	Unusual clinical manifestation of illness	0	0	0	3
6.	Unusual strain	6	6	0	3
7.	Motivation (risk)	6	6	0	0
8.	Accessibility to the strain	6	6	0	0
9.	Accessibility to the food vehicle	4	4	4	4
	Total[a]	37	34	13	19

[a]Evaluation of scores: 0–14, Low probability of an intentional outbreak; 15–29, medium probability of an intentional outbreak; 30–46, high probability of an intentional outbreak

each criterion: 0-criterion has been ruled out; 1-peculiarities or suspicions exist but are uncertain and indistinct; and 2- peculiarities or indications obviously exist with causes yet to be clarified for certain. The weighing factor (1–3) is included to reflect on the Impact of each criterion on the decision. Although an unusually large number of cases and multiple epidemics were both reported in the intentional outbreak in The Dalles, such characteristics were also observed in two "natural" large outbreak of foodborne illness, the 2008 outbreak of *Salmonella* Saintpaul infections in the US and the 2011 outbreak of EHEC O104:H4 infections in Germany (described in the following sections in this chapter), therefore, weighing factors of 1 and 2 were given to these criteria. Evaluation class boundaries were set up at 15 and 30, reflecting approximate thirds of the total available points. A summary of this scoring system is presented in Table 9.9.

9.6.2 New Model Based on Bayesian Network Analysis

A Bayesian network incorporates probability distributions for model components and combines these according to transparent logical rules to develop an overall outcome distribution (Holt et al. 2012). Rather than producing a single outcome score, as in the scoring model, the overall outcome distribution represents the cumulative uncertainties inherent in the combination of model criteria. Bayesian networks are more complex to create and are dependent on the rules used for combining distributions. However, they are explicit in dealing with uncertainty in the values for each component and can be interpreted to indicate the overall strength of the conclusion.

The Bayesian network model uses the same nine non-conclusive criteria as the proposed scoring model. These are combined into four intermediate criteria, and two of these are combined into the final outcome distribution. Examples of the model are demonstrated in Figs. 9.2, 9.3, 9.4, and 9.5.

Intermediate criteria are created by combination rules. These combinations are logical rules with subjective assignment of probabilities of outcomes (the values in the output for that combination) for each of the possible sets of input values, which are expressed as conditional probability tables.

These show the probability associated with each combined node given the values of the immediate input nodes. For example, if the answers to Questions 1 and 4 are both 'False' the result is 'False' and conversely, 'True' if both are 'True'. Mixtures of 'True' and 'False' are judged to give an outcome somewhere in between, in this case symmetrical according to the probabilities in the table. The values selected reflect the relative influence each input node is judged to have on the result in the combined node (each column must sum to 1). The combinations used in the illustrations for the four intermediate and the final combination nodes have been chosen to be symmetrical to demonstrate the principle, and are as follows:

Epidemiology nodes:

1. An unusually large number of cases have been found		False		True	
4. The epidemic has an unusually high attack rate		False	True	False	True
Scale of epidemic is indicative of deliberate cause	False	1	0.5	0.5	0
	True	0	0.5	0.5	1

2. Point-source epidemic curve		False		True	
3. The epidemic curve is very compressed		False	True	False	True
Spatial pattern of epidemic is indicative of deliberate cause	False	1	0.5	0.5	0
	True	0	0.5	0.5	1

5. The epidemic has unusual illness symptoms		False		True	
6. The pathogen is of an unusual or laboratory strain		False	True	False	True
Unusual nature indicates deliberate cause	False	1	0.5	0.5	0
	True	0	0.5	0.5	1

Epidemiology aggregate node:

Scale of epidemic is indicative of deliberate cause		False				True			
Spatial pattern of epidemic is indicative of deliberate cause		False		True		False		True	
Unusual nature indicates deliberate cause		False	True	False	True	False	True	False	True
Epidemiology and agent suggest deliberate cause	False	1	.67	.67	.33	.67	.33	.33	0
	True	0	.33	.33	.67	.33	.67	.67	1

Motivation and Capability nodes:

7a. Personal motivating factors have been identified		False		True	
7b. Political and/or social motivating factors are identified		False	True	False	True
Motivation could result in criminal act	False	1	0.5	0.5	0
	True	0	0.5	0.5	1

8. Accessibility to the strain		False		True	
9. Accessibility to the food vehicle		False	True	False	True
Accessibility indicates deliberate cause	False	1	0.5	0.5	0
	True	0	0.5	0.5	1

Perpetrator aggregate node:

Accessibility indicates deliberate cause		False		True	
Motivation indicates deliberate cause		False	True	False	True
Perpetrator motivation and capability	False	1	0.5	0.5	0
	True	0	0.5	0.5	1

Final node:

Epidemiology and agent suggest deliberate cause	False		True	
Perpetrator motivation and capability could result in criminal act	False	True	False	True
Epidemic only	1	0.5	0	0
Epidemic or Motivated	0	0.5	0.5	0
Motivated	0	0	0.5	1

The relative potential of the three outcome results by this model in the final node are in the ratio 1.5 : 1.0 : 1.5, by adding the values in the three rows in the Final node table above. So there is less likelihood of an intermediate outcome than either of the two extremes. This comes from the symmetrical approach to the inputs. We assume two False inputs leads to a False outcome; two True inputs leads to a True outcome; and either mixed set of inputs leads to a mixed 0.5 outcome of the extreme or intermediate case. The columns in the table must add to 1.0.

9.7 Demonstrations/Validation of New Models with Natural and Intentional Outbreaks of Foodborne Illnesses

Two confirmed intentional outbreak cases (The Dalles 1984 and Dallas 1996) and two additional natural/accidental foodborne illness outbreaks (German 2011 *E. coli* outbreak and St Paul 2008) are used to demonstrate the discrimination achieved by the proposed scoring and Bayesian network models. Information and data related to the two additional natural/accidental outbreaks are presented in the following paragraphs.

9.7.1 2008 Outbreak of Salmonella Saintpaul Infections in the US

On May 22, 2008, the Centers for Disease Control and Prevention (CDC) were notified by the New Mexico Department of Health about 19 cases of *Salmonella* infection in their state (Behravesh et al. 2011). Seven clinical isolates were serotyped as

Salmonella Saintpaul and four of them had indistinguishable patterns by pulsed-field gel electrophoresis (PFGE). The next day, three additional isolates having the same PFGE pattern, from Colorado and Texas, were identified by CDC staff at PulseNet (a national molecular-subtyping network in the US). Epidemiological investigation was initiated with a case–control study for May 2008. In the meantime, more isolates of *Salmonella* Saintpaul having the same, indistinguishable PFGE pattern were uploaded (identified) in PulseNet from additional states.

Additional case–control studies and investigation of clusters linked to restaurants or events studies were carried out via a hypothesis-generating questionnaire about consumption of numerous food items. Although the early case–control study yielded a strong association between illness and consumption of raw tomatoes, the results of multiple investigations indicated that jalapeno and serrano peppers were the major vehicles for transmission (in homes and restaurants). State and local health and agricultural departments, the Food and Drug Administration (FDA), and the CDC conducted traceback investigations of the distribution pathways for food items associated with several ill persons and restaurant clusters, as well as for the distribution pathway including distribution centers, packing facilities, and farms. All tracebacks led to distributors in Texas and Mexico that had received jalapeno peppers from Mexico. The outbreak strain was isolated from a jalapeno pepper sample from one importer in Texas. The FDA investigated two farms (Farm A and Farm B), located approximately 100 miles from each other in Mexico, which had supplied peppers to the packing facility. Farm A grew jalapeno and serrano peppers and Roma tomatoes. All three crops had been harvested between late April and late July. Salmonella (but not the outbreak strain) was detected in agricultural water samples from Farm A. Farm B only grew jalapeno and serrano peppers, which they harvested from mid-April to mid-June. The outbreak strain of Salmonella was isolated from two environmental samples, agricultural water, and serrano peppers on Farm B. Between April 16 and August 26, 2008, 1500 infections with the outbreak strain of *Salmonella* Saintpaul were identified in the US involving 43 states and the District of Columbia. Among these case subjects, 21 % were hospitalized, and two died (Behravesh et al. 2011). This outbreak of foodborne illness in the US was one of the largest salmonellosis outbreaks ever identified (Behravesh et al. 2011).

9.7.2 2011 Outbreak of E. Coli O104:H4 Infections in Germany

From May 1 to July 26, 2011, there was a large outbreak of foodborne illness associated with infections by enterohemorrhagic *E. coli* (EHEC) O104:H4. The illnesses were characterized by acute gastroenteritis (including bloody diarrhea) and haemolytic-uremic syndrome (HUS). A total of 3842 cases attributed to this outbreak including 2987 cases of acute gastroenteritis and 855 cases of HUS (Fig. 9.1).

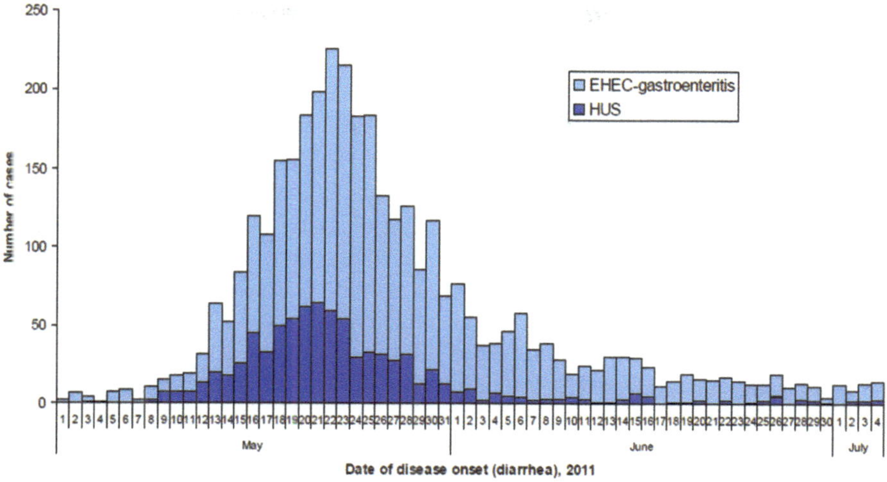

Fig. 9.1 Epidemiological curve for HUS and EHEC outbreak cases in *E. coli* O104:H4 infections in Germany, 2011 (RKI 2011)

Between May 19 and 20, 2011, investigation of a cluster of three cases of HUS by the Health and Consumer Protection Agency in Hamburg and Robert Koch Institute (RKI) led to the epidemiological investigation of this outbreak. Several case–control studies and investigation of clusters linked to restaurants or events studies were carried out via a hypothesis-generating questionnaire about consumption of numerous food items. Although the early case–control study yielded a strong association between illness and consumption of raw tomatoes, cucumbers, and lettuce, the results of multiple investigations, including recipe-based restaurant cohort studies, revealed significant association between the consumption of sprouts and risk of disease. Forty one outbreak clusters involved more than 300 cases were identified by the investigation task force and provided link to sprouts produced by Company A in Lower Saxony. In addition, two individuals infected with the outbreak strain had consumed sprouts grown by them (self-sprouters) from a sprout seed mix, suggesting the contaminated sprout seeds as the source of this outbreak. Cases of illness were reported from all federal states of Germany with 5 northern states being most affected. The majority of cases involved adults with 53 deaths: 35 (4.1 %) of the HUS patients and 18 (0.6 %) of the EHEC gastroenteritis patients. The outbreak strain was a rare EHEC serotype O104:H4 which had only been described rarely in humans (Germany 2001, France 2004, Korea 2006, and Georgia 2009, Finland 2010). It had the virulence characteristics of enteroaggregative *E. coli* (EAEC) (which is different from the majority of EHEC strains) and unique antibiotic resistance profile (RKI 2011).

9.7.3 Demonstration/Validation of Scoring Model

Assessment of the two intentional (Salmonellosis outbreak in The Dalles, Oregon, 1984; Shigellosis outbreak in Dallas, Texas, 1996) and two natural (*Salmonella* Saintpaul infections, USA, 2008; *E. coli* O104:H4 infections, Germany, 2011) outbreaks of foodborne illness by the new scoring model is summarized in Table 9.10.

An unusually large number of cases would apply to all outbreaks except Shigellosis Dallas TX. Point-source epi curve were reported for the two intentional outbreaks. Since the cases in the two large natural outbreaks were reported from multiple geographic locations/states in the affected country, with clusters of cases as well as satellite individual cases reported in a relatively extended period of time, an overall point-source epi curves couldn't not be concluded. However, multiple epidemics (different locations) were observed in one intentional (Salmonellosis outbreak in The Dalles, Oregon, 1984) and both natural outbreaks. The exact attack rates were not reported for the outbreaks expect the Shigellosis outbreak in Dallas TX 1996; however, given the large number of cases in these outbreaks, peculiarities or suspicions do exist. Unusual clinical manifestation of illness was not observed for all the outbreaks except *E. coli* O104:H4 infections, Germany, 2011. In this outbreak, a longer incubation period and different antibiotic resistant profile than those of normal EHEC infections was reported. Although these observations could be explained by the rare outbreak strain as EAEC, a half of the score is given to this outbreak. The strains involved in both intentional outbreaks could be traced to their laboratory origins, therefore, they were unusual. The strain involved in *E. coli* O104:H4 infections, Germany, 2011 was reported rarely, so a half score was given whereas the S. Saintpaul strain implicated in the outbreak of Salmonella Saintpaul infections, USA, 2008 was not as unusual.

From the potential perpetrator(s) aspects, evidence for the motive, the perpetrator's access to the outbreak strain and implicated food vehicle (s) were reported in both of the outbreaks that were intentionally incited. Although there was no evidence for the potential perpetrator(s) in the two natural outbreaks, a full score was given to the accessibility to the food vehicle(s).

According to the new scoring model, 37 and 34 points were assigned to the outbreaks of Salmonellosis outbreak in The Dalles, Oregon, 1984 and Shigellosis outbreak in Dallas, Texas, 1996, respectively. These points would place these outbreaks in the category of high probability of an intentional outbreak. A total of 13 points was generated for *Salmonella* Saintpaul infection, US, 2008, placing it into the group of low probability of an intentional outbreak. The *E. coli* O104:H4 infections, Germany, 2011 received a total of 19 points, positioning this outbreak in the category of medium probability of an intentional outbreak.

9.7.4 Demonstration/Validation of Bayesian Network Model

All variables are entered as No, Possibly or Yes, in a spreadsheet as below (Table 9.11). This is uploaded to the case manager in the Bayesian network software and puts values in the rating nodes (white boxes in Figs. 9.2, 9.3, 9.4, 9.5 examples). These are interpreted as probabilities (pale blue/grey). The conditional probability nodes (green) encode a reasoning which reflects uncertainty in what can be implied from the risk factor values. For example, just because no motivating factors have been found does not mean that there is zero probability that they do actually exist. The values in the green nodes (Figs. 9.2, 9.3, 9.4 and 9.5) need further consideration. The final or target node (at the bottom of the networks in Figs. 9.2, 9.3, 9.4 and 9.5) shows the nature of the evidence for an intentional outbreak: how likely that it is just epidemiological, just motivational or could be both. Bars in the final outcome node of each Bayesian network model indicate the relative likelihood of each conclusion.

9.8 Discussion

Assessing the likelihood that an outbreak of foodborne illness might have been the result of intentional, criminal activity is of great importance to public health management and to the attribution needed for a criminal conviction. Forensic evidence

Table 9.11 Data input file for a demonstration Bayesian network model to differentiate natural and intentional outbreaks of foodborne illnesses

Scenario		*Salmonella* Saintpaul infections US	*E. coli* O104:H4 infections Germany	Salmonellosis The Dalles, OR	Shigellosis Dallas, TX
1.	Cases	Yes	Yes	Yes	Possibly
2.	Point source	No	No	Yes	Yes
3.	Multiple epidemics	No	Possibly	Possibly	No
4.	High attack	Possibly	Possibly	No	Yes
5.	Unusual illness	No	Possibly	Yes	No
6.	Unusual strain	No	Possibly	No	Yes
7a.	Personal motivation	No	No	Yes	Yes
7b.	Political or social motivation	No	No	Yes	No
8.	Accessible strain	No	No	Yes	Yes
9.	Accessible vehicle	Yes	Yes	Yes	Yes

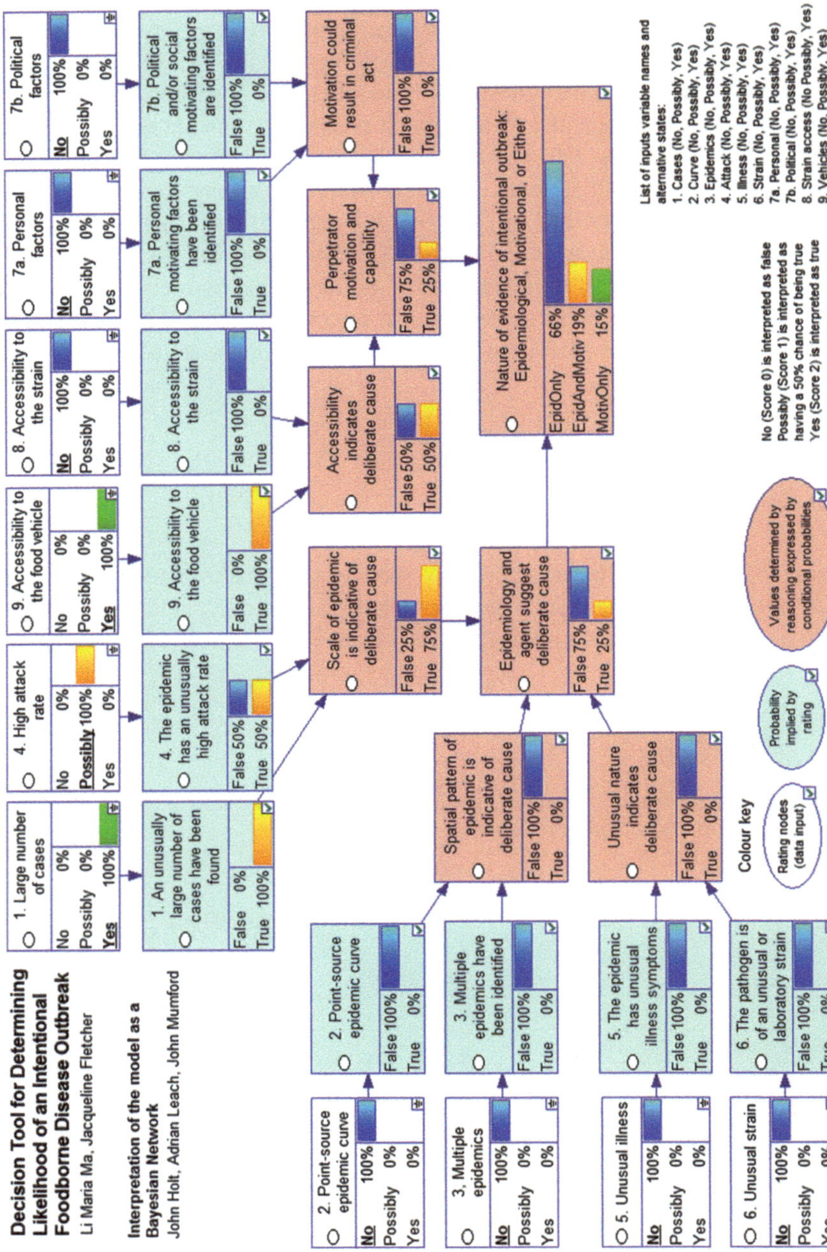

Fig. 9.2 *Salmonella* Saintpaul Infections US (accidental) outbreak in the Bayesian network model

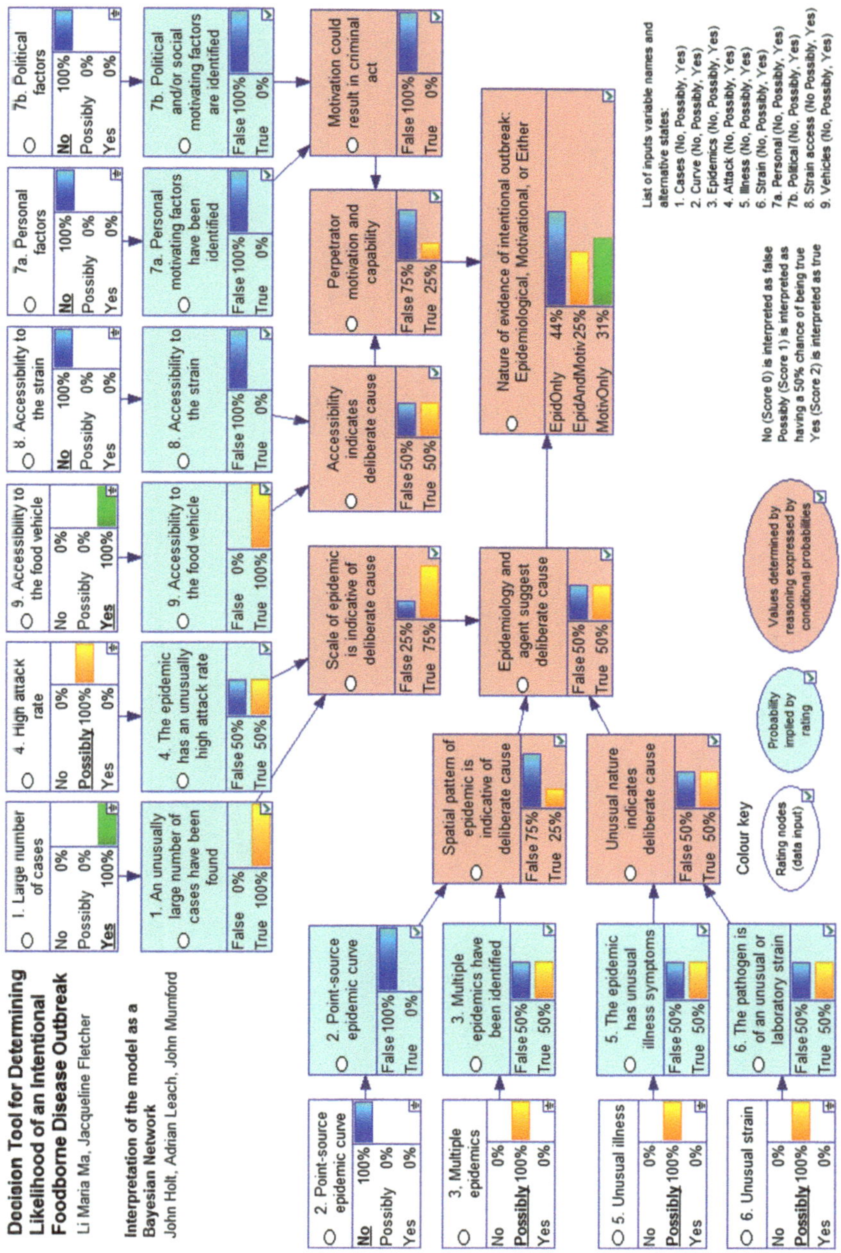

Fig. 9.3 *E. coli* O104:H4 infections Germany (accidental) outbreak in the Bayesian network model

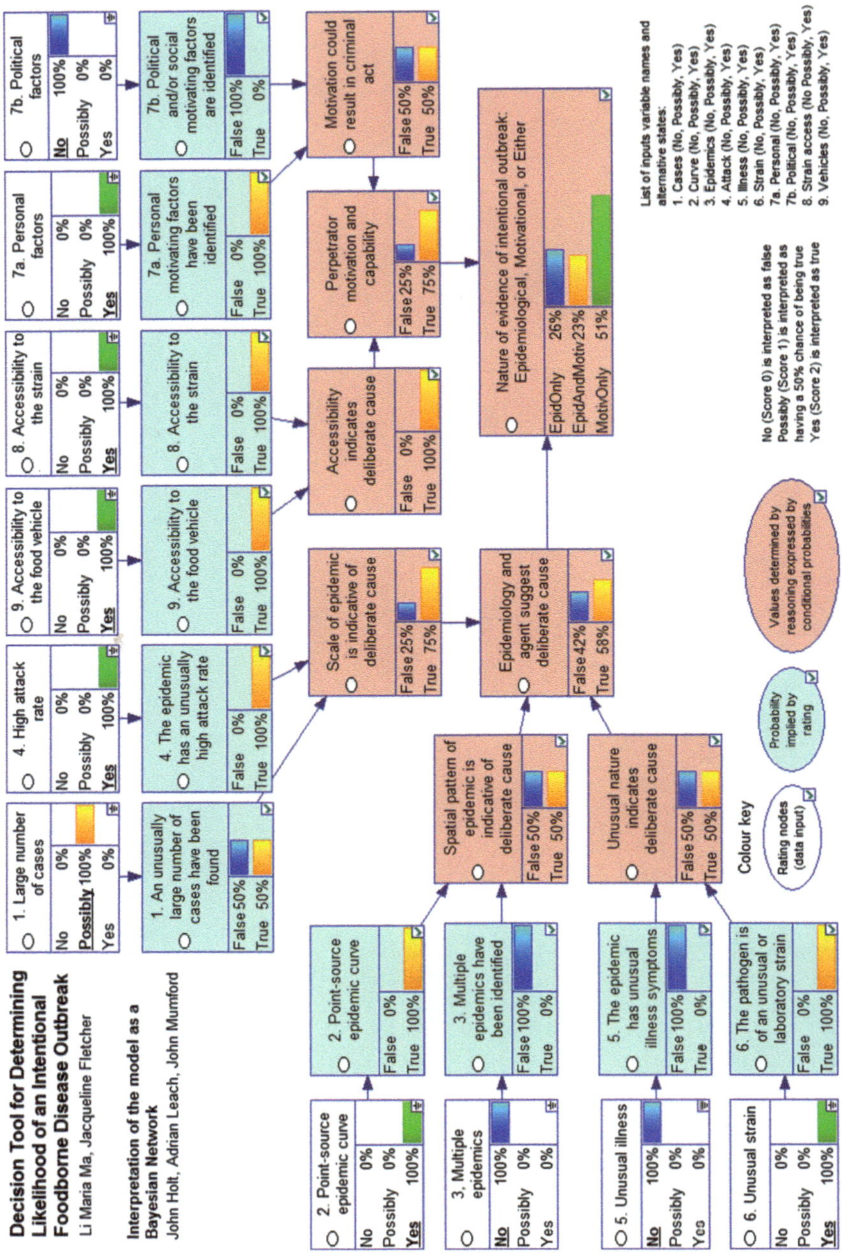

Fig. 9.4 Dallas, Texas Shigellosis (intentional) outbreak in the Bayesian network model

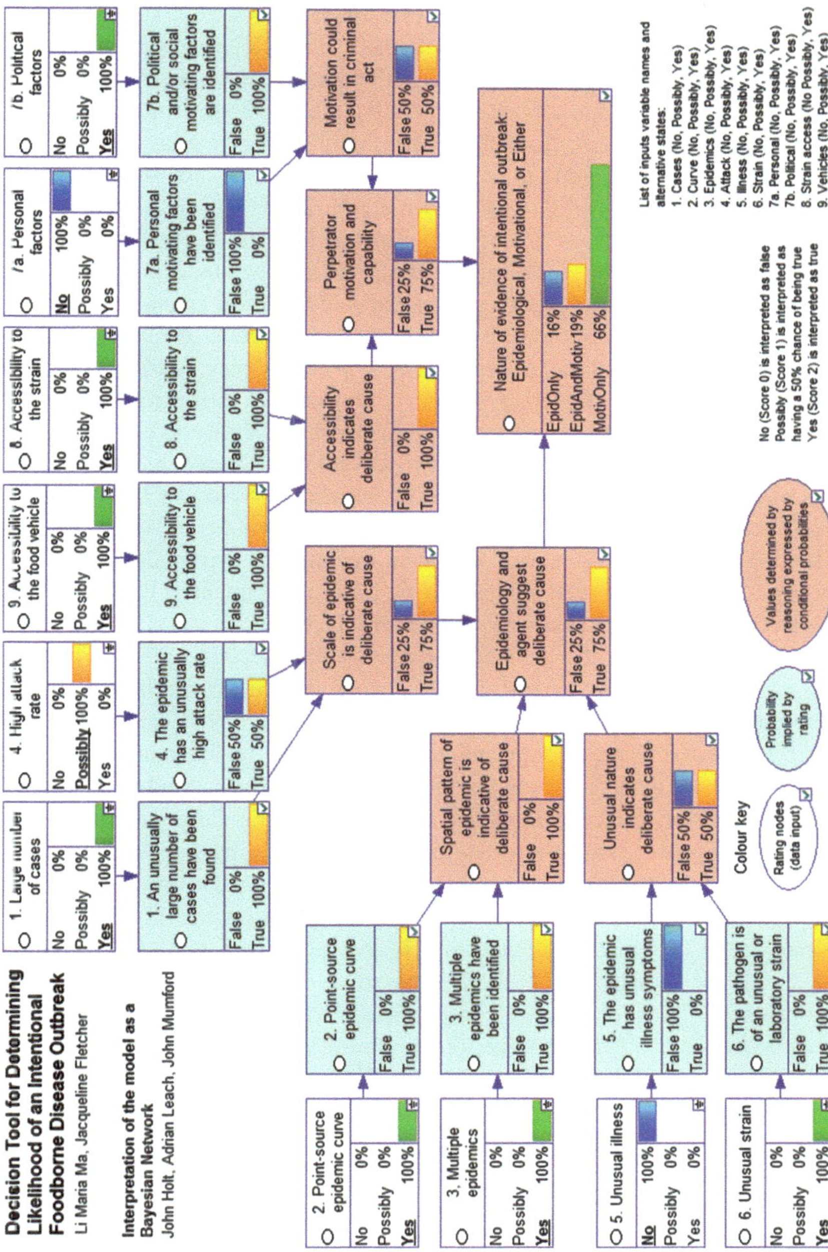

Fig. 9.5 The Dalles, Oregon Salmonellosis (intentional) outbreak in the Bayesian network model

needed for convictions should be gathered as early as possible, so recognizing possible indications of criminal intent early on will help to ensure that such evidence can be effectively collected. Two assessment models were developed previously by others (Grunow and Finke 2002; Rodosavljevic and Belojevic 2012) for differentiating between natural and deliberate outbreaks of biological warfare agents. However, these models are not appropriate for the assessment of outbreaks of foodborne illness, as illustrated in this chapter. Unlike most biological warfare agents, foodborne pathogens or toxins are often "ubiquitous" in nature and "natural" outbreaks of foodborne illness, due to negligence or mishandling, are relatively common events around the world (CDC 2014; CSPI 2014; EFSA 2015). Typical epidemiological investigations are often performed in these outbreaks, with the goals of identifying the source of the infectious agent and then reducing or stopping the further spread of the outbreak. However, certain features of an outbreak, generated from the routine epidemiological investigation, could serve as clues or indicators for potential intentional acts that may lie behind an outbreak. In these cases, the possibility of intentional tampering should be considered in regard to potential motivation, access to the pathogen or toxin involved, the ability to culture/handle these agents, and access to the implicated food vehicles. Therefore, based on the analysis of two previous, well-documented intentional outbreaks of foodborne illness and the typical nature and scope of natural foodborne illness outbreaks, we have developed a set of criteria and used them to construct two models (a scoring system and a Bayesian network model) for use in the assessment of likelihood that an outbreak of foodborne illness was intentionally incited.

The nature and scope of a deliberate outbreak will depend on many factors. Because a wide range of hypothetical biological attack scenarios related to intentional outbreaks of foodborne illness can be imagined, it is impossible to anticipate all possible potential parameters or indicators. A rational model should start from the minimum of components necessary to differentiate a natural outbreak from an intentional attack as early as possible, while minimizing false positive cases. Examples of this process have been demonstrated in the scoring and Bayesian models illustrated in this chapter. However, because only two criminal cases were available to us for testing our models, we suggest that these models should be applied to additional known criminal cases that arise in the future, and that our evaluation criteria should be adjusted as appropriate based on new evidence. Although the number of known intentional cases is too limited to validate such models robustly, they may be valuable in the meantime for demonstrating the models' potential, and can serve as a useful guide for investigations of foodborne illness.

9.9 Conclusion

The two general biowarfare determination models are too broad to be applied to foodborne illness outbreaks. Each has categories that are not applicable to outbreaks of foodborne illness and the weightings may not be relevant to such scenarios. Two new models are proposed, a scoring system and a Bayesian network model. Both use categories derived from the wider models but with more applicability to outbreaks of foodborne illness. Both models appear to be able to differentiate intentional and accidental outbreaks, based on the limited set of cases for analysis. A Bayesian network model is more complicated to build and interpret, but takes account of uncertainty in inputs explicitly. Specific values for weighting in the scoring model, and the evaluation classes, are subjective and need additional cases for further validation. This is also the situation with combination rules in the Bayesian network model. However, both FBIO models demonstrate the potential for more specific discrimination tools to help determine whether a specific case should be considered as intentional.

References

Behravesh CB, Mody RK, Jungk J, Gaul L, Redd JT, Chen SI (2011) 2008 outbreak of salmonella saintpaul infections associated with raw produce. N Engl J Med 364:918–927

Carus WS (2001) Bioterrrism and biocrimes: the Illicit use of biological agents since 1900. Center for Counterprofiration, National Defence University, Washington, DC

Center for Science in the Public Interest (CSPI) (2014) Outbreak alert! 2014. A review of foodborne illness in America from 2002–2012. Center for Science in the Public Interest. Washington, DC

Centers for Disease Control and Prevention (CDC) (2014) Surveillance for foodborne disease outbreaks, United States, 2012, Annual Report. Atlanta, Georgia: US Department of Health and Human Services, CDC

European Food Safety Authority (EFSA) (2015) The european union summary report on trends and sources of Zoonotic agents and foodborne outbreaks in 2013. EFSA J 13(1):3991

Gould LH, Walsh K, Vieira A, Herman K, Williams I, Hall A, Cole D (2013) Surveillance for foodborne disease outbreaks – United States, 1998–2008. Morb Mortal Wkly 62:SS2

Grunow R, Finke EJ (2002) A procedure for differentiating between the intentional release of biological warfare agents and natural outbreaks of disease: its use in analyzing the tularemia outbreak in Kosovo in 1999 and 2000. Clin Microbiol Infect 8:510–521

Holt J, Leach AW, Knight JD, Griessinger D, MacLeod A, van der Gaag DJ, Schrader G, Mumford JD (2012) Tools for visualizing and integrating pest risk assessment ratings and uncertainties. EPPO Bulletin 42:35–41. doi:10.1111/epp.2548

Kolavic SA, Kimura A, Simons SL, Slutsker L, Barth S (1997) An outbreak of Shigella dysenteriae type 2 among laboratory workers due to intentional food contamination. JAMA 278:396–398

Radosavljevic V, Belojevic G (2012) Unusual epidemic events: a new method of early orientation and differentiation between natural and deliberate epidemics. Public Health 126:77–81

Riedel S (2004) Biological warfare and bioterrorism: a historical review. Proc (Baylor Univ Med Cent) 17(4):400–406

Robert Koch Institute (2011) Report: final presentation and evaluation of epidemiological findings in the EHEC O104:H4 outbreak, Germany 2011. Berlin 2011

Sjodin A, Broman T, Melefors O, Andersson G, Rasmusson B, Knutsson R, Forsman M (2013) The need for high-quality whole-genome sequence dtabases in microbil forensics. Biosecurity Bioterrorism-Biodefense Strategy Pract Sci 11:S78–S86

Torok TJ, Tauxe RV, Wise RP, Livengood JR, Sokolow R, Mauvais S, Foster LR (1997) A large community outbreak of salmonellosis caused by intentional contamination of restaurant salad bars. J Am Med Assoc 278:389–395

Chapter 10
Diagnostic Tools for Plant Biosecurity

Jane E. Thomas, Thomas A. Wood, Maria Lodovica Gullino, and Giuseppe Ortu

Abstract There is now a wide range of diagnostic tools in the armoury to help prevent or control damaging disease outbreaks. When applied in the context of biosecurity, they have immense power to protect the plants on which food, feed, fuel and fibre supplies rely. Diagnoses which used to rely on culturing organisms, examining spores, or testing viruses on indicator plants, often taking many weeks to complete, can now be achieved in a matter of hours. Moreover, the advent of in-field diagnostic tests allows growers, agronomists or plant health and seeds inspectors to get a reliable test result without sending a sample to a laboratory. Remote sensing, using ground vehicles, unmanned aerial vehicles, or satellite technology, can bring a new dimension to surveillance, detection and diagnostic systems. Pathogen variation can be characterised rapidly by molecular marker techniques, potentially accelerating the process of identifying new pathotypes or fungicide resistant strains which threaten plant productivity. Metagenomic methods will undoubtedly play a part in non-targeted diagnostics, and identifying new threats to biosecurity. While diagnostic methods have advanced rapidly, their use in disease management in the field must be supported by robust sampling methods, treatment thresholds, and in depth understanding of disease risks.

Keywords Direct and indirect diagnostics • Molecular techniques • Pathotypes • Next generation sequencing • Remote sensing systems

J E. Thomas (✉) • T.A. Wood
NIAB, Huntingdon Road, Cambridge CB 3 0LE, UK
e-mail: jane.thomas@niab.com

M.L. Gullino • G. Ortu
AGROINNOVA, Centre of Competence for the Innovation in the agro-environmental sector, University of Turin, Grugliasco (TO), Italy

M.L. Gullino et al. (eds.), *Practical Tools for Plant and Food Biosecurity*, Plant Pathology in the 21st Century 8, DOI 10.1007/978-3-319-46897-6_10

10.1 Introduction

Approximately 37 % of the earth's land surface is classified as agricultural land, that is land used for food production, including pastures and plantation crops. Natural or planted forests account for a further 30 % of land area. It is a major challenge to protect this area of vegetation from pests and pathogens in an era of global trade, commonplace international travel, and frequent severe weather events which can move disease propagules many hundreds of miles (Schmale and Ross 2015). The longer term effects of climate change will alter the distribution profile of many pathogens, creating new risks and reducing others (Pautasso et al. 2012) Traditional skills in plant pathology have declined in many developed countries, and surveillance resources have been severely stretched. New, or newly emerging, and established pathogens threaten productivity in many countries, affecting the food supply, livelihoods, and landscapes (Anderson et al. 2004). About 30 % of potential productivity is lost worldwide, pre-harvest, to the effects of pests and diseases, and a further 20 % may be lost post-harvest (Oerke et al. 1994). Pesticides, often the most effective method of control, are under pressure, both from regulatory requirements and rapid development of pathogen resistance. Plant breeding for resistance to pathogens can be very effective, but is relatively long term, and for many organisms there is rapid selection for the variants which render resistance factors ineffective. Prevention of new pathogen incursions, whether they are accidental, natural, or deliberate, is thus by far the most preferred approach, and rapid, accurate diagnostics are an essential component of preventative action. Plant biosecurity, defined by Waage and Mumford (2008) relates specifically to the protection of national boundaries against the introduction of alien pests and diseases. These organisms will normally be regulated and subject to statutory actions. However, other pest and diseases can be regarded as "emerging" threats, and have probably been present in a country or region for several years before first detection. For example, *Verticillium longisporum* was first confirmed on oilseed rape in the United Kingdom in 2007, (Gladders et al. 2011) and has now been found in over 20 % of the surveyed crop area (www.cropmonitor.co.uk). Early blight of potato, caused by *Alternaria solani*, once a rarity in Europe, has become prevalent in many European countries (Hausladen and Leiminger 2007), often necessitating specific spray programmes. The control of these, as well as long term established diseases, requires ever more sophisticated approaches to maintain profitable production, avoid unnecessary pesticide use, and still meet exacting quality standards, and the use of diagnostics in precision disease management will be increasingly required.

In the context of this chapter, the term biosecurity is used to refer to the prevention of incursions of alien pests or pathogens, by natural pathways or otherwise. However, since sophisticated diagnostic tools have advanced rapidly in the context of protecting crops and plants from plant pathogens in general, examples which relate to the management of indigenous diseases will also be discussed. The processes surrounding the application of diagnostics for biosecurity purposes, and some of the demands placed on a diagnostic method, *versus* those for disease management may differ considerably. For instance, if a deliberate introduction is suspected, and forensic evidence for prosecution is needed, such as traceability to a source, appropriate accreditation standards will be needed. In some instances, precise strain or isolate discrimination will be needed, which may not always be required for indigenous disease management. For biosecurity purposes, the ability to detect extremely low levels of an organism is usually necessary, and quantification may not be needed. However, for disease management, accurate quantification is usually required, linked to treatment thresholds and economic relevance to a grower.

Diagnostic tools have to be deployed in a wide range of situations, and by a wide range of operators. These may include inspection services, growers, agronomists or staff in diagnostic service labs. In the context of regulated pathogens, training in the diagnosis of specific disease symptoms is usually available for inspection staff, and highly developed protocols, often using molecular techniques, are used for confirmation. Emerging or new, non- regulated threats in field crops are much more dependent on the observation skills of growers and agronomists, backed by "plant clinic" services, though a number of techniques such as remote or proximal sensing devices are increasing capability for wide area surveillance linked to a diagnostic imaging system of one type or another. Newer developments in molecular diagnostics such as loop mediated isothermal PCR (LAMP) and recombinase polymerase amplication (RPA) methods have brought specific and sensitive techniques for direct analysis of a sample within reach of field operations, giving rapid answers so that management actions can be put in place.

Diagnosis is the process of identifying a disease, and diagnostics describe the tools and tests used for this. Diagnostic tools may be considered in two broad categories. Firstly, those techniques where a sample can be viewed or analysed directly, and secondly, techniques where some remote or indirect analysis is involved. The first category is still the most common, and can cover everything from symptom recognition, immunological methods, and advanced molecular methods. The second category is less well established, but sophisticated technologies are increasingly offering a new dimension to diagnostics, incorporating wide area surveillance in the methodologies. This chapter will review some recent diagnostic approaches and show how they can contribute to plant and food biosecurity, and to the management of pathogens which continually threaten productivity.

10.2 Direct Sample Analysis

10.2.1 Symptom Recognition and Culturing

Diagnosis of a large number of diseases can often be achieved by visual inspection, either by eye on symptom type alone, or by observation of fruiting bodies with the aid of a hand lens or binocular microscope, or by culturing if appropriate and examining mycelium and spore characteristics using compound microscopes. There are many high quality image libraries and descriptions available, such as the APS Press Compendium Series, and many on-line resources. Informative though these resources are, they usually show highly typical, clear and distinct symptoms. These can be very different to those a practitioner may see on a sample which has travelled in the mail, or which is at a different stage of development, or on a different part of the plant, or where other organisms or stress symptoms may be involved. Pattern analysis and machine learning of disease symptoms are techniques which are being developed (Camargo-Rodriguez et al. 2012, and Camargo-Rodriguez and Smith 2009) and provided symptom variability is incorporated, such systems could prove valuable in identifying diseases in areas where diagnostic support is not readily available and, importantly, perhaps to indicate when an expressed symptom does not match the range of known problems, and might be a new incursion. Web-enabled microscopy, allowing inter-laboratory examination of a specimen or culture, is available in several diagnostic networks (for example, the National Plant Diagnostic Network in the USA, www.npdn.org). Real-time consultation with experts while viewing a sample is a useful training mechanism as well as aiding diagnosis.

Simple techniques such as incubation in a damp chamber are useful for accelerating spore production, and can be used by growers, agronomists or laboratory staff. Early diagnosis of *Pyrenopeziza brassicae*, the causal agent of light leaf spot on oilseed rape, is essential to inform effective fungicide timings. The pathogen has a long latent period (Gilles et al. 2000) during which symptoms are absent or very unclear, and can easily be confused with frost damage, fertiliser scorch or possibly other diseases. However, where infections are present, incubation of a leaf sample in a polythene bag at about 18 °C for 3–4 days, will reveal the presence of the spore producing acervuli, which are easily visible to the naked eye.

Culturing from an infected sample to obtain putative causal organisms for re-inoculation and the proof of Koch's Postulates is of course a standard procedure, and there is a multitude of modified substrates, selective agars, surface sterilisation and incubation conditions which can be used to try and identify primary causal agents. In the Plant and Food Biosecurity programme, a long term study of the onion- *Fusarium proliferatum* pathosystem was initiated after the pathogen was identified in cultures from diseased onion bulbs in Israel in 2008. A selective medium was developed which encouraged growth of *F. proliferatum* over non-target fungi present in the sampling environment, and which enhanced the development of distinctive spore types which differentiate *F. proliferatum* from other Fusaria (Isack et al. 2014). Highly selective diagnostic media such as this are particularly valuable in labs where molecular diagnostic facilities may be limited.

The Plant and Food Biosecurity project has developed a virtual diagnostic network, described in detail elsewhere in this book, which can act as a repository for the types of information described above, including methodologies and outbreak mapping, expert finding, the potential for web-enabled microscopy as an add-on facility, community pages for protocols, and diagnostic training.

10.2.2 Molecular Diagnostic Methods

Despite the effectiveness of traditional diagnostic skills in many instances, they are usually lengthy and while they are ongoing, a potential disease threat may be increasing in the environment. Some pathogens are obligate, and some threats will occur with sub-species, pathotypes, or other variants which cannot be distinguished from an established problem visually or in culture. Viruses are particularly difficult to diagnose visually and symptoms may vary according to cultivars of the host species as well as virus strain. Enzyme Linked Immunosorbant Assays (ELISA) have long been the basis for plant virus diagnosis. Boonham et al. (2014) suggested that high levels of reproducibility, repeatability. the availability of an industry standard format and easy access for most laboratories were all reasons why the technique remained successful and widely used.

Immunological assays have been developed into lateral flow devices for on-site pathogen detection, for viruses, bacteria and some fungi. The kits are of use to both regulatory officials and growers or agronomists, and give rapid answers, though for testing multiple samples, their cost is relatively high. As with other targeted test methods, LFDs can only answer the specific question of whether a known organism is present, and low sensitivity can limit their value. However, they are increasingly used as on-site diagnostics by growers and agronomists, and have significant value for early detection of disease threats, and are available for virus, bacteria, oomycete and fungal targets. LFDs have been developed to diagnose the presence of *Rhizoctonia solani* in soil (Thornton et al. 2004). Wakeham (2015) developed an LFD for the detection of the club-root pathogen in commercial brassica growing soils. Though the accuracy of the LFD was lower than a qPCR test for disease prediction, it could be used for rapid indexing of soils, in 10 min, by growers to indicate whether the risk of club-root was zero or low, or medium to high. Kennedy and Wakeham (2008) developed monoclonal antibodies against the onion downy mildew pathogen, *Peronospora destructor*, and devised a lateral flow format for the detection of sporangia. Lane et al. (2007) evaluated LFD assays for two Phytophthora species associated with tree diseases and concluded that they compared favourably to a PCR diagnostic and far exceeded visual accuracy. They were considered suitable for phytosanitary purposes, and identifying samples for more stringent analyses. The Phytophthora kits may also be used for detection of late blight on potatoes and tomatoes. LFD reader systems have also been developed, allowing for semi-quantitative interpretation (Faulstich et al. 2009), thus adding value to the assays in terms of the degree of risk which may be predicted. LFDs have the advantages of being easy to use, rapid, and can be applied to many types of plant tissue which can be extracted on site, including woody plants. However, cross reaction with non-target species, delayed or weak reactions can result in misleading information, and potentially wrong management decisions being made. Unless specificity is very high, such as with LFDs developed for detecting potato viruses, the use of the devices as on-site diagnostic may be limited. The majority of commercially available

LFDs are for viruses, with some bacterial and relatively few fungal or oomycete targets. Multiplexing is rare, and available assays target one organism only.

10.2.3 Air Spora Diagnostics

The use of spore traps for detecting the presence of a pathogen can play a major role in monitoring the environment for new incursions and also management of established pathogens. Types of trap samplers have been extensively reviewed by West and Kimber (2015). In the past, spore traps have required a lengthy period between time of sampling and the point at which a quantitative figure for the presence of a target pathogen in the air could be given. Microscopic examination of tapes and strips required a very high level of mycological expertise. However, grower-operated Burkard spore traps can now be used in conjunction with LFDs for analysis of the sample. The Brassica Alert system (www.syngenta-crop.co.uk/brassica-alert) offers the option of an LFD assay for the brassica ringspot fungus (*Mycosphaerella brassicicola*). Early detection of the spores offers the opportunity for early intervention with fungicide and prevention of lesion development, since the retail standards for blemish in vegetable brassicas can be particularly stringent in many European countries. In the context of biosecurity, the analysis of spore trap catches could be very valuable for tracking incursions. However, Jackson and Bayliss (2011) reviewed the use of spore traps for biosecurity objectives, and concluded that their application was constrained by the relatively small volume of air that could be sampled, and the difficulty of achieving adequate representation of the air spora of a region. Analysis time was considered to be less of an issue given new diagnostic approaches.

Direct analysis of a spore trap sample by qPCR is possible, though this still requires a laboratory test. It has the advantage that tests can be multiplexed, and a range of targets from a single spore trap sample can be quantified. Isothermal assays could be conducted in the field, with answers available in a few minutes. Thiessen et al. (2016) used an isothermal assay to detect spores of grape powdery mildew from impaction spore traps placed in vineyards. The data was used to alert growers on spray schedules, and by using the system they were able to reduce spray number compared to calendar driven spray schedules, without affecting disease severity. West et al. (2013) developed an assay to quantify Sclerotinia risk for oilseed rape crops. Spores from an impactor trap were captured in vials containing a semi-selective medium. This was then testing for the presence of oxalic acid, secreted by ascospores of *S. sclerotiorum*, using a biosensor, and the results wirelessly transmitted to a server. Together with on site weather recording, the risk of Sclerotinia was evaluated and sent to the grower *via* text message. Fully integrated trap and diagnostic systems, using LFDs, isothermal PCR methods or other analyses are being developed further, and potentially can give local, regional or national risk indications depending on siting, and type of trap used. The concept of mounting spore traps on lower atmosphere UAVs, or even higher altitude systems, in combination with a diagnostic system, could increase sampling capacity and give new insights into mass spore movements, and the possibility of predicting pathogen outbreaks or incursions well in advance of disease symptoms being discovered.

10.2.4 DNA and RNA Diagnostics on Plant Material and Soil

The rapid development of nucleic acid detection systems in plant pathogen diagnostics has given rise to many standardised and validated protocols. For regulated pathogens, the European and Mediteranean Plant Protection Organisation (EPPO) publishes new diagnostic procedures, and several of these now contain PCR based methods. The International Seed Testing Association (ISTA), and the International Seed Health Initiative (ISHI) have also validated a number of molecular diagnostic protocols to establish seed health status. Apart from regulated materials and quarantine pathogens, many other studies have resulted in the production of nucleic acid based methods which can be used to diagnose pathogens rapidly and accurately, and contribute to more effective disease management. PCR, qPCR and RT-PCR are now standard laboratory diagnostic techniques, and have been reviewed extensively over the last 10 years (see Atkins and Clark 2004; Bradshaw et al. 2006; Martinelli et al. 2014). Signifcant advances in the use of PCR diagnostics for soil-borne pathogens, and other problematic substrates, have been made. Woodall et al. (2012) developed a quantitative PCR test to detect sclerotia of the onion white rot pathogen (*Sclerotium cepivorum*) in soil. Bilodeau et al. (2012) developed an assay to detect microsclerotia of *Verticillium dahliae* from soil samples in strawberry growing fields. Deora et al. (2015) used a Taqman PCR assay to detect and quantify club-root (*Plamodiophora brassicae*) from soil samples. These, and other examples, typify situations where conventional mycological tests take extended periods to produce results, or require complex extraction and processing, with the need for highly experienced staff to count propagules. They also represent serious pathogens of high value crops, where avoidance of infected land is the principle means of control. Several commercial soil extraction kits are now available and the cost of diagnostic services using them should be acceptable to growers.

10.2.5 Early Detection of Emerging Diseases in High Value Crops

Vegetable and salad crops are usually produced in relatively small, specific areas of a country, in close rotation, and often with two or more crops produced in a single season. These intensive growing conditions are very favourable for the rapid increase of disease. If pathogens are carried on or in seeds, even very low levels of contamination can lead to the rapid emergence of new disease problems which threaten crop quality and saleability (Gullino et al. 2014). The development of diagnostic tools for early detection and investigation of causal agents is essential to support prevention and control measures.

Case Study 1 – Wild Rocket and Lettuce Diseases
Two recently occurring diseases on wild rocket and lettuce in Italy were investigated as part of the Plant and Food Biosecurity project. *Plectosphaerella cucumerina* was described as the causal agent of a leaf spot on wild rocket (*Diplotaxis tenuifolia*) in Italy in 2012 (Garibaldi et al. 2012). The disease was severe under glasshouse conditions and caused significant production losses, with rotting of material occurring after processing and packing. A real-time PCR assay for the detection of *Pa. cucumerina* was first developed by Atkins et al. (2003). However, there are four *Plectosphaerella* species, and analysis of the sequence data suggested that the assay was not specific enough to distinguish between these. Specific real time Taqman probes were therefore designed targeting areas of the ITS sequence with most divergence between species to detect and quantify *Pa. cucumerina*. This assay was able to detect *Pa. cucumerina* DNA in symptomatic rocket leaf tissues and was shown to be highly specific, with no amplification of non-target species of *Plectosphaerella*.

Fusarium. oxysporyum f. sp. *lactucae* is the causal agent of wilt on lettuce (*Lactuca sativa*). It has spread through many countries and was first detected in Europe in Italy in 2002 (Garibaldi et al. 2002) causing up to 70 % losses in summer production. Three different races (1, 2, and 3) have been identified within *F. oxysporum* f.sp. *lactucae.* Currently only race 1 has been reported in European countries (Garibaldi et al. 2002; Gullino et al. 2004) while races 2 and 3 have been only reported in Japan (Fujinaga et al. 2001, 2003) and Taiwan (Lin et al. 2014). Pasquali et al. (2007) developed an assay based on inter-retroelement amplified polymorphism (IRAP) PCR which differentiated *F. oxysporum* f.sp. *lactucae* Race 1 from other isolates of the pathogen, and other *F. oxysporum formae speciales* This assay could be applied to both plants and seeds for rapid detection of Race 1 of the pathogen, thus supporting disease management and use, or not, of seed stocks.

10.2.6 Multiplex Diagnostics

Multiplex detection of several pathogens in a single assay is still relatively limited, though many target organisms are either part of a disease complex, or cause distinct diseases while existing in the same sample substrate. Cullen et al. (2000), developed a qualitative multiplex PCR test for three potato pathogens in soil and on tubers. Qu et al. (2010) used a TaqMan PCR for simultaneous detection of powdery scab and common scab on potato tubers. Visual discrimination of symptoms of these two diseases can be difficult, but is necessary for effective management of seed tubers. Recently, the Luminex® platform has provided a new mechanism for multiplexing nucleic acid or protein based assays. In the case of nucleic acids, there is specific hybridisation between DNA fragments on colour coded paramagnetic beads.

Kostov et al. (2015) used Luminex® technology to detect and identify 26 species of Phytophthora simultaneously. A further 22 samples were identified to clade or sub-clade levels. Discrimination at this level provides a significant advance for detecting closely related organisms which have major phytosanitary implications throughout Europe and elsewhere. Charlermroj et al. (2013) used Luminex® technology for the simultaneous detection of three plant viruses and a seed- borne bacteria. The assay detected all pathogens from infected plant material, and could be adapted to detect up to 50 targets at the same time. Though laboratory based, these multiplex systems offer considerable efficiencies which will assist the safe movement of plants and seeds across national borders.

10.2.7 On Site, Field Deployable Diagnostics

Despite the speed and sophistication of new laboratory based molecular diagnostics, on-site diagnosis is still a practical requirement for both regulated and non-regulated pathogens (De Boer and Lopez 2012). For regulatory organisms, rapid diagnosis is usually needed at the point of inspection, so that consignments can be moved onwards or held back (Brasier 2008). For non-regulated pathogens, where effective disease management is at stake, growers and agronomists frequently need a rapid diagnosis to select an optimal spray programme, including appropriate active ingredients and rates. The use of LFDs as previously discussed has already made an impact, and on-site DNA diagnostics is now a rapidly developing area. Loop mediated isothermal PCR (LAMP) operates at between 60 – 65 ° C and does not require a thermal cycler. Isothermal assays can be carried out on crude extracts from target tissues, with no DNA purification steps, and can operate effectively in the presence of many inhibitors which are usually present in plant, seed or soil samples. Developed assays and closed tube systems are available (see www. optigene.co.uk) marketed with Genie II or Genie III instrumentation. These units are portable and robust, running for a day on battery power, and delivering easy to interpret visual assay results within 15–30 min. Other isothermal diagnostic systems have been developed, and are now being investigated in the plant health arena. Miles et al. 2015 used recombinase polymerase amplification (RPA) to detect Phytophthora species, and suggested its use as field deployed diagnostics for Phytophthora infections of forest trees. The RPA assays were very rapid, and as with LAMP did not require DNA purification. Doan et al. (2014) used RPA to diagnose *Fusarium oxysporum* f.sp. *vasinfectum* race 4 in cotton plant tissues. The assay was race specific, and was carried out at a constant 39 °C, giving a result in 30 min. Identifying Race 4 infected fields is important for the prevention of spread of the race *via* seed and soil. In the growing region of California, where the work was carried out, seed and ware production of cotton takes place in the same area, so restricting infection to as few fields as possible is necessary. Ultimately, highly sensitive techniques may be able to detect target regions of DNA without any amplification step. These approaches are already being developed for medical applications (Cai et al. 2015) as point of care diagnoses.

The on-site diagnostic systems currently available in the plant health sector are single target assays, so several tests may need to be performed on a single sample. Nevertheless, they are providing much needed efficiencies, and new information to feed into disease prevention and control strategies.

10.2.8 Biosensors

Biosensors offer a novel and potentially very efficient method of diagnosing plant pathogens. They consist of a biological molecule (protein, nucleic acid, enzyme etc) which can specifically recognise a target analyte. The reaction is then converted to a signal by a physiochemical mechanism (the transducer). Biosensors could be deployed in large numbers in agricultural environments, or in produce storage systems, to detect the presence of pathogens or rots. Signals could be captured wirelessly and transmitted to a server in the farm office or to a store manager, or integrated with a disease forecasting system, to create alerts and prompt action. Fang and Ramasamy (2015) reviewed the potential for biosensors in plant disease detection. Though biosensors for fungi, bacteria, oomycetes and viruses had been developed in the laboratory, their practical use as an in- field detection systems has not yet been realised. They concluded that enzyme based biosensors probably had the greatest potential as they were stable, relatively low cost, easy to use and specificity could be very high.

10.2.9 Non-targeted Diagnostics

A major challenge facing diagnostics is the presence of an "unknown" threat. Next generation sequencing technologies, and the bioinformatic systems which accompany them, are being deployed to detect new causal agents without any prior knowledge of their identity. Simply, sequencing extracts from a plant expressing symptoms and comparing it to sequence data from a plant with a healthy appearance, can reveal differences which point to potential causal agents. Adams et al. (2009) identified a previously unknown cucumovirus using this approach, and Barba et al. (2014) in a review of the techniques involved, itemised the discovery of viruses or viroids in tissues of many plant species that exhibited unknown disease etiologies. They suggested that NGS technologies were a future tool in quarantine and certification of high value fruit species and in woody plants where virus titres may be very low.

Nanopore sequencing is a third generation sequencing technology generating long read lengths. While relatively new to plant pathology, it is already being used in metagenomic studies to identify human bacterial and viral infections

(Greninger et al. 2015). The technology identifies nucleotides in a DNA molecule by measuring conductivity as DNA molecules pass through pores in a membrane. The MinION™ device is a portable pocket sized unit containing disposable flow cells and which can be plugged into a PC *via* a USB port and results read in real time on the screen. Multiple re-sequencing of pathogen isolates and comparison with reference genomes could aid understanding of genetic variation and virulence changes, and metagenomic sequencing of plant, soil or air samples could identify emergent problems before they become established. While the technology is still in very early stages of research for plant pathogens, it offers great potential for future precise understanding of microbial diversity, and the new threats that may exist.

10.2.10 Diagnostic Techniques for Detection of Novel Pathotypes

The discrimination of pathogen strains or pathotypes is a major factor in plant biosecurity. Though pathogens may be well established threats in particular areas, and the diseases caused easy to recognise by symptoms alone, the evolution and selection of new virulences which defeat previously effective host resistance factors, can frequently cause sudden and serious disease outbreaks. One of the most common examples are the rust pathogens of cereals and other crops. Long distance transport of spores, either by extreme weather events or inadvertently by humans during international travel, is probably responsible for the movement of pathotypes common in one region to a different area. However, rust pathogens have also been weaponised in the past (Rogers et al. 1999) so the potential for deliberate introduction exists.

The diagnosis of pathotypes of common, established diseases is still achieved largely by field sampling and testing on host differential lines which reveal variation in virulence matching host resistance genes or factors. For wheat rust pathogens, many countries worldwide carry out regular surveys of field outbreaks from commercial crops, or selected sentinel plots which remain unsprayed and expose different resistances to try and detect early occurrences of new virulences. Despite these efforts, there is often a considerable time lag before confirmation of new pathotypes, particularly when field testing on adult plants is required as well as seedling tests. Extensive facilities and staff resources are also needed, and the sampling intensity of outbreaks is inevitably restricted. Frequently, new pathotypes are not detected at very early stages of establishment, with the result that previously resistant cultivars can suddenly "break down" in the field. Marker techniques such as simple sequence repeats (SSRs) have proved effective in tracking programmes for rust populations and associated pathotypes, and understanding global pathogen evolution (Hovmoller et al. 2015). Nevertheless, such systems still require careful increase of these obligate pathogens on host plants, which can take many weeks.

Case Study 2 – *Puccinia striiformis* **f.sp.** *tritici*
Recently, the concept of field pathogenomics has been developed to obtain more rapid information on the occurrence of new virulence phenotypes (Hubbard et al. 2015). Using the wheat yellow rust pathogen (*Puccinia striiformis* f.sp. *tritici*) in the U.K. as a model, the technique employed transcriptome analysis of field samples collected in tubes of RNAlater. A section of the same infected leaf was reserved to increase the rust spores and carry out conventional pathotyping on host differentials. Analysis of the genetic structure of the samples showed that four distinct lineages were present, which corresponded to distinctive virulence phenotypes. The method was able to show that the yellow rust population had recently undergone a major change genotypically, with a highly diverse structure compared to historical isolates. Together with field observations of a higher than usual incidence of teleutosori, the evidence pointed to an exotic incursion, probably from a region where the pathogen underwent a full heteroecious cycle on its alternate host. The technique also created the potential for greatly increased sampling intensities at a relatively low cost, and prediction of which samples appear unusual genotypically, and should be taken forward for standard phenotypic analysis. Rapid genotyping techniques for diagnosing new races of stem rust (*Puccinia graminis* f.sp. *tritici*) have also been developed using a SNP chip (Singh et al. 2015). The field pathogenomics concept, though laboratory based, offers very considerable scope for the rapid characterisation of emergent pathotypes, particularly for obligate organisms, and so contributes to enhanced biosecurity through speed of response and re-direction of breeding programmes.

10.3 Indirect Diagnostic Methods

10.3.1 Disease Diagnostics by Remote or Proximal Sensing

The vast areas of cultivated plants that may be at risk of pathogen attack cannot be adequately monitored by the standard method of crop walking and visual inspection, though these tried and tested methods will always be needed, both for identifying regulated pests at ports of entry, and for the management of indigenous diseases. Halting epidemics or spread of a pathogen threat is most effective at very early stages of development, and for this different approaches are needed. Aerial photography has long been available for locating foci of a range of diseases, though confirming identity of the specific problem has been achieved by site visits. Now, ground vehicles, UAVs, and satellite platforms, can all be equipped with various types of sensing equipment which can identify unhealthy plants, creating the potential for wide area monitoring. Mahlein (2016) reviewed the use of disease detection by imaging sensors and concluded that there was as yet no commercially available system which could be used for the specific detection of plant disease. The recognition of specific diseases, and discriminating between them, and other causes of plant stress, with sensing technologies, remains a major research challenge, though some experimental progress has been made.

Measuring reflectance from plants in the field in the elcctromagnetic spectrum from multispectral or hyperspectral cameras holds significant promise for disease detection. The most informative spectral bands range from visible though to infrared and thermal wavelengths. Passive remote sensing devices measure reflectance of incident radiation from the plant canopy, while active sensors emit radiation to the target and measure returned reflectance. The collection of multiple spectral information, at as high a resolution as possible, may start to build up "signature" information for specific disease or pest infections, but extensive ground truthing is a critical aspect for research. There is also a significant image processing and computational requirement before positional data on "affected" versus "healthy" plants can be mapped onto field layouts.

Information from different areas of the electromagnetic spectrum will have different functionalities. Visual wavelengths are unlikely to detect pre-symptomatic infection, but will be important in tracking disease epidemics and providing new insights into disease dynamics. Early detection or pre-symptomatic disease is more likely to be achieved by multispectral or hyperspectral images. Bauriegel et al. (2011) identified Fusarium head blight infection in wheat using hyperspectral imaging under laboratory conditions. However, they further identified two specific and narrow wavelength ranges which could be used under field conditions and defined the optimal growth stage for predicting infection levels. Huang et al. (2007)) used air-borne hyperspectral imaging to detect wheat yellow rust, and obtained a high correlation with ground assessment of a disease severity index. Specific spectral features were associated with yellow rust infection, but not with water or nutrient stress.

Wahabzada et al. (2015) used hyperspectral reflectance to identify specific signatures of foliar diseases of barley at different stages of development, and coined the term "metromaps" to describe them. Though the material was inoculated and assessed under laboratory conditions, such specific hyperspectral signatures could possibly be used to identify field outbreaks of diseases before complete symptom development, enabling precision application of eradicant fungicides.

UAVs, either fixed wing or rotary platforms, have the ability to survey large areas, though suffer from several drawbacks. Licences to operate may be restricted in some areas, and payload will be limited for rotary type systems. Though more can be carried by fixed wing UAVs, these are more expensive and operators require more training, with higher standards and more restrictions. The frequency of flights can thus be limited, and insufficient to enable the build up of images which might indicate a spreading disease. Satellite technology has the advantage of more frequent image collection, continuing improved resolution, and relatively inexpensive access. Cloud cover and other environmental conditions can restrict operational periods. However, to date, applications are mostly in the surveillance of crop type and planted areas rather than within field diagnosis of potentially diseased plants and there are few examples of remote sensing of specific diseases from satellite platforms. Mirik et al. (2013) mapped outbreaks of wheat streak mosaic virus in

Texas using reflectance data gathered by the Landsat 5 satellite. Yuan et al. (2014) detected wheat infected with powdery mildew using multispectral imaging from the SPOT 6 satellite with a maximum of 89 % accuracy. For the medium term however, it is probable that satellite sensing for disease diagnostics is more likely to be used to identify areas of stressed plants, where ground confirmation of specific causal agents is needed.

The detection of volatile organic compounds (VOCs) is also an indirect diagnostic method, aiming to detect specific signatures of volatiles that a host plant emits when challenged with a pathogen. Many factors may affect the volatile signature of a plant-pathogen combination, including the age of the plant, non-disease related stress, nutrient status etc. A wide range of control plants must therefore be investigated in order to separate disease associated volatiles reliably. A number of electronic nose (EN) systems have been developed. Ghaffari et al. 2012 used 13 gas array sensors to discriminate successfully between healthy and pathogen or pest challenged leaves of tomato, cucumber and pepper, and concluded that the EN system would be effective for diagnosing disease in commercial glasshouses. FAIMS (field asymmetric ion mobility spectroscopy) is an alternative to EN technology. Both methods use fingerprinting of volatile biomarkers, but FAIMS uses mobility of molecules rather than a chemical detection of compounds. It is thought to be more sensitive, and Rutolo et al. (2014) reported successful application of FAIMS profiling to detect bacterial soft rots in potatoes under laboratory conditions. Aksenov et al. (2014) recently identified a specific VOC profile of citrus plants infected with the citrus greening disease (*Candidatus Liberibacter*). The assay was very accurate, even at early pre-symptomatic stages, and could be deployed in the field for growers as a portable device with further refinement. VOC profiling in this case has great potential for critical early identification of a devastating disease, so that infected trees can be removed.

Obtaining pathogen specific profiles for VOCs remains a research challenge in the majority of cases (Sankaran et al. 2010), though detecting healthy versus generically challenged plants may suffice in many situations, such as removing rots from storage environments.

10.4 Conclusions

The range of pathogen diagnostic tools, and the level of their sophistication, has never been greater, and with the advent of new sequencing techniques and novel, field deployable diagnostic systems, their potential can only increase. Maintaining plant biosecurity in the area of regulated pathogens will be a clear beneficiary of these developments. Recent occurrences of new forest and amenity tree pathogens in many countries has brought an unprecedented level of awareness of plant pathology to the general public. In crop plants, cereal rust fungi, citrus greening disease, downy mildew of basil, and the *Xyllela fastidiosa* infections of olive trees have made recent national headlines in different countries. This new awareness brings advantages in that well informed crowd-sourced observations can help to monitor and track disease outbreaks, and potentially discover unusual occurrences, at a level that research organisations and inspection services alone could not achieve. It has also brought investment, both public or private, in developing diagnostic systems. To some extent, it has revitalised interest in plant pathology, and the existence of a skills gap has been acknowledged in many countries, though there is still significant effort needed to ensure that skills can be taught and acquired by new generations of pathologists.

In the regulated pathogen sector, the simple aim of diagnostics must be to identify the causal agents of disease accurately even when levels are very low, and thus contribute to their exclusion, or containment. Detecting and diagnosing "unknown" threats with metagenomic analyses brings a new level of protection against threats to biosecurity. Strain, or pathotype discrimination is important to detect continually evolving variants of pathogen species. Diagnostic tools themselves require validation, and service providers need ongoing training, participation in reference testing and maintenance of international accreditation. Where indigenous organisms are concerned, diagnostic systems face different challenges. While accurate identification is still an obvious pre-requisite, the whole disease management system has to be taken into account. Sampling techniques, weather conditions, treatment thresholds, product availability, rotational decisions, cultivation options, seed bed conditions, cultivar choice, end use requirements, likely price for a crop, are all part of the decisions that need to be made when particular diseases are diagnosed, and a diagnostic tool needs to give results which are readily interpreted in the overall context of a crop situation. Practicalities need to be considered as well. For instance, growers frequently need to apply agrochemical for disease control regardless of risk, or at times which are not optimum, because of farm workloads, contract sprayer schedules, and weather. In some cases, prophylactic spraying is preferred rather than risk determined sprays because quality requirements are so high, and failure to meet them can result in significant loss of income. Countering this however are increasing restrictions on agrochemical use, loss of active ingredients, and low to zero residue level requirements. Diagnostics has a role to play in the effective targeting of control, by integrating with smart spray systems linked to GPS which only apply product where needed. Precision farming, and the drive for sustainable intensification, will place more demands on diagnostic systems, both field deployable units and rapid, sophisticated laboratory methods, all linked with a thorough understanding of the epidemiology of disease causing organisms.

References

Adams IP, Glover RH, Monger WA, Mumford R, Jackeviciene E, Navalinskiene M, Samuitiene M, Boonham N (2009) Next-generation sequencing and metagenomic analysis: a universal diagnostic tool in plant virology. Mol Plant Pathol 10:537–545

Aksenov AA, Pasamontes A, Peirano DJ, Zhao W, Dandekar AM, Fiehn O, Ehsani R, Davis CE (2014) Detection of Huanglongbing disease using differential mobility spectrometry. Anal Chem 86:2481–2488

Anderson PK, Cunningham AA, Patel NG, Morales FJ, Epstein PR, Daszak P (2004) Emerging infectious diseases of plants: pathogen pollution, climate change and agrotechnology drivers. Trends Ecol Evol 19:535–544

Atkins SD, Clark IM (2004) Fungal molecular diagnostics – a mini review. J Appl Genet 45:3–15

Atkins SD, Clark IM, Sosnowska D, Hirsch PR, Kerry BR (2003) Detection and quantification of *Plectosphaerella cucumerina*, a potential biological control agent of potato cyst nematodes, by using conventional PCR, Real-Time PCR, selective media, and baiting. Appl Environ Microbiol 69:4788–4793

Barba M, Czosnek H, Hadidi A (2014) Historical perspective, development and applications of next-generation sequencing in plant virology. Viruses 6:106–136

Bauriegel E, Giebel A, Geyer M, Schmidt U, Herppich WB (2011) Early detection of Fusarium infection in wheat using hyper-spectral imaging. Comput Electron Agric 75:304–312

Bilodeau GJ, Koike ST, Uribe P, Martin FN (2012) Development of an assay for rapid detection and quantification of *Verticillium dahliae* in soil. Phytopathology 102:331–343

Boonham N, Kreuze J, Winter S, van der Vlugt R, Bergervoet J, Tomlinson J, Mumford R (2014) Methods in virus diagnostics: from ELISA to next generation sequencing. Virus Res 186:20–31

Bradshaw RE, Foster SJ, Monahan BJ (2006) Molecular diagnostic tools for detection of plant pathogen fungi. In: Rao JR, Fleming CC, Moore JE, (eds) Molecular diagnostics – current technology and application, Horizon Biosciences, p 47–66

Brasier CM (2008) The biosecurity threat to the UK and global environment from international trade in plants. Plant Pathol 57:792–808

Cai H, Parks JW, Wall TA, Stott MA, Stambaugh A, Alfons K, Griffiths A, Mathies RA, Carrion R, Patterson JL, Hawkins AR, Schimdt H (2015) Optofluidic analysis system for amplification-free, direct detection of Ebola infection Nature Science Reports 5, doi:10.1038/srep14494

Camargo-Rodriguez AV, Smith J (2009) An image-processing based algorithm to automatically identify plant disease visual symptoms. Biosyst Eng 102:9–21

Camargo-Rodriguez AV, Molina JP, Cadena-Torres J, Jimenez N, Kim JT (2012) Intelligent systems for the assessment of crop disorders. Comput Electron Agric 85:1–7

Charlermroj R, Himananto O, Seepiban C, Kumpoosiri M, Warin N, Oplatowska M (2013) Multiplex detection of plant pathogens using a microsphere immunoassay technology. PLoS One 8(4):e62344. doi:10.1371/journal.pone.0062344

Cullen DW, Lees AK, Toth IK, Bell KS, Duncan JM (2000) Detection and quantification of fungal and bacterial potato pathogens in plants and soil. EPPO Bull 30:485–488

De Boer SH, Lopez MM (2012) New grower friendly methods for plant pathogen monitoring. Annu Rev Phytopathol 50:197–218

Deora A, Gossen BD, Amirsadeghi S, McDonald MR (2015) A multiplex qPCR assay for detection and quantification of *Plasmodiophora brassicae* in soil. Plant Dis 99:1002–1009

Doan H, Zhang S, Davis RM (2014) Development and evaluation of AmplifyRP Acceler8 diagnostic assay for the detection of *Fusarium oxysporum* f. sp. *vasinfectum* race 4 in cotton. Plant Health Prog doi:10.1094/PHP-RS-13-0115

Fang Y, Ramasamy RP (2015) Current and prospective methods for plant disease detection. Biosensors 5:537–561

Faulstich K, Gruler R, Eberhard M, Lentzsch D, Haberstroh K (2009) Handheld and portable reader devices for lateral flow immunoassays. In: Wong RC, Tse HY (eds) Lateral flow immunoassay. Humana, New York, pp 157–183

Fujinaga M, Ogiso H, Tsuchiya N, Saito H (2001) Physiological specialization of *Fusarium oxysporum* f. sp. *lactucae*, a causal organism of Fusarium root Rot of crisp head lettuce in Japan. J Gen Plant Pathol 67:205–206

Fujinaga M, Ogiso H, Tuchiya N, Saito H, Yamanaka S, Nozue M, Kojima M (2003) Race 3, a new race of *Fusarium oxysporum* f. sp. *lactucae* determined by a differential system with commercial cultivars. J Gen Plant Pathol 69:23–28

Garibaldi A, Gilardi G, Gullino ML (2002) First report of *Fusarium oxysporum* on lettuce in Europe. Plant Dis 86:1052

Garibaldi A, Gilardi G, Ortu G, Gullino ML (2012) First report of *Plectosphaerella cucumerina* on greenhouse cultured wild rocket (*Diplotaxis tenuifolia*) in Italy. Plant Dis 96:1825

Ghaffari R, Laothawornkitkul J, Iliescu D, Hines E, Leeson M, Napier R, Moore JP, Paul ND, Hewitt CN (2012) Plant pest and disease diagnosis using electronic nose and support vector machine approach. J Plant Dis Prot 119:200–220

Gilles T, Evans N, Fitt BDL, Jeger MJ (2000) Epidemiology in relation to methods for forecasting light leaf spot (*Pyrenopeziza brassicae*) severity on winter oilseed rape (*Brassica napus*) in the UK. Eur J Plant Pathol 106:593–605

Gladders P, Smith JA, Kirkpatrick L, Clewes E, Grant C, Barbara D, Barnes AV, Lane CR (2011) First record of verticillium wilt (*Verticillium longisporum*) in winter oilseed rape in the UK New Disease Reports 23: 8

Greninger AL, Naccache SN, Federman S, Yu G, Mbala P, Bres V, Bouquet J, Stryke D, Somasekar S, Linnen J, Dodd R, Mulembakani P, Schneider B, Muyembe J, Stramer D, Chiu CY (2015) Genome Med 7:99. doi:10.1186/s13073-015-0220-9

Gullino ML, Gilardi G, Garibaldi A (2004) New *Fusarium* wilts on vegetable crops in Italy. Commun Agric Appl Biol Sci 69:405–413

Gullino ML, Gilardi G, Garibaldi A (2014) Seed-borne fungal pathogens of leafy vegetables. In: Gulliono ML, Munkvold G (eds) Global perspectives on the health of seeds and plant propogaton material. Springer, p 47–56

Hausladen H, Leiminger J (2007) Potato early blight in Germany (*Alternaria solani* – *Alternaria alternata*), PPO Special report no. 12, Westerdijk CE, Schepers HTAM, (eds) Applied Plant Research BV, Wageningen.

Hovmøller MS, Walter S, Bayles RA, Hubbard A, Flath K, Sommerfeldt N, Leconte M, Czembor P, Rodriguez-Algaba J, Thach T, Hansen JG, Lassen P, Justesen AF, Ali S, de Vallavieille-Pope C (2015) Replacement of the European wheat yellow rust population by new races from the centre of diversity in the near-Himalayan region. Plant Pathol 65(3):402–411. doi:10.1111/ppa.12433

Huang W, Lamb D, Niu Z, Liu L, Wang J (2007) Identification of yellow rust in wheat using in situ spectral reflectance measurements and airborne hyperspectral imaging. Precis Agric 8:187–197

Hubbard A, Lewis CM, Yoshida K, Ramirez-Gonzalez RH, de Vallavielle-Pope C, Thomas J, Kamoun S, Bayles R, Uauy C, Saunders DGO (2015) Field pathogenomics reveals the emergence of a diverse wheat yellow rust population. Genome Biology 16:23. doi:10.1186/s13059-015-0590-8

Isack Y, Benichis M, Gillet D, Gamliel A (2014) A selective agar medium for isolation, enumeration and morphological identification of *Fusarium proliferatum*. Phytoparasitica 4:541–547

Jackson SL, Bayliss KL (2011) Spore traps need improvement to fulfil plant biosecurity requirements. Plant Pathol 60:801–810

Kennedy R, Wakeham AJ (2008) Development of detection systems for the sporangia of *Peronospora destructor*. Eur J Plant Pathol 122:147–155

Kostov K, Verstappen ECP, Bergervoet JHW, de Weerdt M, Schoen CD, Slavov S, Bonants PJM (2015) Multiplex detection and identification of *Phytophthora* spp. using target-specific primer extension and Luminex xTAG technology. Plant Pathol 65(6):1008–1021. doi:10.1111/ppa.12481

Lane CR, Hobden E, Walker L, Barton VC, Inman AJ, Hughes KJD, Swan H, Colyer A, Barker I (2007) Evaluation of a rapid diagnostic field test kit for identification of *Phytophthora* species, including *P. ramorum* and *P. kernoviae* at the point of inspection. Plant Pathol 56:828–835

Lin YH, Lai PJ, Chang TH, Wan YL, Huang JW, Huang JH, Chang PFL (2014) Genetic diversity and identification of race 3 of *Fusarium oxysporum* f. sp. *lactucae* in Taiwan. Eur J Plant Pathol 140:721–733

Mahlein A (2016) Plant disease detection by imaging sensors – parallels and specific demands for precision agriculture and plant phenotyping. Plant Dis 100:241–251

Martinelli F, Scalenghe R, Salvatore D, Panno S, Scuderi G, Ruisi P, Villa P, Stroppiana D, Boschetti M, Goulart LR, Davis CE, Dandekar AM (2014) Advanced methods of plant disease detection. A review. Agron Sustain Dev 35:1–25. doi:10.1007/s13593-014-0246-1

Miles TD, Martin FN, Coffey MD (2015) Development of rapid isothermal amplification assays for detection of Phytophthora spp. In plant tissue. Phytopathology 105:265–278

Mirik M, Ansley RJ, Price JA, Workneh F, Rush CM (2013) Remote monitoring of wheat streak mosaic progression using Sub-pixel classification of landsat 5 TM imagery for site specific disease management in winter wheat. Adv Remote Sens 2:16–28

Oerke EC, Dehne HW, Schnbeck F, Weber A (1994) Crop production and crop protection: estimated losses in major food and cash crops. Elsevier, Amsterdam

Pasquali M, Dematheis F, Gullino ML, Garibaldi A (2007) Identification of race 1 of *Fusarium oxysporum* f. sp. *lactucae* on lettuce by inter-retrotransposon sequence-characterized amplified region technique. Phytopathology 97:987–996

Pautasso M, Doring TF, Garbelotto M, Pellis L, Jeger MJ (2012) Impacts of climate change on plant diseases – opinions and trends. Eur J Plant Pathol 133:295–313

Qu XS, Wanner LA, Christ BJ (2010) Multiplex real-time PCR (TaqMan) assay for the simultaneous detection and discrimination of potato powdery and common scab diseases and pathogens. J Appl Microbiol 110:769–777

Rogers P, Whitby S, Dando M (1999) Biological warfare against crops. Sci Am 280:70–75

Rutolo M, Covington JA, Clarkson J, Iliescu D (2014) Detection of potato storage disease via Gas analysis: a pilot study using field asymmetric Ion mobility spectrometry. Sensors 14:15939–15952

Sankaran S, Mishra A, Ehsani R, Davis C (2010) A review of advanced techniques for detecting plant diseases. Comput Electron Agric 72:1–13

Schmale DG, Ross SD (2015) Highways in the Sky: scales of atmospheric transport of plant pathogens. Annu Rev Phytopathol 53:591–611

Singh RP, Hodson DP, Jin Y, Lagudah ES, Ayliffe MA, Bhavani MN, Rouse S, Pretorius SA, Szabo LJ, Huerta Espinosa J, Basnet BR, Lan C, Hovmoller MS (2015) Emergence and spread of new races wheat of stem rust: continued threat to food security and prospects of genetic control. Phytopathology 105:872–884

Thiessen LD, Keune JA, Neill TM, Turechek WW, Grove GG, Mahaffee WF (2016) Development of a grower conducted inoculum detection assay for management of grape powdery mildew. Plant Pathol 65:238–249

Thornton CR, Groenhof AC, Forrest R, Lamotte R (2004) A one-step, immunochromatographic lateral flow device specific to *Rhizoctonia solani* and certain related species, and its use to detect and quantify *R. solani* in soil. Phytopathology 94:280–288

Waage JK, Mumford JD (2008) Agricultural biosecurity. Philos Trans R Soc Lond B Biol Sci 363:863–876

Wahabzada M, Mahlein A-K, Bauckhage C, Steiner U, Oerke E-C, Kersting K (2015) Metro maps of plant disease dynamics – automated mining of differences using hyperspectral images. PLoS One 10(1):e0116902. doi:10.1371/journal.pone.0116902

Wakeham A (2015) Validation of the clubroot lateral flow in UK commercial Brassica cropping systems. Report CP 099a, AHDB Horticultural Development Council

West JS, Kimber RBE (2015) Innovations in air sampling to detect plant pathogens. Ann Appl Biol 166:4–17

West JS, Canning GGM, Heard S, Wili SG (2013) Proceedings of future IPM in Europe. development of the miniature virtual impactor – MVI – for long-term and automated air sampling to detect plant pathogen spores., 19–21 March 2013, Riva del Garda, Italy. (Abstract)

Woodall JW, Webb KM, Giltrap PM, Adams IP, Peters JC, Budge GE, Boonham N (2012) A new large scale soil DNA extraction procedure and real time PCR assay for the detection of *Sclerotium cepivorum* in soil. Eur J Plant Pathol 134:467–473

Yuan L, Zhang J, Shi Y, Nie C, Wei L, Wang J (2014) Damage mapping of powdery mildew in winter wheat with high-resolution satellite image. Remote Sens 6:3611–3623

Chapter 11
The European Union Plant Diagnostic Information System (EUPDIS): A Platform for Collaborative Diagnostics and a Tool for Early Detection of Plant Pathogens

Paul J. Verrier, Jane E. Thomas, and James P. Stack

Abstract Agricultural and natural plant systems within the European Union are diverse, widely distributed, and vulnerable to the introduction of new pests and pathogens. Detection, surveillance and diagnosis are all necessary to ensure effective protection of crops and foods from pathogen incursions resulting from natural or intentional introductions. Each EU nation relies upon a cadre of specialized plant health experts operating within an administrative framework that is unique to that country. However, since the movement of pathogens and pests is influenced more by weather patterns and human activities such as trade and travel than by politics and boundaries, communication among plant health specialists across EU national borders would facilitate informed preparations and decision-making to minimize the impacts of invasive pathogens. A web-based plant diagnostic information system and plant disease and pest database were designed and implemented for deployment within the European Union. This system, designated the EU Plant Diagnostic Information System (EUPDIS), evolved from consideration of the features of other systems, particularly the National Plant Diagnostic Network (NPDN) deployed in the USA, from considerations of the needs of EU plant health practitioners and plant protection officers, and the recognition of the need for a system that alerts neighbouring countries of emerging threat situations outside their own borders. Increasing free trade agreements will ensure the continued movement of pests and pathogens across borders and regions. Early detection and rapid response are essential to minimize the economic and environmental impacts from these introductions. A virtual EU Plant Diagnostic Network supported by a comprehensive diagnostic information system such as EUPDIS will provide the platform necessary to facilitate

P.J. Verrier (✉) • J.E. Thomas
NIAB, Huntingdon Road, Cambridge, CB3 0LE, UK
e-mail: paul.verrier@niab.com

J.P. Stack
Department of Plant Pathology, Kansas State University, Manhattan, KS, USA

M.L. Gullino et al. (eds.), *Practical Tools for Plant and Food Biosecurity*, Plant Pathology in the 21st Century 8, DOI 10.1007/978-3-319-46897-6_11

the collaboration and cooperation required to protect plant systems. The main features of the system, data collection, data reporting and disease identification assistance are described along with the rationale for the data collection and dissemination deployed.

Keywords Virtual diagnostic network • Plant disease information system

11.1 Introduction

The correct identification of organisms causing diseases of the world's plants, in both cropping systems and natural landscapes, has become an increasingly critical need as nations and continents strive to protect their agricultural and horticultural plant systems. International travel, weather extremes and global trade in seeds, propagating material and produce has meant that the opportunity for movement and introduction of plant pathogens has never been greater. While sophisticated diagnostic techniques are now available and being enhanced rapidly, their success still relies on the primary observations of first responders and the training and support services available to them. Suspect samples may be sent to a wide range of organisations for further analysis. Sometimes these will be staffed by experts in pathology, entomology and nematology, but these skills are generally decreasing, rather than increasing, in society (Howie 2012). As a means of providing easy access to high quality resources, diagnostic networks and information systems have been developed in many countries to provide information and support to both first responders in the field and testing laboratories. While the structure and specific objectives of different networks are varied, the overall principles are to facilitate access to expertise and information, allow the exchange of disease outbreak intelligence, permit analyses of recorded data to aid understanding of pathogen movement, and ultimately to improve control strategies.

Miller et al. (2009) reviewed diagnostic networks and their place in improving surveillance systems and the application of diagnostic tools. The major diagnostic networks include the National Plant Diagnostic Network (NPDN: http://www.npdn.org) in the USA (Stack et al. 2014) and the National Plant Biosecurity network in Australia (www.planthealthaustralia.com.au/NPBS). The NPDN consists of five linked sub-networks, each representing a group of states; e.g., the Great Plains Diagnostic Network comprised of nine states in the Great Plains Region of the U.S. (http://www.gpdn.org) and a central data repository. The European Union funded project, "Plant and Food Biosecurity," proposed the development of a virtual diagnostic network for Europe having the capability to link plant pathogen intelligence from all member states.

Increasing free trade agreements among nations will result in the continued movement of pests and pathogens across borders and regions. Climate change is altering the habitats of insect vectors of pathogens (Cannon 1998), often adding to the potential and the uncertainty of outbreaks. Early detection of introductions and rapid response to detections and outbreaks are essential to minimize the economic and environmental impacts from these introductions. A virtual EU Plant Diagnostic Network supported by a comprehensive diagnostic information system such as EUPDIS will provide the platform necessary to facilitate the collaboration and cooperation required to protect plant systems and to ensure the safe trade of plants and plant products.

This chapter reviews the main components that may be part of diagnostic networks, and describes a proposed model for construction of a diagnostics recording system for the EU, designated the EU Plant Disease Information System (EUPDIS), and a web based diagnostic network, and considers how its uptake by participating organisations could improve communications on disease surveillance and diagnosis in member states.

For the purposes of this paper, the following definitions apply: (1) a database is a repository of organized data; (2) an information system is a database with a user interface that allows access to the database and the analytical tools to process data; and (3) a network is a group of nodes (e.g., people, institutions, computers, etc.) connected virtually or physically to enable interaction and the digital and/or physical exchange of materials (e.g., information, goods, etc.).

11.2 Components of an Information System to Support Diagnostic Networks

11.2.1 Administrative

As well as supporting a scientific function, an information system (IS) can be constructed to provide sample tracking, reporting, and invoicing for commercial laboratories, the latter has not been implemented in the EUPDIS as the existing laboratories have their own mechanisms to handle this. The inclusion of this aspect is of most benefit where designated laboratories are already part of an existing coordinated diagnostic delivery system with national mandates. However, in countries where there are many extant private organisations providing services, internal systems are usually well established already and the rate of uptake of a new, international administrative function may well be slow.

11.2.2 Diagnostic Protocol Repository

Novel diagnostic procedures are being developed rapidly as outcomes of research projects or targeted protocol development for high interest regulated organisms. Validated protocols for the latter are available at national plant protection websites. For non-regulated organisms access to protocols is usually via published literature or research reports, requiring potential users to undertake some level of adaptation in their own laboratories. However, use of standardised, validated procedures that facilitate inter-laboratory comparison is clearly advantageous to generate confidence in outcomes and improve efficiency and quality for service-providers. Providing access to, and updates on, diagnostic protocols is a key function of diagnostic networks.

11.2.3 Community News, Training and Accreditation

A diagnostic network can provide a forum for alerting diagnosticians, growers and agronomists to new, changing levels of virulence or emerging disease occurrences. While numerous mechanisms already exist for alerting stakeholders to disease and pest outbreaks in near real-time, many are subscription services and these are frequently related to specific crops or pests. In contrast, the concept of an EU Plant Diagnostic Network (EUPDN) and an EU Plant Diagnostic Information System (EUPDIS), is the provision of free information, the creation of a discussion forum, and the training of first responders and diagnosticians. In the context of the EU, the confirmation of a disease outbreak in one region can alert others that a potential problem may arrive, and could prompt early action, thus minimizing production losses. Training of diagnosticians and first responder crop agronomists, along with the accreditation of diagnostic laboratories, are major elements of the diagnostic networks and information systems developed in the USA. These functions may be delivered through workshops, formal courses or on-line mechanisms. Accreditation systems, to include annual proficiency testing and adherence to international quality certificate standards or other audited mechanisms, can also be coordinated via a diagnostic network.

11.2.4 Sample Data Repository

Searchable records of plant disease outbreaks from confirmed diagnostic outcomes provide both a valuable research tool for tracking disease development and a key mechanism for identifying new or unusual outbreak patterns that may suggest accidental incursion, deliberate introduction, or a previously unknown means of pathogen dispersal. Establishing appropriate levels of access to different sample records is challenging since there is tension between preserving the confidentiality of sample details. avoiding the generation of negative impacts on individual growers, and responding rapidly to a damaging new pathogen introduction to protect an industry. In the NPDN, access to records is strictly limited and there is high security for the data repository (Stack and Baldwin 2008). However, certain authorised personnel have access to all data records and can take appropriate actions based on the information deposited if necessary.

11.2.5 Alert System and Outbreak Tracking

Networks may act as alert systems for regulated organisms. In the USA, the NPDN has no regulatory or enforcement function, but has developed a strong relationship with the statutory body (the United States Department of Agriculture (USDA) Animal and Plant Health Inspection Services (APHIS) Plant Protection and Quarantine (PPQ)); it is an integral part of the plant biosecurity infrastructure and has a formal method of handling suspected regulated pathogens or pests (Stack et al. 2014). In the case of established pests and diseases, organisations belonging to a network may allow each other access to summary sample data or to more detailed sample records. Sharing of diagnostic information from a plant clinic during the growing season for particular crops can be a valuable way of alerting growers and agronomists of disease pressures. The concept may extend to in-field disease diagnosis by growers, crop inspectors, and agronomists via mobile phone image capture; depending on the disease, diagnoses may need to be flagged as unconfirmed.

11.2.6 Expertise Database

For first responders or disease clinicians, knowing where to send suspect samples for expert analysis can save considerable time. It is rare for a diagnostic laboratory to have in-depth skills across all pest and pathogen types, and a searchable database of the expertise residing in various laboratories is an important component of a diagnostic network. It may be used by laboratories seeking expert advice, or by agronomists wanting to direct a sample to an appropriate specialist.

11.2.7 Live Sample Analysis

Web-enabled microscopy, if established as an additional functionality in a network, may be used for training purposes, to refer a sample from a remote field clinic to an expert, or to facilitate expert-to-expert consultation. It could be especially beneficial in less developed countries, where expertise is generally localized to one or a few centralised laboratories. The Australian Centre for International Agricultural Research has supported the development of a network with 58 web-enabled microscopes in Australia, New Zealand and Eastern Asian countries, supported by the development of an image library (Thompson et al. 2011). CABI's Plantwise remote diagnosis system (http://www.plantwise.org/diagnostic-and-advisory-service/) is being deployed globally to developing nations to provide support to rural farmers. While web-enabled microscopy systems can be a feature of diagnostic networks, their value in the EU, where samples can reliably be sent in cool boxes between laboratories for next-day delivery, was considered to be limited. Consequently, this capability was not included in the EUPDN/EUPDIS network structure proposed in this project. Nevertheless, stand-alone systems can easily be set up between laboratories if desired.

11.3 A European Plant Diagnostic Network

National Plant Protection Organisations (NPPOs) in European countries are members of the European and Mediterranean Plant Protection Organisation (EPPO), which coordinates the means for prevention of introduction of harmful pests and pathogens and to harmonize phytosanitary regulations and diagnostic procedures. Its activities relate to organisms that are regulated by one or more member countries: i.e., those on the A1 and A2 EPPO Alert Lists (www.eppo.int). The EPPO website provides information on diagnostic protocols, NPPO laboratories and named experts, pest and pathogen distribution maps, image libraries, and news concerning workshops, meetings, and publications. However, its remit is quarantine-designated, regulated organisms.

The proposed virtual EUPDN and the EUPDIS information system developed in the Plant and Food Biosecurity project, described here, was designed to complement the EPPO system and provide a means for diagnostic laboratories that are not designated NPPOs to access disease intelligence, protocol details, training opportunities, and to share information. In the European region, many former government supported organisations have become private commercial entities and a large number of diagnostic laboratories now provide services to seed merchants, agronomists, distributor companies, and direct to growers.

EUPDIS has been aimed at these entities, and at filling gaps in data sharing and interactions between diagnosticians in EU member countries. Though there are multiple sources of high quality information to support diagnostic services, there is no mechanism, other than by individual laboratory agreement, for sharing information or coordinating mutually beneficial activities across national borders. In addition, laboratories handling plant clinic samples from growing crops also may be the first to see new or unusual symptoms that could be indicative of the arrival, emergence or establishment of a regulated pest or pathogen that has not yet been detected by official inspection services. Ensuring that network diagnosticians are knowledgeable about regulated organisms, and the procedures to immediately alert the designated authorities, is critical for the rapid and effective containment of an outbreak, greatly increasing the probability that an exotic organism can be eradicated. Utilisation of EUPDIS could therefore provide extra "eyes and ears" for the early identification of regulated organisms. Where multiple outbreaks occur, and have been detected during the course of routine sample diagnosis, knowledge of the pattern and timing of occurrence can help to track potential sources and provide evidence that can assist in judging whether the introduction might have been deliberate rather than natural or accidental. The EUPDN and EUPDIS proposed in this project can thus support regulatory functions as well as contribute to intelligence on occurrence and severity of indigenous organisms. However, it was not designed to be a statutory system and will not generate official alerts on plant biosecurity events. Individual network laboratories would be responsible for notifying relevant authorities if a regulated organism were to be detected in a sample. However, the proposed system indicates the need to notify the appropriate authorities, providing the contact details and optionally forward the details of the sample recording to the specific countries NPPO.

Detection of a plant disease outbreak depends on receiving reports of pathogen occurrences. While mundane common pathogen reports may seem tedious to add to the system, the presence of reports on a specific pathogen enables agronomists and pathologists to observe the pattern of outbreaks, and may form the basis of future investigations, if sample sizes are adequate.

If diagnosticians are to invest time and effort in registering and using EUPDIS for diagnostic record capture in place of, or as an adjunct to, any internal system that has already been established, they must receive value in return. The establishment of incentives for reporting disease outbreaks by laboratories, diagnosticians, crop consultants, and farmers is likely to enhance the development, robustness and utility of the EUPDIS database. Ensuring that EUPDIS creates value for the user was a primary focus during development of the system. The EUPDIS model offers several benefits for users, including: (a) the ability to visualise and interrogate the full detail of their own data in a variety of ways; (b) complete confidentiality of their data, unless users decide to share it; (c) the ability for the sender of a sample to create sample details and then view the diagnostic outcome on a map that geo-locates a sample or set of samples; (d) the ability to see summary records in map format, but without exact location detail, so that the confidentiality is preserved but regional disease patterns can be observed; (e) the ability to locate a subject matter expert on a specific pest or disease; and (f) access to a community information page on pest and pathogen issues for diagnosticians, researchers and crop management professionals.

11.4 System Construction

11.4.1 User Registration and Access

EUPDIS was constructed using the PHP language and databases were constructed with the MySQL package. It is hosted on a server at the National Institute of Agricultural Botany (NIAB), in Cambridge, UK (https://www.niab.com/pfs). Users must first register, which provides an opportunity to assign local administrative and access rights and provides a mechanism to secure sensitive information from public view.

EUPDIS registration is relatively simple, but is monitored and requires an administrator to review the provided information and to establish the appropriate access level. Users can be registered directly by a laboratory administrator using pre-set roles with defined access rights.

Four main user levels have been proposed:

1. Crop advisor, inspector or agronomist
2. Provider of diagnosis or diagnostic expert
3. Provider of laboratory facilities for diagnosis
4. None of above (e.g., regulatory officials and researchers may register)

Differentiation of the user role has several important consequences. Firstly, it enables the system to assign different levels of confidence to any diagnosis logged onto the system. For instance, an agronomist may wish to enter an unconfirmed diagnosis to flag an issue of interest while a diagnostic laboratory will want to enter confirmed diagnostic outcomes. Secondly, it enables EUPDIS to store user expertise information as a summary for other users wishing to access particular knowledge and expertise, and thirdly, to filter the diagnosis maps and to enable users to see contact details and location of experts according to the role of the user seeking the contact. In practice, an expert will be able to find the direct contact details for another expert, but a non-expert will find only the diagnostic laboratory details where an expert can be contacted.

11.4.2 Expertise and Laboratories

While it is relatively simple to locate NPPOs within the EU via the EPPO database, it is more difficult to identify laboratories that have capability for diagnosing plant diseases and offer services to plant industries, and even harder to locate expertise on specific organisms within these laboratories. In addition, some plant pathogens generate toxins that have adverse effects on human and animal health, and knowledge of where toxins can be analysed is also difficult to find. A survey was conducted within the Plant and Food Biosecurity project to collect information on the main plant diagnostic laboratories, their locations and contact details in each EU country. Much of this information has been placed in EUPDIS. In addition information on available expertise on specific organisms or classes of organisms was collected. The resulting database is fully searchable and provides a valuable resource to the plant health community not available elsewhere. It will need to be regularly reviewed and updated to remain valuable. At present, public health laboratories in the UK authorized to identify food-borne organisms causing human illness have been included in EUPDIS as an example (for the UK only). The potential exists to expand EUPDIS to include such laboratories throughout Europe.

Maintaining a database can be a challenge once the funding used to create the database has ended. EUPDIS was designed to enable experts to update their areas of expertise and the capabilities of the laboratory to which they are attached, or for an organisation to modify and update registration information. Any laboratory EUPDIS designated administrator has permission to do this and to add an expert within their organisation.

The user-update mechanism will still require verification and moderating, but avoids the need for central data entry via a central administrator. It is envisaged that the main contact of each laboratory and individual experts will be sent an automatic reminder annually to check and update their data. The automatic reminders require a positive response of 'checked' or 'no update required'. A system administrator would follow up if no response.

Laboratory locations have been tagged on a Google-based map, enabling users to rapidly view information within each country and quickly establish the level of information available for each area. The current distribution of diagnostic laboratories in Europe was determined from responses received during the information gathering phase of the project (note that public health laboratories responsible for identifying human pathogens on plants were added for the UK only).

EUPDIS allows a user to navigate through the list or the map to a specific laboratory to obtain the contact details and list of attached experts. Similarly, by clicking on an expert, the details of that expert and his/her areas of expertise will be presented.

Various search functions are provided to locate a laboratory, an area of expertise or available equipment. Searchable terms include country, type of laboratory (e.g., whether it is an advisory, or an official NPPO), facilities available (e.g., containment level, molecular biology facilities, toxin analysis capability, or standard mycological equipment). Similarly, search functions (list or map) are provided to seek expertise by selecting a combination of expert area (pathogen, organism group, etc.), crop and country.

11.4.3 The Pathogen Database

EUPDIS has a pathogen database with accepted pathogen names and variants, including different common disease/pathogen names in different countries; diagnoses and pathogen identifications entered into EUPDIS are reconciled with the accepted names. A similar database of crop names that accounts for the variation in names across countries is also needed. EUPDIS treats all entities (pathogens or their host plants) as organisms. It is envisioned that experts will provide updates to pathogen and crop names, including common names in the local language(s) of each country; any changes should be vetted by a system administrator.

While some 2500 organisms are listed in the USA's NPDN database, currently there are 859 organisms present in the EUPDIS database; the numbers will increase with adoption of EUPDIS. If EUPDIS is to retain its value to the EU community, funding will be necessary to support the system administrator, country administrators as well as required infrastructure maintenance and updates. EUPDIS was intended to include a limited description of each pathogen, crop, and crop-pathogens associations. Many high quality information sources and publications are available to support diagnosticians, some freely available. It is not intended that EUPDIS will contain complete documentation of every pathogen or crop, but rather will point users to detailed information. However, for pathogens of specific interest, such as those on local alert lists or new and emerging threats, the detail will be as comprehensive as possible to aid advisors to identify the pathogen's potential presence in the field. Further details of a pathogen may be included within articles embedded within the system that might, for example, provide a case study of a specific organism, an update on different pathotypes that have been detected, or a review of molecular markers to assist in pathotype or strain identification. As stated above, this will require funding to maintain and update EUPDIS.

While EUPDIS will contain information on those regulated organisms for which statutory action is required, it will also provide an automated warning if a diagnosis of such an organism is entered. An "alert if found" message will warn a laboratory that the finding must be reported to their national plant health organisation, which will then confirm the diagnosis and take appropriate action.

The EUPDIS user interfaceprovides access to the "Full list (alternative names)" of Latin and common names of the organisms. The "Full list (nested)" provides a hierarchical list of pathogens and crops/hosts, the "Crops" list shows all crops, the "Pathogens" list shows all pathogens and the "Alerts" tab shows all pathogens on the EPPO A1 list.

Selecting any organism from the list will take the user to a page providing many properties of the organism, including alternative names (in different languages), parentage in the organism hierarchy (all linked), hosts (with links to their descriptions), images (if available), description of the pathogen and its control (if available).

The pathogen report provides a link ("Overview of samples" tab) to show any registered diagnoses (see below) for this pathogen. Also available, though not shown, is a list of experts that have associated themselves with this disease, microscopic and field images of disease symptoms and signs, notes specific to the pathogen and links to further information.

11.4.4 Registering a Diagnosis

A diagnosis, whether made in the field or in the laboratory, can be entered into EUPDIS. A laboratory diagnosis entered into EUPDIS is considered accurate, a 'confirmed diagnosis.' A field diagnosis may be entered as a 'tentative diagnosis'. This designation may be particularly useful to agronomists or crop advisors who enter sample details, images and observations in the field or from the farm office from a mobile device. An initial field diagnosis can be followed by sending a sample to an expert laboratory for confirmation, the details of which can be entered when the confirmation is made. Details on cultivar, soil type, and previous cropping, all of which may be useful in any subsequent analysis of field data can be entered into EUPDIS; a "sample notes" text box also is available. Laboratories may submit many samples of the same crop with the same disease, entered individually or uploaded in bulk from a spreadsheet template devised for this purpose.

The basic critical information to be entered about any diagnosis includes the pathogen, host, the sample date and the location (address or nearest town). Diagnoses are automatically geo-referenced from the supplied information. Any that fail automatic geo-location are flagged for the laboratory administrator to examine and perhaps contact the person entering the diagnosis to obtain clarifying information to enable geo-referencing. Samples and diagnostic outcomes may then be viewed in complete detail by the organisation entering the data. They can be filtered by crop and disease, and displayed as a list or in a variety of visual representations with designated start and end dates.

11.4.5 Viewing an Outbreak

Within EUPDIS, various levels of permissions can be set by users, enabling regulation of access to data at different levels of detail. Summary data can be viewed by all users, however all sample detail, including precise locations and diagnostic details, can only be accessed with appropriate permissions. Data for a particular disease can be viewed as a heat map, which shows the intensity of disease (e.g., prevalence and severity) reported within a region. Thus, a user in France may be able to see the summary heat map showing records of downy mildew on oilseed rape in the UK; however, access to specific data records would require the appropriate system permission. Animated versions of the heat map show progression over time, with timings adjustable to daily, weekly or monthly intervals. The number of samples and the identity of laboratories providing the diagnoses would remain hidden. A query for a specific pathogen or disease can select diagnoses that were "confirmed", "tentative", or "all", depending on the interest of the user.

EUPDIS users within a country, or across countries, may share more extensive diagnostic detail, allowing for much more informed use of diagnostic data and closer cooperation among diagnostic laboratories. Secure and flexible group access capabilities are built into EUPDIS to permit cooperation among multiple laboratories. Such arrangements are private, voluntary, and provide specific visualisations and summaries securely to the group. A more detailed reporting feature, would allow a user or group of users to view all disease records for their countries in tabular form over a season at monthly intervals. Flags can be set to show the numbers of increases and decreases in diagnostic records over a growing period. Initial occurrences, rapid increases, and declines of disease or pest outbreaks can thus be easily pinpointed.

The features described above provide a rapid method of viewing the occurrence of disease and enables monitoring disease progress in neighbouring countries. Future development of EUPDIS may enable linkage of the time and location of an outbreak with local actual or interpolated weather data, creating the potential for users to carry out epidemiological investigations using their own data, or with others by agreement. If viewing arrangements are agreed upon between laboratories in different countries, early warnings of diseases in one location could trigger responses in other areas on disease "pathways."

11.4.6 Community Information

EUPDIS contains a section open to all users providing information on relevant events (such as training courses, summer schools, workshops, conferences, webinars, field meetings, etc). Useful articles on disease outbreaks, new or unusual occurrences, high impact threats and links to other sites and information sources are also included. This section could also house documentation, or links to it, for validated protocols, protocol updates or comments on protocol issues. An accreditation system for diagnostic laboratories could be administered via EUPDIS. In the United States' NPDN, a System for True, Accurate and Reliable Diagnostics (STAR-D) accreditation system has been launched for member laboratories, with the objective of maintaining and enhancing the credibility of NPDN laboratory diagnoses (Stack et al. 2014). While the scope for accrediting laboratories within the EU clearly exists, the number of independent countries involved, and the complexities that this introduces, precluded the development of a formal system within the Plant and Food Biosecurity project. Nevertheless, the community section of EUPDIS may still be used for the organisation of periodic reference testing and the comparison of informal inter-laboratory ring tests.

11.5 Utilisation of EUPDIS

A critical need for a diagnostic network that is to help monitor disease progression and the incidence of new diseases is the input of current data on the diagnoses being performed throughout the EU. The Plant and Food Biosecurity project recognised the need for early disease detection and the need for greater collaboration among all plant pathologists across the EU. EUPDIS has, to date, been tested by only a relatively small number of users, and will continue to be optimized and improved with continued use and feedback.

The various informational elements, functions, and capabilities of EUPDIS in support of a virtual diagnostic network will add value and convenience to users, and provide a vehicle to foster cooperation and communication among the plant disease diagnostic communities in the EU. The success of EUPDIS depends on data acquisition. As diagnostic records accumulate, the network's capacity to visualise and manage outbreaks will increase, and this in turn should enhance value and attract more users. The closer to "real-time" that data visualisation becomes, the more useful it is likely to be as disease intelligence that can inform management practices. There are obvious limitations to the use of unconfirmed, or tentative, diagnoses.

However, for some obvious and well known diseases, direct field observation and submission of a diagnostic record could produce crowd-sourced data providing new insights into disease patterns. The future development of remote systems for detection and diagnosis of plant disease, though still far from commercial application (Mahlein 2016), may render some laboratory diagnoses unnecessary, but the value of capturing data in a diagnostic database will remain. EUPDIS will need to evolve as user experience increases and diagnostic strategies and technologies advance. For this reason, flexibility has been built into EUPDIS to allow modifications and additions according to the needs and vision of users.

11.6 Confidentiality

Confidentiality is a critical feature of the network system. EUPDIS is designed to be secure and only registered users may access data. Full details of any sample diagnosis may be viewed only by the submitter or their associated laboratory, and only summary, anonymous data, as described earlier, are available to all users. If a laboratory indicates detection of a regulated organism, detailed or follow up information can be accessed only by plant health officials in the country of origin of the sample.

It is essential to understand that diagnostic network and information system such as EUPDIS should be considered as an enabling partner with regulatory agencies and their confirmatory laboratories. Through effective partnerships among regulatory and non-regulatory plant diagnostic laboratories, the capacity for early detection is enhanced thus increasing the effectiveness of rapid response.

11.7 Opportunities for Outbreak Surveillance

In the event of an incursion from a regulated organism or an unusual occurrence of a non-regulated organism, ongoing monitoring is usually required to determine its rate of spread or whether containment or eradication processes have been effective. EUPDIS implementation by an EU virtual diagnostic network could readily be used to support surveillance during outbreak response. It would be relatively simple to assign a geographical area to a field worker for a disease surveillance operation. Diagnosis reports could be associated with a given field worker and negative as well as positive diagnoses collated and visualised. Observations of diseases from sentinel plants or plots could also be captured in EUPDIS, which provides a ready-made repository for such targeted surveillance records. The Asian Soybean Rust sentinel plot system in the USA, and training of first responders/field personnel in its recognition, has involved both the NPDN and regulatory agencies, and is an excellent exemplar of how diagnostic networks become a key part of the process of surveillance for high threat organisms. The concept of triage, in which non-statutory but highly skilled, well equipped laboratories undertake elimination of clearly negative samples, while relaying all suspect positive samples to official laboratories, was also developed as a function of the NPDN (Stack 2010), and could become another function of the EU diagnostic network and its user organisations in the event of a significant threat.

11.8 Conclusions and Future Development

The Plant and Food Biosecurity project that funded the development of EUPDIS and established the diagnostic network has concluded. However, much thought is being given to additional features which would provide increased user benefit: Registered EUPDIS users can provide suggestions within EUPDIS.

The main areas currently targeted for further development are (a) to develop more response layouts so that EUPDIS can be viewed easily on mobile devices, (b) to construct the means by which a sample record or tentative diagnosis, can be submitted from a mobile device, with location information and (c) to add further pathogen information to the pathogen database.

During the Plant and Food Biosecurity project, discussions with EPPO officials considered the contribution that a well-utilised diagnostic network system could make to the regulatory sector. The development of productive interactions between diagnostic laboratories having no regulatory function and NPPOs should start to flow with system uptake, and potentially be realised in the same way as the NPDN has achieved by its interactions with the USDA APHIS PPQ in the USA.

Developing a plant diagnostic network and an information system to support that network for the EU, with its 27 independent nations, and endeavouring to apply the best principles from single country networks such as the USA and Australia, has inevitably resulted in compromises and differences in approach. Uptake of the EUPDIS network system will demonstrate its value, and data sharing will steadily increase this value. However, even with minimal sharing, the diagnostic network system still offers private laboratories a stand-alone mechanism for managing their diagnostic records, and extracting value from them as they address the needs of their plant growing and crop production industries.

References

Cannon RJC (1998) The implications of predicted climate change for insect pests in the UK, with emphasis on non-indigenous species, 1998. Glob Chang Biol 4:785–796

Howie B (2012) Plant pathology and entomology capability study, 2012. CQual Agritelligence Pty Ltd report

Mahlein A (2016) Plant disease detection by imaging sensors – parallels and specific demands for precision agriculture and plant phenotyping. Plant Dis 100:241–251

Miller SA, Beed FD, Harmon CL (2009) Plant disease diagnostics and networks. Ann Rev Phytopathol 47:15–38
Stack JP (2010) Diagnostic networks for plant biosecurity. In: Hardwick NV, Gullino ML (eds) Knowledge and technology transfer for plant pathology, plant pathology in the 21st century. Springer Science + Business Media B.V., Dordrecht, p 4. doi:10.1007/978-1-4020-8934-3_6
Stack JP, Baldwin W (2008) The need for secure communications networks and global connectivity. In: Gullino ML, Fletcher J, Gamliel A, Stack JP (eds) Crop biosecurity: assuring our global food supply. Springer, Dordrecht, pp 103–109
Stack JP, Bostock RM, Hammerschmidt R, Jones JB, Luke E (2014) The national plant diagnostic network: partnering to protect plant systems. Plant Dis 98:708–715
Thompson M, Lyons A, Kumarasinghe L, Peck D, Kong G, Shattuck S, La Salle J (2011) Remote microscopy: a success story in Australian and New Zealand plant biosecurity. Aust J Entomol 50:1–6

Chapter 12
Containment and Eradication of Invasive Pathogens

Abraham Gamliel and Jacqueline Fletcher

Abstract A plant disease outbreak resulting from an invasive pathogen can threaten a country's agricultural enterprise, economy and trade, and pose a threat to human food and animal feed. Therefore, following the detection of a new disease, the preferred response objective is elimination of invading pathogen(s). Invasive pathogen eradication requires a well prepared infrastructure and a coordinated process of early and rapid detection, identification of the pathogen, and the adoption and careful execution of an appropriate strategy. Since selection of the best approach in a given situation depends upon a realistic assessment of the effectiveness of various available approaches, and the feasibility for their use and success, a quantitative assessment of all the factors influencing the eradication process is recommended.

Keywords Emerging infectious disease • Emerging infectious pests • Quarantine • Management • Pest control

12.1 Introduction

Introduction of an invasive pathogen is a threat to a country's agriculture, economy and trade (Anderson et al. 2004). Furthermore, certain pathogens can threaten the safety of human food and animal feed. Therefore, response objectives following a disease outbreak due to an invasive pathogen are targeted to pathogen containment and eradication from the invaded area, emphasizing the goal of zero inoculum and disease left in the invaded area. These goals differ from those used in managing an endemic pathogen, which, driven mainly by economic considerations, emphasize

A. Gamliel (✉)
Laboratory for Pest Management Research, Institute for Agricultural Engineering, ARO Volcani Center, P.O. Box 6, 50250 Bet Dagan, Israel
e-mail: agamliel@volcani.agri.gov.il

J. Fletcher
Department of Entomology and Plant Pathology, Oklahoma State University, Stillwater, OK, USA

M.L. Gullino et al. (eds.), *Practical Tools for Plant and Food Biosecurity*, Plant Pathology in the 21st Century 8, DOI 10.1007/978-3-319-46897-6_12

reduction of pathogen development and limitation of disease impact below an economical threshold. Eradication of invasive pathogens should begin with risk assessment of short and long-term consequences from the establishment of a new pathogen and implementation of procedures to effectively eliminate the pathway for such establishment. It involves a number of interrelated or repeated management measures combined with surveys and inspections to validate the efficacy of the process. Moreover, successful pest eradication requires coordinated action among regulatory authorities, plant pathologists, extension personnel and, ultimately, the farmers. Individual tasks of an eradication program may have little impact unless all tasks relate, logically and temporally, to each other (ISPM 9).

Eradication of an invasive pest begins with the assessment of various approaches and measures for the elimination of a particular pathogen and their effectiveness in interrupting the disease cycle and dynamics. Validated means for quantitative assessment of the effects of the control measures, singly and in combination, should be implemented. All of these actions are essential for an integrated and practical strategy to eliminate the pathogen, and in some cases the relevant host, from the outbreak area. The choice of inappropriate measures, their inefficient application, or omission of key factors from consideration may lead to failure of eradication (Gottwald and Irey 2007).

Pathogen eradication may not always be practical or achievable. In certain cases it may be evident early on that eradication is a far-reaching goal. Nevertheless, if the possible consequences of an unmanaged plant disease epidemic are highly significant, an eradication strategy may be justifiable as the first line of action.

Criteria for characterizing the impact of an invasive pathogen were numerically rated in the "Effective Pathogen Index – EPI" by Schaad et al. (1999)). These ratings reflect parameters of the pathogen, i.e. survival, establishment and spread, which can increase its potential damage when introduced to a new area.

Approaches to prevent the establishment of an introduced invasive begin with preventing the invasion, and then, following detection of an outbreak, eradication. Plant biosecurity principles are designed to achieve rapid mitigation, and eventual eradication or management of invasive pathogens. Eradication, the preferred goal in any invasion of a new pest, depends upon preparedness and rapid response.

The following terminology further explains the various approaches and mitigation concepts.

- ***Prevention*** – The first set of precautionary measures is intended to prevent the introduction of pathogens in any form. Prevention strategies begins with priority setting; i.e. conducting regular surveillance of high-threat exotic plant pathogens, assessing the probability that one of those pathogens could move into the nation or region of concern and the possible impact of such movement, and taking precautionary measures to prevent its entry. Most countries have developed prioritized lists of quarantine pests, those agents considered to be of greatest threat to that nation's biological resources. Quarantine pest lists form the basis for border inspections and controls to stop the entry of any items (propagation material, plant parts, food) infested with listed agents. The prevention

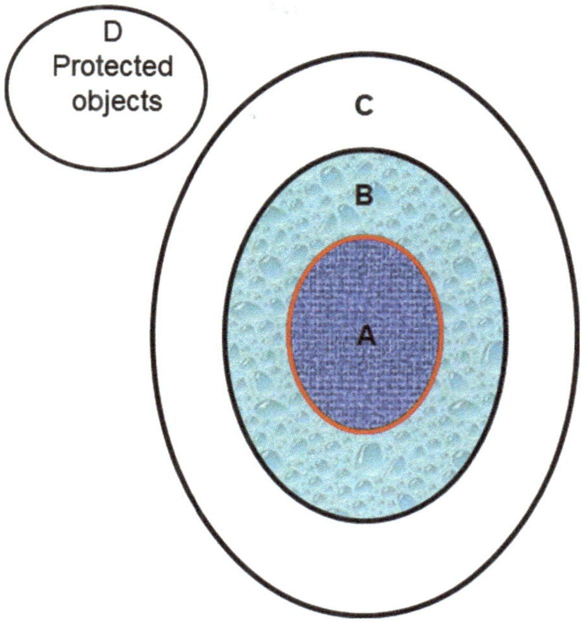

Fig. 12.1 Spatial delineation of the area of an outbreak for the deployment of containment and eradication procedures. *A* – outbreak area; *B* – buffer zone; *A* + *B* – quarantine area; *C* – area outside quarantine zone; *D* – protected object or area (e.g. nurseries)

strategy also should include contingency plans, including emergency regulations and special unit training, to mitigate a possible outbreak should it occur. Appropriate, dedicated administrative and physical infrastructures are essential to achieve the goals of prevention.

- ***Outbreak and quarantine setup*** – After an outbreak is detected and its causal pathogen identified, response begins with delineation of the outbreak area as a quarantine zone, which is under rigorous control to prevent any means by which the invasive pathogen could be moved into or out of the quarantine area (Fig. 12.1). Restricted items often include plants (plant parts and fruits), machinery, equipment, and packing materials.
- ***Containment*** – In the attempt to prevent the movement of an invasive pathogen from the outbreak area to new loci, measures are taken to suppress the initial inoculum and to prevent new infection and spread beyond the detected outbreak (quarantine) zone.
- ***Eradication*** – Executed following, and in parallel to, the containment process, eradication is the application of available and effective phytosanitary measures to eliminate all existing and potential inoculum, including infected host plants, possible vectors and alternate hosts, from a contained area (ISPM 5).
- ***Management*** – Following successful execution of the containment and eradication process, or in cases in which eradication strategies are deemed unlikely to be effective, management actions are applied. Strategies include prevention of new infections, introduction of resistant plant cultivars, application of pesticides and more. Management should continue for an extended period to ensure the continued success of the eradication process. Ref?

12.2 Aspects of Pathogen and Disease with Relevance to Containment and Eradication

12.2.1 Pathogen Traits

Successful eradication of invasive pathogens requires a concerted complex of detection, risk assessment, adoption of appropriate strategies, and careful execution of management procedures. Early disease detection in a confined outbreak area is followed by adoption of appropriate strategies to terminate the outbreak. The latter depends on realistic assessment of the effectiveness of available approaches, and the feasibility for their success. Quantitative assessment of all the factors that influence the eradication process can assist in the selection of the most effective eradication approach and eventually to pathogen elimination. The following factors are important for evaluating the threat and assessing its relevance to the selection of appropriate response measures.

- ***Type of threat and its possible impact***. Threats posed by an invasive pathogen include food poisoning, trade interruption, economic damage (loss of crop yield or quality), and loss of biodiversity in natural habitats. The potential impact of each threat in terms of time and severity dictates the urgency and the priorities for response. For example, a threat of food toxicity should draw the highest attention and response, since human and animal health are at the top of the priority ladder.
- ***The pathogen***. The systematic classification of pathogens (*i.e.* viruses, bacteria, fungi, and their taxonomic ranks) indicates types of potential disease and can therefore help investigators to assess the magnitude of the threat (Gamliel 2008). Moreover, it also suggests what types of response may be effective. For example, strategies to eradicate myctoxin-producing fungi target both the fungus and the contaminated products. For viral diseases, in contrast, most measures are directed towards eradication of insect vectors and destruction of the infected crop.
- ***Pathogen biology and disease epidemiology***. The ultimate objective of containment is to suppress new infectious inoculum. To apply appropriate countermeasures and accomplish this goal requires knowledge of the pathogen and host biology, life cycle and disease progress (Jeger 2004). For example, soilborne fungi, which have a relatively slow, spatial distribution pattern, can be contained if the inoculum is suppressed. In contrast, it is much more difficult to contain foliar fungal diseases, such as Karnal bunt of wheat (*Tilletia indica*), in which large masses of spores are produced.
- ***Vectors***. The involvement of vectors, usually (but not always) insects, in a disease cycle introduces complexity in several ways. Some plant pathogens move from plant to plant only through the actions of vectors, while for others insects may disseminate propagules to greater distances and more quickly than they would move on their own. Vector transmission also introduces new elements of host and geographical specificity that are characteristic of the vector rather than

of the plant or pathogen. Therefore, it is critical to prevent vector entry to the outbreak area, or to eradicate vectors already present. For example, *Xyllela fastidiosa*, the bacterium that causes Pierce's disease of grapevine, has recently detected in Italy, but not in other EU countries (EPPO 2016). Since, numerous species of *Cicadellidae* and *Cercopidae* known to be vectors of *X. fastidiosa* (Hopkins and Purcell 2002) reside in these areas, vector management should play an important role in any preparedness and eradication program if and when this pathogen invades that territory.

- ***Other hosts***. Many pathogens can infect, survive on and spread to hosts other than an economically important crop (primary host). The range of pathogen hosts can include cultured or wild plants that are taxonomically close (or not) to the primary host, and a wide spectrum of weeds. Failure to identify and eradicate all host species from the invaded area can result in failure of the overall eradication process. For example, because *Phytophthora ramorum*, the causal agent of sudden oak death, colonizes at least 97 host species (USDA-APHIS 2006), containment and eradication must include all the possible hosts.
- ***Size and location of the outbreak area***. The success of containment and eradication measures is inversely correlated with the size of the outbreak area. In a small and confined area, a rapid response could be successful. However, if the disease is present over a wide area, or in multiple sites, pathogen distribution may have occurred beyond the detected location. In such cases the chances for successful containment and eradication are lower. introduction of a tree pathogen into an urban area or forest could be much more difficult to handle than one in an open agricultural field setting, since other factors may dominate the response approach. For example, the fact that the Florida citrus canker outbreak was initially localized within an urban area with many back yard citrus trees, prompting vigorous opposition to the tree eradication strategy that had been adopted,was one of the main reasons for eradication failure during 1995–2001 (Gottwald et al. 2001, 2002; Graham et al. 2004).
- ***Extreme climatic events***. Unusual climatic conditions can induce, spread or suppress epidemics. For example, hurricanes were a significant factor in the spread of citrus canker in Florida in 2005 (Gottwald and Irey 2007), and in the introduction of soybean rust to the southern U.S. in 2004 (Rupe and Sconyers 2008).
- ***The lag time from infection to detection***. Early detection and accurate diagnosis are crucial to prevent the establishment and dispersal of introduced pests and pathogens and to minimize subsequent impact. Once an invading pathogen species becomes established in an area it can be difficult or impossible to eradicate. A good example of effective and quick detection is the case of pathogens in propagation material that are detected before their introduction into the soil. In contrast, symptoms of citrus greening (caused by *Liberobacter* sp.) were expressed in a period of 2.5–3.5 months after leaves emerged from buds on diseased trees (Su and Huang 1990). Furthermore, detection of citrus greening pathogens in asymptomatic tissue is inconsistent by any known method Molecular detection assays may be complicated, and results are not always reliable. The incubation period (i.e. the time from infection to disease), and the latent period

(the time from infection to production of an infectious propagule) further extends the time from invasion to detection, possibly beyond the threshold timing for effective containment and eradication.

- ***Available measures and time of response***. Two critical steps in emerging infectious diseases (EIDs) are pathogen establishment in a new area and spread to other loci. Because preventing these events is time dependent, the success or failure will depend on the rapidity of the response as well as to the specific measures taken.

12.2.2 Clustering Pathogens by Recommended Eradication Strategies

Quarantine pathogens and pests can be grouped into categories to facilitate appropriate selection of "containment and eradication" approaches. In this chapter we describe five such pathogen clusters (Table 12.1). The containment-eradication approach for each aims at addressing both general and specific traits of the cluster's members. A list of representative or example pathogens for each cluster, which are relevant for EU countries, is also shown in Table 12.1.

Table 12.1 Pathogen clusters for selection of containment-eradication protocols

	Pathogen group	Representative pathogens of concern to EU nations	Pathogen type
1.	Pathogens that contaminate edible plant parts with toxins/byproducts harmful to human consumers	*Tilletia indica*	Fungus
		Fusarium proliferatum	Fungus
2.	Viral pathogens	*Andean potato latent virus*	Virus
		Beet leaf curl virus	Virus
		Pepino mosaic potexvirus (PeMV),	Virus
		Plum pox potyvirus	Virus
		Potato spindle tuber viroid (PSTV)	Virus
3.	Foliar pathogens	*Xanthomonas axonopodis* pv. *citri*	Bacteria
		Xyllela fastidiosa	Bacteria
		Magnaporthe grisea/Pyricularia oryzae	Fungus
		Phakopsora pachyrhizi	
4.	Soilborne pathogens	*Aphanomyces euteiches*	Fungus
		Fusarium oxysporum f.sp. *albedinis*	Fungus
		Ralstonia solanacearum race 3 biovar 2	Bacteria
		Synchytrium endobioticum	Fungus
5.	Forest tree pathogens	*Ceratocystis fagacearum*	Fungus
		Mycosphaerella poplorum	Fungus
		Microcyclus ulei	Fungus

12.3 Containment and Eradication Procedures

Successful eradication of invasive pathogens can be accomplished through rapid response and the use of the appropriate strategy, including an accurate delineation of the outbreak area and a well-structured and interdisciplinary coordinated set of activities for containment and eradication. Adoption of the appropriate strategies and their effective application are the key to outbreak termination. The temporal chain of procedures and activities in the containment and eradication process are discussed in this section.

12.3.1 Delineation of the Quarantine Zone

Once an outbreak is reported, responders conduct a spatial delineation of the area to establish the quarantine zone and areas to which the appropriate containment and eradication procedures will be applied (Fig. 12.1). The following definitions describe the zones involved in the process:

- **Outbreak area** – The area in which the pathogen was detected originally
- **Buffer zone** – An area surrounding or adjacent to the outbreak area, officially delimited for phytosanitary purposes in order to minimize the probability of spread of the target pest into or out of the delimited area, and subject to phytosanitary or other control measures, as appropriate
- **Quarantine area** – An area within which a quarantine pest is present and is being officially controlled
- **Protected objects** – areas such as nurseries, seed production fields, etc., which are remote and outside the quarantine but areof significant importance for potential spread of the pest

The zone delineation can follow man-made boundaries (e.g. roads, large buildings, walls), or natural boundaries (e.g. rivers, valleys) to support management within the relevant zone.

The initial focus in setting a quarantine zone is to block every possible pathway of pathogen escape from the contained area, and restrict any entry and exit of machinery, equipment, farm materials and products that may contain the pathogen. An outstanding example is the spread of *Synchytrium endobioticum*, the causal agent of potato wart, from infected fields to other fields by contaminated automobile wheels (Jennings et al. 1997). *Erwinia amylovora*, the causal agent of fire blight of pome fruits, is spread by the movement of infected fruits that are ready for market (Roberts et al. 1998). Furthermore, it is essential to clean and disinfect vehicles, machinery, commodities and any products that can potentially carry contaminants within the quarantine area. These measures are important to apply, especially with regard to accidental transfer of pathogens such as those causing potato wart and fire blight of pears and apples. These two examples demonstrate that care should be taken to address any possible pathogen exit pathway. Measures which are applied in the quarantine area to assure the success of quarantine process include:

12.3.2 Containment

Containment procedures in the outbreak area should cover the agricultural, rural and the urban sectors. Specific procedures include:

- **Sanitation** . The main goal of sanitation is to reduce and suppress the spread of the pathogen within and beyond the quarantine zone. It includes disinfection of large and small equipment, machinery and tools. On large scale farms this procedure applies to heavy machinery, which can transfer inoculum by moving soil particles, infected grains and more. Examples of pathogens that can be transmitted by tools include a wide spectrum of viruses, bacteria and fungi.
- **Physical barriers** . Barriers can be positioned to contain the inoculum within the outbreak area. Such practice is especially important with soilborne pathogens, which can spread by root to root contact. For example, trenching to disrupt grafted root systems is an effective control strategy for oak trees infected by the fungus *Ceratocystis fagacearum*, the causal agent of oak wilt (Wilson and Lester 2002).
- **Vector control**. Intensive insect control and monitoring is directed at eliminating any vector that transmits the pathogen, and preventing further infection within and outside the outbreak area. Vector elimination is important especially with certain insect transmitted viruses, phytoplasmas and spiroplasmas, and fastidious walled bacteria, but a few plant pathogenic fungi and other bacteria are insect-transmitted as well.
- **Intensive chemical treatment of the plants before or as a part of removal of infected plants**. This strategy is used to suppress the epidemic, prevent further infection and possible pathogen spread outside the outbreak area, and suppress generation of new inoculum. It also prevents the spread of inoculum during the process of removal and destruction of infected plants or plant parts. An appropriate pesticide should be applied to infected plants to reduce the pathogen and vector populations and to prevent inoculum spread during the consecutive activities of plant removal and destruction. For example, USDA-APHIS guidelines for the eradication of citrus greening disease, caused by *Candidatus* Liberibacter sp., indicate that the psyllid vectors *Trioza erytreae* and *Diaphorina citri* should be controlled prior to tree removal to minimize pathogen spread (USDA-APHIS 2016).
- **Destruction and removal of infected plants** . Infected plants or plant parts can harbor infectious inoculum internally, but the removal and destruction (by burning, composting, etc.) of the entire plant or plant part can eliminate the majority of the inoculum. Whether a whole plant, or only the infected part, is uprooted or destroyed depends on the type of infection (systemic or localized), the area and the size of infection, the crop type and many other considerations. Because of the time lag between infection and symptom development, the procedure should cover all the cultivated plants regarded as potential hosts in the outlined area, not just those that are visibly infected. Root diseases and soilborne pathogens can be controlled by destruction of host root systems by soil fumigation and herbicide

applications. These procedures can suppress both existing and new pathogen inoculum and are especially important with annual crops, as they also will minimize any viable inoculum left in soil after uprooting the infected plants.

- **Intensive foliar treatment program (perennial crops)**. When it is impractical or impossible to remove whole plants efforts are made to suppress the internal inoculum, arrest further development of the epidemic, and prevent further infection and spread. Such approaches may be successful when an infected area is small and spread is limited. However, this approach may be a weak link in the eradication chain; Hopkins and Purcell (2002) noted that the decision to not remove Pierce's disease affected grapevines may have led to the failure to control that disease despite intensive management of vineyards in California to control the vectors of *Xyllela fastidiosa*, the causal agent.
- **Volunteer cultivated hosts and wild weeds** . Elimination and eradication of other plant species that can serve as volunteer hosts may be very helpful in assuring containment success. For example, because *X. fastidiosa* has a wide host plant range in California vineyards (Wistrom and Purcell 2005), eradication of all the possible host species from the outbreak area will be crucial for containment.
- **Water reservoir management**. Treatment in and around water reservoirs to prevent pathogen contamination can be followed by suppressing possible movement of inoculum into and through water. Eradication of volunteer hosts and pathogen vectors around water areas is recommended. However, this procedure may not be feasible in cases where the pathogen has already invaded large water reservoirs or in areas where access for treatment is limited (forests and rangelands).

12.3.3 *Eradication*

Although eradication is the main pillar in the chain of steps toward ultimate pathogen elimination, it is interlinked with containment and can be successful only if containment procedures are also fulfilled. In fact, measures relevant to and part of the containment strategy serve also as initial steps for eradication. Eradication procedures are performed both within the affected area and in the outlying buffer zone. Additional measures of eradication are:

- **Removal of infected parts**. It is generally preferable to remove and destroy entire plants, often by burning (Schubert et al. 2001), although composting can be effective also (Termorshuizen et al. 2003). It is important to delineate an area larger than that containing visibly infected plants, and to remove all the plants within it, because symptomless plants may be infected but in a latent period during which the pathogen population is increasing. The appropriate dimensions of the eradication perimeter must be determined through epidemiological study and risk assessment (Gottwald et al. 2001). In practice, to suppress any inoculum remaining after the containment treatments and to prevent new infections, the area should be surveyed several times over the following days and weeks to identify new infections and remove additional plants.

- **Intensive foliar treatment program (perennial crops)**. Efforts are made to apply the relevant pesticides in perennial crops and susceptible plants that were not removed in order to suppress any existing internal inoculum, and prevent further infection.
- **Soil disinfestation** . Soil borne pathogens can survive in soil in many forms, including dormant and chemical-resistant resting structures. Eradication of such pathogens from the soil requires, therefore, a robust treatment such as soil disinfestation using highly toxic soil fumigants having non-selective activity. Effective disinfestation depends upon establishing proper application conditions (e.g. appropriate soil cultivation, moisture levels, etc.). So as to be effective at deep soil levels, disinfestation should be repeated to assure removal of pathogens that may survive the first application. Until 2005, methyl bromide was an effective and recommended soil disinfestation fumigant. However, due to its ozone depletion potential, this chemical is no longer available. Other fumigants (MBTOC 2007) are available; however, their performance is currently inferior to that of methyl bromide.
- **Destruction of new emerging plants from a treated area**. The use of soil fumigants or herbicides to kill new emerging plants or offshoots from destroyed perennial plants prevents reestablishment of inoculum left in soil after the containment treatment and soil disinfestation.

12.3.4 Management

Management follows successful execution of containment and eradication actions in the outbreak area. One objective is to assure the elimination of any new emerging inoculum and to prevent conditions suitable for the beginning of a new epidemic. Alternatively, management strategies may be employed when an eradication strategy failed or is regarded as not feasible. An effective management program should include all the above mentioned procedures for quarantine and containment practices. Additionally, measures performed during the eradication process (e.g. removal of infected plants, pesticide applications against the pathogens and/or their possible vectors, and destruction of weeds and wild hosts) should continue. Intensive pesticide applications are most important in tree crops, if trees were not removed during eradication. Additional specific practices relevant to the management stage include:

- **Cultural practices** . The cropping system may be modified to create conditions that hamper reemergence of the pathogen. For example, to suppress new infections of *Erwinia amylovora* (the causal agent of fire blight of pome fruit trees), recommendations include reducing fertilization to slow the growth rate of the trees, withholding irrigation water and nitrogen fertilizer, and cultivation (Brunner 1994). Similarly, practices that reduce tree wounding and bacterial

movement can reduce the risk of infection. Other cultural procedures may include changes in planting dates and the establishment of windbreak rows of trees as mechanical barriers to pathogen movement.
- **Resistant cultivars**. Planting of crop varieties bred for resistance to specific pathogens, thereby eliminating susceptible hosts for a period beyond the survival of the pathogen, can reduce the likelihood or rate of new infection.
- **Pathogen-free propagation material** . The use of certified planting material, and disinfection of seed and other propagation material by means of chemicals, thermal treatment or combination of approaches, may be recommended.

12.4 Selection and Adoption of the Appropriate Strategy Against Invasive Pathogens

Although eradication is generally the preferred goal following the introduction of a new pathogen, it may not always be feasible. Eradication steps are often very expensive and there is no guarantee of success. Therefore, in addition to having knowledge of all the factors described above, it is helpful to understand the impact level of each specific tactic on each disease element (e.g. the initial inoculum, the infection rate or the vector, etc.), and the implications of the strategy on the possible outcome. Failure to adopt a quantitative approach can lead in many cases to establishment and spread of the pathogen over a wide area. In many cases eradication is sought to protect trade, particularly if a pathogen is new (exotic) to a country or region (Gamliel et al. 2008). After the invasion, a rapid assessment of the potential for pathogen spread and disease epidemiology should be made. If key elements of the disease are not known, eradication plans may be ineffective and the pathogen may become established and distributed in spite of efforts to prevent it. If eradication is not a reachable goal, its pursuit will only waste resources. Therefore the following factors should serve as guidelines when selecting the strategy to mitigate an invasive pathogen:

- The impact of the pathogen and its potential to disrupt the economy and the stability of the society should be assessed in order to gauge an appropriate level of response. Various aspects of plant disease impacts were discussed in previous sections.
- The extent of the affected area plays an important role in the decision. Eradication is more likely to be effective if the outbreak area is relatively small and/or located in a remote and isolated place. It contrast, eradication of a pathogen from an outbreak that spans several locations or covers a huge area, is much less likely to be successful.
- Specific characteristics of the causal pathogens (type of organism, vector, epidemiology, etc.) that influence pathogen spread and establishment, regardless of the initial size and location of the introduction, will impact the likelihood of successful eradication.

- An appropriate regulatory framework for delineating the quarantine zone and for supporting decisions related to crop destruction are crucial for rapid response. The lack of regulation will result in delay of containment and eradication procedures, hampering the success of these activities. Such considerations are particularly relevant in urban areas in which plant removal and destruction should be made in backyards and home gardens.
- Response measures should be available and able to be applied by appropriate responders. Relevant issues include the previous registration of the relevant pesticide, and the availability of the appropriate technology as well as personnel trained to apply them.
- The cost of the eradication process is also a factor to consider.

After considering these aspects, the selection of a strategy must meet the capabilities and resources of the responders. Since the weights of these factors vary among pathogens, and location types, it is useful to make a quantitative assessment of the probability of successful eradication in a given situation. Previously, we suggested the value of a "successful eradication probability" (SEP), calculated from various elements of the pathogen's characteristics and the specific disease situation (Gamliel and Fletcher 2008). SEP is a cumulative value based on a hierarchy of criteria specific to the relevant event. It can give a weighted assessment for the probability of eradication success, and indirectly suggest an appropriate strategy. Important to any of SEP assessment is knowledge of previous documented eradication efforts for the relevant pathogen in other locations and situations.

SEP is not a mathematical model, but rather a practical tool for simple and quick assessment of the probability of eradication success, and indirect suggestion of appropriate strategies. In a simple arithmetic calculation of all assessed factors, the weight of each variable in overall SEP scoring is identical (although in practice their influence on disease eruption and spread may differ).

A practical plant disease eradication "manual" should be available as part of each country's preparedness for crop biosecurity, in order to facilitate the identification and execution of an appropriate management approach (USDA-APHIS 2016). Such a document should include analyses of all factors relevant to the biology and epidemiology of pathogens of high priority in that country, as well as recommended response plans.

12.5 Conclusions

Although eradication of an invasive pathogen involves many uncertainties, it is usually the strategy of top priority. Therefore, to successfully eliminate an introduced pathogen a concerted series of simultaneous as well as sequential procedures should be planned and executed. Clearly, developing a successful strategy against invasive pathogens requires knowledge of the factors described above, as well as estimates of the level of the impact of each specific tactic on each factor (e.g. on the initial inoculum or, the infection rate, the vector, etc.), and understanding of the implications of strategy choice on possible results. If eradication fails, the pathogen is likely to become established and to spread further, and the input made in the eradication program will have been wasted. Hence, preparedness and cooperation within the agricultural community are critical for successful eradication. A robust preparedness plan depends upon having an organized and effective agricultural management infrastructure, and reliable and sensitive detection and diagnosis tools. Effective eradication requires the availability of the appropriate measures and a cadre of well-trained plant health specialists to implement them. Finally, because a high percentage of invasive pathogen incursions occur across national boundaries, international cooperation and collaborations are also crucial to the establishment of optimal practices.

References

Anderson PK, Cunningham AA, Patel NG, Morales FJ, Epstein PR, Daszak P (2004) Emerging infectious diseases of plants: pathogen pollution, climate change and agrotechnology drivers. Trends Ecol Evol 19:535–544

Brunner JF (1994) Integrated pest management in tree fruit crops. Food Rev Int 10:135–157

European and Mediterranean Plant Protection Organization (2016) EPPO A1 and A2 Lists of pests recommended for regulation as quarantine pests. http://www.eppo.org/QUARANTINE/quarantine.htm Accessed 20 Jan 2016

Gamliel A (2008) High consequence plant pathogens. In: Gullino ML, Fletcher J, Gamliel A, Stack JP (eds) Crop biosecurity: assuring our global food supply.. Springer, Dordrecht, pp 25–36

Gamliel A, Fletcher J (2008) Crop biosecurity: containment and eradication of invasive pathogens. In: Gullino ML, Fletcher J, Gamliel A, Stack JP (eds) Crop biosecurity: assuring our global food supply. Springer, Dordrecht, pp 73–92

Gottwald TR, Irey M (2007) Post-hurricane analysis of citrus canker II: predictive model estimation of disease spread and area potentially impacted by various eradication protocols following catastrophic weather events. Online. Plant Health Prog. doi: 10.1094/PHP-2007-0405-01-RS

Gottwald TR, Hughes G, Graham JH, Sun X, Riley T (2001) Citrus canker epidemic in Florida: the scientific basis of eradication policy for an invasive species. Phytopathology 91:30–34. doi:10.1094/PHYTO.2001.91.1.30

Gottwald TR, Graham JH, Schubert TS (2002) Citrus canker: the pathogen and its impact. Online. Plant Health Prog. doi:10.1094/PHP-2002-0812-01-RV

Graham JH, Gottwald TR, Cubero J, Achor D (2004) *Xanthomonas axonopodus* pv. *citri*: factors affecting successful eradication of citrus canker. Molec Plant Pathol 5:1–15

Gamliel A, Gullino ML, Stack JP (2008) Crop biosecurity: local, national, regional and global perspective. In: Gullino ML, Fletcher J, Gamliel A, Stack JP (eds) Crop biosecurity: assuring our global food supply. Springer, Dordrecht, pp 37–61

Hopkins DL, Purcell AH (2002) *Xylella fastidiosa*: cause of Pierces's disease of grapevine and other emergent diseases. Plant Dis 82:1056–1066

Jeger MI (2004) Analysis of disease progress as a basis for evaluation disease management practices. Annu Rev Phytopathol 42:61–82

Jennings D, Hampson MC, Wood SL (1997) Contamination of automobiles by *Synchytrium endobioticum* in relations to home gardens in Newfoundland. Can J Plant Pathol 19:52–56

MBTOC (2007) Assessment report of the Methyl Bromide Technical Options Committee (MBTOC). 2006 Assessment of the montreal protocol on substances that deplete the Ozone layer. UNEP, Nairobi, Kenya

Roberts RG, Hale CN, van der Zweff T, Miller CE, Redlin SC (1998) The potential for spread of *Erwinia amylovora* and fire blight via commercial apple fruit; a critical review and risk assessment. Crop Prot 17:19–28

Rupe J, Sconyers L (2008) Soybean rust. The plant health instructor. Am Phys Soc. doi:10.1094/PHI-I-2008-0401-01

Schaad NW, Shaw JJ, Vidaver A, Leach J, Erlick BJ (1999) Crop biosecurity. APS feature. http://www.Apsnet.org/online/feature/Biosecurity

Schubert TS, Rizvl SA, Sun X, Gottwald TR, Graham JH, Dixon WN (2001) Meeting the challenge of eradicating citrus canker – again. Plant Dis 85:340–350

Su, H.-J. and R.-S Huang. 1990. The nature of likubin organism, life cycle, morphology and possible strains, pp. 106-110 In B. Aubert, S. Tontyaporn, and D. Buangsuwon [eds.], Rehabilitation of Citrus Industry in the Asia Pacific Region. Proc. Asia Pacific International Conference on Citriculture. Chiang Mai, Thailand, 4-10 February 1990. UNDP-FAO, Rome

Termorshuizen AJ, Volker D, Blok WJ, ten Brummeler E, Hartog BJ, Janse JD, Knol W, Wenneker M (2003) Survival of human and plant pathogens during anaerobic mesophilic digestion of vegetable, fruit, and garden waste. Eur J Soil Biol 39:165–171

United States Department of Agriculture, Animal and Plant Health Inspection Service (2016) online: http://www.aphis.usda.gov/programs/ag_selectagent/ag_bioterr_toxinslist.html Accessed Jan 2016

USDA-APHIS (2006) New pest response guidelines Candidatus Liberobacter africanus, candidatus liberobacter americanus, Honglongbing, citrus greening disease. Version 1.1

USDA-APHIS (2016) Emergency response manual https://www.aphis.usda.gov/import_export/plants/manuals/emergency/downloads/epm.pdf Accessed Jan 2016

Wilson AD, Lester DG (2002) Trench inserts as long-term barriers to root transmission for control of oak wilt. Plant Dis 86:1067–1074

Wistrom C, Purcell AH (2005) The fate of *Xylella fastidiosa* in vineyard weeds and other alternate hosts in California. Plant Dis 89:994–999

Chapter 13
Applications and Assessment of Microbial Forensics in a Field Outbreak of Salmon Blotch of Onion in Israel

Jacqueline Fletcher, Abraham Gamliel, James P. Stack, Heinz W. Dehne, Yochai Isack, and Ian Moncrief

Abstract Microbial forensics is a scientific discipline devoted to analyzing evidence from a bioterrorist act, biocrime, or inadvertent release of a microorganism/toxin with a goal of attribution (Breeze et al. 2005), linking a pathogen and/or a perpetrator to a specific biocrime or bioterrorst act. Attribution includes identifying the microbe(s) involved (Breeze et al. 2005) as well as those responsible. The components of microbial forensics described by Breeze et al. include (1) detection and identification of a pathogen; (2) bioinformatics, including genome sequencing and genetic databases; (3) strain repositories for pathogens or microbes of interest as well as their near-neighbors; (4) validation and standardization of forensic methods; and (5) rigorous attention to quality assurance steps. Because forensic casework is subject to vigorous challenge in a court of law, the rigor of standardization and validation of experimental, analytical and application methods goes beyond levels that are normal for typical research and management activities.

J. Fletcher (✉) • I. Moncrief
National Institute for Microbial Forensics & Food and Agricultural Biosecurity, Department of Entomology & Plant Pathology, Oklahoma State University, Stillwater 74078, OK, USA
e-mail: jacqueline.fletcher@okstate.edu

A. Gamliel • Y. Isack
Laboratory for Pest Management Research, Institute for Agricultural Engineering, ARO Volcani Center, P.O. Box 6, 50250 Bet Dagan, Israel

J.P. Stack
Department of Plant Pathology, Kansas State University, Manhattan, KS, USA

H.W. Dehne
INRES – Institute for Crop Sciences and Resource Conservation, Department of Phytomedicine, University of Bonn, Bonn D-53115, Germany

M.L. Gullino et al. (eds.), *Practical Tools for Plant and Food Biosecurity*, Plant Pathology in the 21st Century 8, DOI 10.1007/978-3-319-46897-6_13

Keywords Analysis of molecular variance (AMOVA) • Assessment • Attribution • BioNumerics minimum spanning tree • Biosecurity • Criteria • Decision tool • Dendrogram • *Fusarium proliferatum* • GeneAlex • Inter-simple sequence repeats (ISSR) • Investigation • Law enforcement • Microbial forensics • Motive • Multilocus sequence typing (MLST) • Mycotoxin • Network of excellence in security • Perimeter • Plant and food security • PLANTFOODSEC • Plant pathogen forensics • Polymerase chain reaction (PCR) • Population biology • Principle component analysis • Salmon blotch of onion • Sampling • Simple sequence repeats (SSR) • Spatial distribution • Strain discrimination • Strain typing • STRUCTURE analysis • Traceback • Unweighted pair group with arithmetic mean (UPGMA) • Weighting factor

Microbial forensics is a scientific discipline devoted to analyzing evidence from a bioterrorist act, biocrime, or inadvertent release of a microorganism/toxin with a goal of attribution (Breeze et al. 2005), linking a pathogen and/or a perpetrator to a specific biocrime or bioterrorst act. Attribution includes identifying the microbe(s) involved (Breeze et al. 2005) as well as those responsible. The components of microbial forensics described by Breeze et al. include (1) detection and identification of a pathogen; (2) bioinformatics, including genome sequencing and genetic databases; (3) strain repositories for pathogens or microbes of interest as well as their near-neighbors; (4) validation and standardization of forensic methods; and (5) rigorous attention to quality assurance steps. Because forensic casework is subject to vigorous challenge in a court of law, the rigor of standardization and validation of experimental, analytical and application methods goes beyond levels that are normal for typical research and management activities.

13.1 Microbial Forensic Technologies Adapted to Plant Pathogens

As plant pathogen forensics continues to emerge as a discipline, the need for establishing standard crime scene practices and evidence handling is needed, and procedures must be adapted and validated for plant pathogens (Fletcher et al. 2006). Technologies such as PCR, DNA sequencing, and mass spectrometry that are used traditionally in plant pathology are also useful for forensics. However, not only must procedures adapted or created for forensic applications in plant disease incident investigation be validated, using standards equivalent to those used for the investigation of more traditional forensic cases, they also should be rigorously evaluated within the context of an agricultural setting. An outbreak of salmon blotch of onions in southern Israel provided the context for a multi-disciplinary and multi-institutional effort to practice, evaluate and document such procedures in a real-life crop setting. To our knowledge, this work is the first to achieve the combination of (1) targeting of forensically relevant goals, (2) development and use of highly stringent and validated protocols, and (3) application and testing within a crop setting.

A European Union research project, "**Plant and Food Biosecurity – Network of Excellence**" (PLANTFOODSEC) program brought together key plant pathology researchers from Europe, Israel, and the United States to work synergistically, in teams, to approach specific, high priority issues in plant and food biosecurity. To our knowledge, this work is the first to achieve the combination of (1) targeting of forensically relevant goals, (2) development and use of highly stringent and validated protocols, and (3) application and testing within a crop setting.

13.2 *Fusarium proliferatum* as a Model Fungal Plant Pathogen for Investigating Microbial Forensic Issues

The *Fusarium proliferatum* – onion "salmon blotch" syndrome in southern Israel was selected as a model system for assessing the utility of plant pathogen forensics approaches. A recent outbreak of the disease in the southern Israel onion production area, familiar to one of us, had been a perplexing case for extension plant pathologists. Elements of the pathogen's biology, disease history and epidemiology and of the Israeli onion production system were rich sources of pertinent research and investigatory questions, clues and challenges. For example, the mycotoxigenic fungal pathogen, *F. proliferatum*, has a very broad plant host range (Proctor et al. 2010). Relevant information on biology, epidemiology, gene sequences, detection tools, and other pertinent features were available for the fungus, as well as for some of its near-neighbors.

13.3 Salmon Blotch – *F. proliferatum* History

Salmon blotch of onion was first observed in onions in commercial fields located at the Southern Arava Research and Development near Yotvata, southern Israel, in 2005–2006. The disease is characterized by salmon-colored blotches, consisting of mycelia and spores, which are clearly visible on the outer scales of white onion cultivars. The fungus also can be isolated from yellow and red onion cultivars, but symptoms are masked by the natural pigmentation in these cultivars. Affected areas extended inward from the bulb surface and the onions eventually rotted. A fungus consistently isolated from symptomatic bulbs, when inoculated onto healthy bulbs, produced identical symptoms. The fungus was identified by PCR as *Fusarium proliferatum* (Isack et al. 2014).

Onion seeds, either imported or produced within Israel, are planted in northern Israel in late January. Small bulbs (sets) are harvested around mid-February and stored until they are sold to production farms in southern Israel. There, toward the end of August or early September, sets are planted directly into the soil, where they grow into mature bulbs and are harvested in January or February and sent to the

Fig. 13.1 Overhead view of Israel. Onion sets are produced near the town of Beit She'an, in northern Israel, while onion bulb production fields are located near Yotvata, in southern Israel

local packinghouses. In this study, onion sets produced near Beit She'an (northern Israel) were planted in fields near the kibbutz towns of Yotvata and Grofit (southern Israel) (Fig. 13.1).

Although there was no reason to postulate that the salmon blotch incidents were intentionally caused, this scenario offered us the opportunity to apply and evaluate many tools and strategies, which we and others had developed in the laboratory for use in plant pathogen forensics, in a real-life field disease investigation.

13.4 *F. proliferatum* Background Information

Fusarium proliferatum (Matushima) Nirenberg 1976, Phylum Ascomycota, was first described as *Cephalosporium proliferatum* (Nirenberg 1976). Prior to 1976 many *F. proliferatum* isolates were identified as *F. moniliforme* (Leslie and Summerell 2006), but new host range and morphological information led to the splitting of the species into seven species, including *F. proliferatum* (Leslie and Summerell 2006; Nelson et al. 1983; Nirenberg and O'Donnell 1998), whose teleomorph (sexual state) was *Gibberella fujikuori* var. *intermedium* (Kuhlman 1982) (later renamed *G. fujikuori* mating population D) (Leslie 1995). Morphological characteristics such as microconidia, chlamydospores, and polyphialides are often used for preliminary identification (Leslie and Summerell 2006), but because such features are very similar in closely related *Fusarium* species molecular diagnostic tools are often required for species discrimination. *Fusarium* species are differentiated also by mating-type tests (Leslie and Summerell 2006).

F. proliferatum causes several types of disease syndromes including rots, die-backs, blights, and wilts (Proctor et al. 2010). It has a very wide host range, including onion, mango, wheat, maize, asparagus, palm, pine, and rice (Proctor et al. 2010). It has been isolated from 75 plant species, including monocots, dicots, and conifers; however, it causes disease in only half of them (Proctor et al. 2010). The fungus occurs worldwide and has been reported in the United States (Leslie et al. 1990; Palermo et al., 2012), the Middle East (Alizadeh et al. 2010; Bayraktar and Dolar 2011; Iqbal et al. 2006); Europe (Gherbawy et al. 2001; Logrieco et al. 1995; Stankovic et al. 2007; Palmero et al. 2010); South America (Sampietro et al. 2010) and Japan (Dissanayake et al. 2009).

A soilborne fungus, *F. proliferatum* poses a threat not only to plants but also to animals, including humans, who consume affected plant products because it produces a variety of mycotoxins including fumonisins, trichothecenes, beauvericin, and moniliformin (Logrieco et al. 1995). Notable trichothecenes, such as deoxynivalenol (DON), nivalenol, and T2 toxin (Bluhm et al. 2002) can lead to growth retardation, reduced ovarian function, immunosuppression, feed refusal, and vomiting (Rocha et al. 2005). Fumonisins are both cytotoxic and carcinogenic to animals and humans, interfering with metabolic functions and disrupting the urea cycle (Hopkins and Adlerz 1988).

13.5 Initiation of a Forensic Investigation

Because forensic investigations are typically costly and disruptive, one will be initiated only if there is reason to think that a crime or misdemeanor might have been committed (Rogers 2011). Suspicion of nefarious activity is rare in agricultural settings, as most growers do not associate plant diseases with intentional acts. Also challenging is the fact that plant disease symptoms do not appear immediately after infection, but can take several weeks (Fletcher et al. 2006). A tool designed to assist investigators in determining whether a disease outbreak was due to natural events or to human involvement could add confidence to early-stage decision-making and reduce response time. For example, the application of such a tool, developed to assess the use of biological warfare agents during an unusual epidemic of tularemia in Kosovo from 1999 to 2000 (Grunow and Finke 2002), helped to rule out the possibility that the epidemic had resulted from an intentional release of the bacterium *Francisella tularensis* (Grunow and Finke 2002).

A similar decision tool, adapted to be suitable for the investigation of an outbreak of a plant disease, was designed and validated by Rogers (2011) using a specific plant disease, wheat streak mosaic, caused by *Wheat streak mosaic virus* (WSMV), as a model. Rating criteria included factors relevant to the pathogen host range, environmental conditions, epidemiology, dissemination, biology, and other disease-relevant elements. This tool was validated in WSMV-affected wheat fields in Oklahoma (Rogers 2011). However, it was not tested for applicability to other types of plant pathogens, such as fungi and oomycetes, bacteria, and nematodes.

In this work, salmon blotch of onions in Israel served as a model for the adaptation of the decision tool to assess the possibility of criminal intervention related to a fungal plant disease.

13.6 *Fusarium proliferatum* Detection and Strain Differentiation for Plant Pathogen Forensics

The goal of a forensic investigation is the attribution of a crime to the perpetrator (s). Generally, when a crime involves a pathogen or other microbial agent, investigators seek to match microbial strains found at the crime scene to strains associated with a suspect. In our study, we also sought to identify the point of origin of strains of *F. proliferatum* that were causing salmon blotch in southern Israel. To accomplish these two goals we needed tools, such as unique, morphological or genetic location-specific signatures to accurately identify and discriminate among *F. proliferatum* isolates collected from a variety of locations in Israel and elsewhere.

13.7 Methods to Distinguish *F. proliferatum* from Other *Fusarium* Species and to Discriminate Among Isolates of *F. proliferatum*

Morphological characteristics of *F. proliferatum* used to identify the fungus include small chains of microconidia formed by polyphialides; however, because several other species of *Fusarium* have similar morphology this approach has insufficient confidence for forensic purposes.

A number of DNA fingerprinting assays have proven effective with *Fusarium* species and strains. Multilocus sequence typing (MLST), which compares microorganisms based on a set of genes usually encoding housekeeping functions (Breeze et al. 2005), is a method of choice in many forensics investigations and is generally reproducible among laboratories; limitations occur when compared organisms have limited genetic variation. MLST was used to differentiate the *F. solani* species complex (Debourgogne et al. 2010) via a strategy involving 25 genes tested in different combinations to yield a 5-locus MLST scheme (Cooke 2005). If MLST is used to distinguish isolates that are genetically very similar, the use of housekeeping genes such as ITS, β-tubulin, and TEF1-α may not be effective. However, if unique regions occur within the species' mtDNA for example, then MLST could provide more insight to the genetic variability of this fungus.

Repetitive genome segments called simple sequence repeats (SSRs), consisting of 2–6 bp repeats occurring in tandem, were used to asses genetic diversity of *Fusarium* species pathogenic to onions in Turkey (Bayraktar and Dolar 2011). A total of 322 isolates belonging to seven species of *Fusarium*, including *F. prolifera-*

tum, were collected from 223 onion fields. ISSR (inter-specific simple sequence repeats) analysis of 70 isolates, representing the seven *Fusarium* species, showed distinct banding patterns (Bayraktar and Dolar 2011). Seventy-seven isolates of *F. proliferatum* from root zone soil of palms (planted 20 m apart) in Finke Gorge National Park, Australia (Neumann et al. 2004), fell into two genetically similar, but separate, clades. The authors speculated that there could have been two separate introductions of *F. proliferatum*, or a single introduction followed by a split over time into two populations (Neumann et al. 2004).

We assessed the spatial and temporal distribution of *F. proliferatum* in Israel, and evaluated the diversity of genotypes within the country, within regions, or within crop fields.

13.8 Hypothesis and Experimental Objectives

We approached this investigation – as far as practicable – from the perspective of a forensic investigator who would not have had any prior knowledge of the disease incident and would have had no input or opportunity to collect samples or data prior to the incident. An investigator also would be unable to transform the disease occurrence into a controlled experiment. At the same time, when relevant and appropriate, we used long-accepted scientific approaches to enhance our work. For example, we formulated an initial testable hypothesis about the source of the fungus causing salmon blotch.

Hypothesis ***Onion sets become infested with F. proliferatum while in the set production fields in northern Israel, and serve as carriers of the fungus during shipment to bulb production fields in southern Israel***

Furthermore, from a scientific perspective, we also established an overall objective and several sub-objectives.

Overall Objective **To Develop, Apply and Validate Forensic Tools to Investigate a 'Real World' Disease Situation, Salmon Blotch of Onions in Southern Israel**

Sub-Objectives

1. To create and validate a decision tool for investigators that will assist in assessing whether the salmon blotch outbreak in onions in southern Israel was due to natural causes or to nefarious actions.
2. To discriminate among *Fusarium proliferatum* strains using inter-simple sequence repeats (ISSRs) and simple sequence repeats (SSRs).
3. To type strains of *Fusarium proliferatum* associated with salmon blotch of onions using simple sequence repeat (SSR) technology.

13.9 Experimental Methods and Results

Sub-Objective 1 **To create and validate a decision tool for investigators that will assist in assessing whether the salmon blotch outbreak in onions in southern israel was due to natural causes or to nefarious actions**

Previously, a self-guiding decision tool, modeled for the plant pathogen *Wheat streak mosaic virus*, was designed to help investigators assess the likelihood that a field outbreak was intentionally caused. In the study reported here, the tool was adapted for use with the plant pathogenic fungus *F. proliferatum* and its efficacy was assessed by applying it to an investigation of the source of the fungus causing an outbreak of salmon blotch of onion in southern Israel.

A commercial onion field (200 m × 400 m) on the Arava Research and Development Experiment Station in Yotvata, Israel, owned and farmed by a local grower, was selected for this study. Three other nearby onion fields were also sampled. In 2012 sets of four onion cultivars, two yellow, Ada (A) and Gobi (G), one white, Milky Way (MW), and one red, Mata Hari (MH) were planted. The field was surrounded by windbreaks of mature salt cedar trees and flanked on two sides by date palm plantations, both of which are hosts of *F. proliferatum.* Within the field, in addition to onions, were a variety of weeds and volunteer plants. The field was drip-irrigated with water from a local well. The water and the sandy soil had a high salinity content of 5 μM and ~3.0 μM respectively (Gamliel, personal communication).

As the bulbs reached maturity in November 2012, salmon blotch symptoms were observed on almost all of the MW cultivar bulbs in all four fields. Disease incidence in the Yotvata field in 2012 was the highest ever seen in the area since the disease was identified there in 2008. Disease symptoms were not obvious on bulbs of the other three cultivars.

13.9.1 Selection of Decision Tool Criteria

A decision tool was adapted from that of Rogers (2011) for assessing the possibility that the salmon blotch outbreak was intentionally incited. Criteria (biological, ecological, situational, or molecular factors relevant to an investigator's judgment of criminal involvement) were adapted for the salmon blotch system. Each criterion was worded in the form of a statement, for which the user inputs an assessment value based on observations made in the field or laboratory, or from interviews with victims, witnesses, and others. An assessment value of 1 indicates that the statement agrees fully with the field situation, 2 indicates partial agreement and 3 indicates no agreement. Further, a weighting factor (WF) of 1, 2, or 3 was assigned based on the degree to which that criterion was judged to impact the assessment (Table 13.1).

Table 13.1 Decision tool criteria, statements and weighting factors

Criteria and statement	WF
I. Geographical distribution	
Fp is commonly found in the area	3
II. Spatial distribution	
Infection pattern typical of Fp	2
III. Weather	
Weather conditions favorable for pathogen survival	3
IV. Temporal	
Usual time of year for outbreak	1
Usual severity of symptoms for time of year	3
V. Field history and cultural practices	
Infection found in field previously	1
VI. Crop rotation	
Onion rotated with host of Fp	1
VII. Human activity	
No unusual human activity present or reported	3
VIII. Physical evidence	
No physical evidence found at scene	3
IX. Surrounding areas	
Nearby fields, volunteer date palms, or weeds, water, fallow fields infected	1
X. Motive	
No motivation to harm the grower	3
No evidence of a national attack	3
XI. Pathogen characteristics	
Fp strain is native to the area	2

WF Weighting factor

13.9.2 Use of the Decision Tool in This Study

To adapt the decision tool for the Yotvata salmon blotch investigation, literature pertaining to the host, the pathogen, the disease, and the Israeli farm production and onion distribution systems was collected. For example, the National Climatic Data Center (http://www.ncdc.noaa.gov/) provided average temperatures near Yotvata at and after the time of onion planting. The tool directs the investigator to interview persons of interest, such as growers, farm employees and their families, extension personnel, and others of interest. In this study, interviews were held with the farmer, an extension specialist and head of vegetable research at the farm, and a plant pathology researcher from the Volcani Institute who conducts research at the experiment station. Many of the tool criteria were assessed at the initial visits to the scene, but others, because of their nature, can be evaluated only after sample collection and lab analyses are completed. A team of investigators/researchers collected soil, plant, and onion samples from the Yotvata/Grofit fields and the surrounding areas. *F. proliferatum* isolates, cultured later at the Volcani Institute, were identified morphologically and molecularly.

Actual law enforcement personnel investigating a suspicious disease outbreak would need to make an initial assessment of potential criminal involvement based only on field observations and witness interviews. However, that initial assessment could change following the collection of laboratory data pertaining to pathogen and strain identities and biology. Accordingly, in our study, the decision tool was applied twice, once during the initial stages of the investigation (criteria I–IX) and a second time after the incorporation of the lab results (criteria I–XI).

13.9.3 Simple Sequence Repeat (SSR) Analyses

SSR markers were identified in *F. proliferatum* (Moncrief 2014; Moncrief et al. 2016) and validated on 10 fungal isolates from Germany, Israel, and North America, and from onion, asparagus, and maize. These SSR primers were used in this study to characterize populations of the fungus from the plant and soil materials collected during the Yotvata field investigation. The population data, analyzed by a variety of statistical programs, provided the basis for establishing groupings of the fungus that informed conclusions about the likelihood that specific *F. proliferatum* isolates were the cause of this outbreak, and that these isolates originated in a particular location.

13.10 Sub-Objective 1: Results

The decision tool was used for assessment of intent at two points in the investigation, first at the initial field visit (criteria I–IX) and again later after laboratory results were obtained (criteria I–XI).

13.10.1 Early Field Assessment

13.10.1.1 Criterion I: Geographical Distribution of *F. proliferatum* in Israel

F. proliferatum was recovered from plant and soil samples collected in southern Israel, in and around the onion field, and from adjacent windbreaks, a nearby date palm orchard, soils, and weeds. Molecular SSR analysis suggests that *F. proliferatum* isolates from the 'Milky Way' sets (grown in the north) are closely related to one another, but differ genetically from isolates recovered from the infected onion bulbs grown in southern Israel. Because *F. proliferatum* had been reported in southern Israel in the past and was detected, in this study, in southern Israel vegetation and soils outside of the Yotvata field, an assessment value of '1' was assigned to this criterion.

13.10.1.2 Criterion II: Spatial Distribution of *F. proliferatum* in the Yotvata Field

Disease incidence in the field and the locations of bulbs from which *F. proliferatum* was isolated were determined. Salmon blotch symptoms were visible only on white onions (cv. Milky Way), for which disease incidence was judged very high; incidence in the yellow onion cultivars, Ada and Gobi, and the red cultivar, Mata Hari, could not be determined visually. *F. proliferatum* was isolated and identified morphologically from 84 % of Milky Way bulbs, 70 % of Gobi bulbs, 56 % of Mata Hari bulbs, and 48 % of Ada bulbs. A 'normal' field distribution for salmon blotch of onions has not been described, but, at the time of early investigation, we assigned an assessment value of 1 because disease symptoms were clearly visible in cv. Milky Way.

13.10.1.3 Criterion III: Weather

The Yotvata area growing season is hot and dry. The average annual rainfall for Yotvata is 2 mm, but the field is drip irrigated as needed. Average 2012 temperatures in July (when the onion sets were planted), August, September and October were 35 °C, 33 °C, 32 °C, and 29 °C, respectively, all of which are conducive to growth of *F. proliferatum*. Only in November 2012, when the bulbs reached maturity, did the average temperature drop to 20 °C, below optimal for *F. proliferatum* microconidia germination. An assessment value of 1 was assigned because July–October weather conditions were conducive for *F. proliferatum*.

13.10.1.4 Criterion IV: Temporal Factors for *F. proliferatum*

Salmon blotch symptoms typically appear late in the growing season as bulbs near maturity (D. Gilette, personal communication). A value of 1 was assigned for the first part of this criterion, because every year since 2008, when the disease was first seen in the Yotvata field, symptoms appeared on white onion cultivars at about this time of year. The 2012 outbreak was more severe than in any other year since 2008 (Gamliel, personal communication), so an assessment value of 3 was given to the second part of this criterion.

13.10.1.5 Criterion V: Field History and Cultural Practices

The Yotvata field grower reported observing salmon blotch symptoms in this field during previous years. In 2012, the year of this study, but before the onion sets were planted, the grower applied a soil solarization regime but skipped his normal fungicide application, possibly impacting plant vulnerability to infection. Since the pathogen was present in this field in previous years, an assessment value of 1 was assigned.

13.10.1.6 Criterion VI: Crop Rotation

The farmer reported that onions were planted in the Yotvata field in 2009, 2010 and 2011 (D. Gilette, personal communication). Thus, it is possible that soil or plant debris remaining in the field from previous years could be the source of the 2012 outbreak fungus because *F. proliferatum* can survive on debris for several years, even though it does not produce overwintering spores. However, solarization of the field in 2012, prior to introduction of the onion sets, likely killed most or all of the fungal propagules. An assessment value of 1 was assigned because, although the grower did not rotate onions with another crop, he planted onions (susceptible to the fungus) in the 3 years leading up to the 2012 outbreak.

13.10.1.7 Criterion VII: Human Activity

As in any farming operation there was significant human activity at the Arava Experiment Station, where the Yotvata field was located; various vehicles, farm machinery and personnel moved in and around the production fields. Staff awareness and the display of vehicle identification logos helps to assure farm security. Any unrecognized individual seen in an unauthorized location would be approached and questioned (D. Gilette, personal communication). However, growers often hire temporary workers to help in the fields during the summer; such short-term employees will be unfamiliar to permanent staff for some time. In our investigation, an interview with the grower established that his workers behaved appropriately, and that there were no known conflicts between them and the grower (O. Mishli, personal communication). The Yotvata field is monitored closely during the day and is equipped with security gates, but unauthorized access cannot be ruled out at night after workers leave. An assessment value of 1 was given for this criterion because interviews with the grower and experiment station manager revealed no unusual activity or motive.

13.10.1.8 Criterion VIII: Physical Evidence

During the initial field investigation a plastic Petri plate and a discarded plastic Petri dish bag were found in the cv. Milky Way section of Yotvata field. Since this onion field was being used also as a research plot by scientists at the Volcani Institute, Bet Dagen, Israel, and since that research team had recently visited the field, using Petri dishes to collect samples, it was deemed highly likely that they were responsible for the found items. This hypothesis was further supported by an interview with the researcher. An assessment value of 1 was given because the physical evidence found in the field was judged not related to the disease outbreak.

13.10.1.9 Criterion IX: Areas Surrounding the Yotvata Field

Attempts were made to culture *F. proliferatum* from samples collected from salt cedar windbreaks (1st perimeter) north, south, and west of the Yotvata field, date palm seedlings (2nd perimeter) east and west of the Yotvata field, and woody shrubs along the highway (3rd perimeter) were collected. No fungus was recovered from salt cedars, but sixteen isolates were recovered from 117 date palm seedlings and one isolate (out of 126 samples) was recovered from vegetation collected along the highway perimeter. An assessment value of 1 was assigned because the fungus was found in vegetation adjacent to the Yotvata field.

13.10.1.10 Criterion X: Motive

An interview with the farmer and the experiment station manager revealed no evidence of motivation to harm the grower. Further, there was no evidence of politically-based grudges or friction, such as news reports of political factions or protest groups. The farmer reported that the local growers know one another well and work to minimize competition. For example, local onion growers purchase their sets from different companies and consult with each other to assure cultivar diversity at market (D. Gillette, personal communication). An assessment value of '1' was given to both subsections of this criterion.

13.10.2 Assessment of the Yotvata Field After the Lab Work

A second assessment of the outbreak was performed after the results of the sample isolations and the molecular analyses were incorporated into the decision tool (criteria X and XI). This assessment was based on all criteria (I–XI).

13.10.2.1 Criterion IX: Areas Surrounding the Yotvata Field

DNAs extracted from *F. proliferatum* isolates obtained from Yotvata onions, both mature and seedling date palms and salt cedars, volunteer field weeds, and highway plant samples were analysed by SSR typing (Moncrief 2014; Moncrief et al. 2016). SSR profiles from the onion bulbs differed from those of the date palm seedlings; phylogenetic analyses indicated that the isolates from these two sources were separate populations and suggested that the date palm plantations are not the source of the outbreak fungus. An assessment value of '1' was given based on the SSR results.

13.10.2.2 Criterion XI: Pathogen Characteristics

The *F. proliferatum* isolates identified morphologically in Israel were shipped to Oklahoma for further analysis (Moncrief 2014). Typically, *F. proliferatum* produces a dark violet pigment on PDA, but the Israel isolates displayed a range of mycelium colors including white, purple, red, green, and yellow. SSR screening of isolates collected from different plant and soil populations in and around the Yotvata field and from northern Israeli set producing farms (Moncrief 2014; Moncrief et al. 2016) established the existence of several distinguishable populations of the fungus in Israel. SSR results clearly discriminated between the set isolates (from the north) and all of the isolates collected from the south. An assessment value of '1' was assigned because the SSR profiles of all of the isolates from the south were highly similar to one another and significantly different from those of the northern population.

The total decision tool point value obtained after the early field assessment (prior to obtaining the lab results) was 33, while that obtained in the later assessment (after the incorporation of the lab results) was 35. Both of these values fall into the range of 'unlikely' for the probability that the outbreak was the result of criminal intent (Rogers 2011).

13.11 Overall Assessment of the Use of the Decision Tool

A decision tool for use in assessing the likelihood that a plant disease was caused intentionally, developed for a plant pathogenic virus (Rogers et al. 2012), was modified in this study to assess its effectiveness when applied to a different plant pathogen and cropping system (the fungus *F. proliferatum* and salmon blotch of onion). Other decision tools have been used to assess, retrospectively, if outbreaks of *Francisella tularensis* in Kosovo and the more recent *Escherichia coli* O104:H4 in Germany were due to natural causes or acts of biocrime or bioterrorism (Grunow and Finke 2002; Radosavljevic et al. 2015). Our decision tool-assisted analysis of the 2012 salmon blotch outbreak in onions in southern Israel suggests that the disease was not the result of an intentional act.

This research provided the opportunity to evaluate the decision tool itself, and to identify areas for improvement. One gap recognized from this study was the limited information about salmon blotch disease on onions in Israel. We observed that different onion cultivars may be impacted differently by the disease. Salmon blotch is usually visible only on white onions even though the fungus may be present also in yellow and red cultivars. In this work, the incidence of fungal presence also differed with the cultivar, being significantly higher in white onions than in bulbs of other colors. Since the Yotvata field had been planted with onions for each of three years prior to 2012, it would be useful to compare SSR profiles of fungal isolates from previous years with those collected in 2012. The field was planted with maize (a host of *F. proliferatum)* prior to 2009, and if the SSR profiles of the maize isolates resembled those from onions then the fungus could have been introduced in the maize and persisted in the field during subsequent years. Probably the most important information we lack is whether *F. proliferatum* is present in the soil of the set fields in the north. However, the fact that *F. proliferatum* was recovered from onion sets grown in that location suggests that the fungus is present in northern field soil, and that it would resemble those recovered by us from the sets. If they are in fact, similar, but if the set field soil isolates do not match the production field soil isolates, then the soil from the set fields could have been the source of the fungus in the sets.

The use of the decision tool for salmon blotch of onion in Israel could be further validated by having other scientists, local growers and law enforcement agents in Israel use the tool during a training exercise, as was done in the *WSMV* study (Rogers 2011). Although our decision tool assessment suggested that this disease outbreak was natural, the tool should be tested also on other onion fields that are naturally infected with *F. proliferatum* (Moncrief 2014) as well as on an onion field that was intentionally inoculated with the fungus for comparison.

The effectiveness of a decision tool to investigate the issue of intentional pathogen introduction related to a disease outbreak is influenced by what information about the pathogen and the disease is available. Even the most basic biological information is helpful when determining which criteria should be chosen for a particular tool, as in a recent paper published by Radosavljevic et al. (2015) describing the development of a decision tool for assessment of the 2011 German *E. coli* O104:H4 outbreak, for which the authors drew their criteria from a variety of literature sources from previous *E. coli* outbreaks in food.

The work described here confirmed the conclusion of Rogers (2011) that a decision tool can be useful for assisting in a forensic investigation of a plant disease. Our study expanded the application of the tool to two very different plant disease systems and pathogens, *Wheat streak mosaic virus* in wheat and *F. proliferatum*, causing salmon blotch, in onions, opening the door for adaptation to other plant pathogens and cropping systems. This tool is not intended to be the sole determinant of whether or not a crime was committed; but rather to assist investigators focus on the criteria most appropriate for making that judgment, increasing the efficiency of their work and providing a systematic framework for determining whether the incident warrants the investments of time, energy and resources necessary for further investigation.

13.12 Sub-Objective 2: To Discriminate Among *Fusarium proliferatum* Strains Using Inter-Simple Sequence Repeats (ISSRs) and Simple Sequence Repeats (SSRs)

F. proliferatum is genetically diverse (Alizadeh et al. 2010), based on the occurrence of vegetative compatibility groups (VCGs), and affects a broad range of plant hosts. The fact that the geographical range of *F. proliferatum* in Israel and the source of the fungus causing salmon blotch in southern Israel were unknown prevented the development of effective management strategies for the disease. The development of a suitable method for typing fungal strains was necessary to conduct a forensic investigation of the source of the causal agent. Several existing molecular fungal strain typing approaches take advantage of characteristic, strain specific genetic elements to categorize isolates and can suggest pathogen relationships, evolutionary history and geographical distributions.

The usefulness of SSR markers in studies of population biology and genetic diversity of plant pathogens has been well established for several oomycetes, such as the common greenhouse pathogens *Pythium aphanidermatum*, *P. irregular*, and *P. cryptoirregulare* (Moorman et al. 2002; Lee and Moorman 2008). They have been identified also in species of *Fusarium* (Chandra et al. 2011), such as the *F. oxysporum* (*Fo*) species complex. For those species having fully sequenced genomes, such as *F. verticillioides* (*Fv*), hundreds of SSR loci can be distinguished (Leyva-Madrigal et al. 2014).

In this work we adapted SSR marker technologies for characterizing *F. proliferatum* isolates collected from four different countries: Germany, Austria, North America, and Israel.

Inter simple sequence repeat (ISSR) markers, short repetitive sequences located between microsatellite loci, are generated by single-primer polymerase chain reaction (PCR) (Wolfe 2005). They can be amplified from a variety of eukaryotes and prokaryotes (Zietkiewicz et al. 1994) for fingerprinting to assess genetic diversity in taxonomic and phylogenetic studies (Bayraktar and Dolar 2011). Simple sequence repeats (SSRs), which offer high reproducibility and high variability among closely related species, have been described for several *Fusarium* species, including *F. verticillioides*, *F. graminearum*, and *F. solani* f. sp. *pisi* (Ren et al. 2012; Singh et al. 2011; Xiang et al. 2012).

Fusarium proliferatum genomic DNAs of seven isolates derived from onions grown in Israel, asparagus grown in Germany and Austria, and maize grown in the United States, were screened using five commercially available, previously published ISSR primers to identify repetitive sequences. Amplicon patterns produced using three of these primers had significant variability, and a commercial mix of oligonucleotides having the same motifs were selected for the development of SSR markers.

Seventeen SSR primers were designed and tested on ten isolates of *F. proliferatum* from the three countries and the six plant hosts. Six primers yielded significant levels of amplicon diversity among isolates from different countries as well as from within a country. Amplicon numbers (which correspond to the number of alleles) ranged from 6 to 9. The other eleven SSR primers tested either amplified the DNA inconsistently or yielded multiple bands per isolate. When DNAs from one isolate each of *F. verticillioides, F. thapsinum, F. subglutinans*, and *F. andiyazi* were used to test the transferability of the SSR markers designed for *F. proliferatum*, two primers amplified all four species while six others amplified some, but not all, of them. The other nine primers did not amplify any of the four species.

To our knowledge this is the first report of SSR primers designed specifically for *F. proliferatum*, and our data reveal the potential for characterizing large numbers of *F. proliferatum* isolates using this tool. Our results are similar to those obtained by others who developed and tested SSR markers for *F. oxysporum* and *F. verticillioides* (Bogale et al. 2005; Leyva-Madrigal et al. 2014). The SSR primers not only allowed discrimination among *F. proliferatum* isolates from different countries and hosts, but also revealed differences among isolates from the same plant host and from the same country, evident from the range of band sizes among the North American and Israeli isolates. The *F. proliferatum* SSR primers were transferable to other species within the genus *Fusarium*, and could be useful for population studies of this genus.

13.13 Sub-Objective 3: To Type Strains of *Fusarium proliferatum* Associated with Salmon Blotch of Onions Using Simple Sequence Repeat (SSR) Technology

Damaging outbreaks of salmon blotch in southern Israel, caused by the mycotoxigenic fungus *F. proliferatum*, provided an opportunity to test and validate, in a field setting, newly developed strategies and technologies for forensic investigation of a plant disease. In this case study, nefarious human activity was not likely or suspected. However, should criminal actions be suspected in a disease outbreak, investigators seek to demonstrate, with high confidence, a "match" between the pathogen strain(s) found at the scene with strain(s) associated in some way with a suspect(s). In the salmon blotch scenario, possible pathogen sources – other than intentional introduction by a perpetrator – include the onion seeds, the onion sets produced in northern Israel, or reservoir plants (onions, other crops, native plants or weeds) and/or soils in onion growing regions of southern Israel. The aim of this objective was to validate the use of SSR primers for (1) the characterization of *F. proliferatum* populations from different locations and hosts within Israel to assess potential sources of the fungus causing salmon blotch of onions, and (2) their application in a forensic investigation within an agricultural setting.

13.13.1 Sampling Sites

Four onion production fields in southern Israel (Figs. 13.1 and 13.2) were planted either by the owners (two commercial fields, designated Yotvata (the case investigation field) and Grofit,) or by the Arava Research and Development Experiment Station (ARDES) Manager (two research plots designated Arava 1 and Arava 2), with sets of white onion, cv. Milky Way, grown from seed produced in northern Israel. At the Yotvata field, additional rows were planted to cvs. Gobi and Ada (yellow) and Mata Hari (red).

Each year, onion sets of all cultivars are purchased by and shipped to growers in southern Israel in June or July, planted immediately into onion production fields and allowed to grow to maturity (October/November). A variety of crops had been planted in these fields in previous years; some of which, like maize, are known hosts of *F. proliferatum*, while others, such as potato, are not. Pre-plant treatments for the selected fields included application of metham sodium to some fields and soil solarization in the Yotvata, Grofit, and Arava 1 fields. Other vegetation in and around these fields included salt cedars, date palms, and weeds. At the Yotvata field, only the road separated the field from surrounding salt cedar windbreaks and date palm plantations just beyond.

13.13.2 Sampling of Onion Bulbs and Soils

Plant and soil samples were collected from each of the four fields (Fig. 13.2) into sterile, individual containers by gloved personnel, as follows:

Yotvata Field Fifty bulbs each of salmon blotch symptomatic white (cv. Milky Way), asymptomatic yellow (cvs. Ada and Gobi), and asymptomatic red (cv. Mata Hari) onions were collected (a total of 200 bulbs). The same number of soil samples were collected, one adjacent to each sampled bulb. Symptomatic cv. Milky Way bulbs had visible salmon blotches on the outer scales. Cv. Gobi, Ada, and Mata Hari bulbs lacked visible salmon blotches, but their yellow or red pigmentation often masks such signs. Individual bulb and soil samples were placed in separate, labeled sterile containers, transported in coolers to the laboratory and stored at 4 C until processed. *Grofit, Arava 1, and Arava 2 fields*: Symptomatic (when present) or asymptomatic white (cv. Milky Way) onion bulbs and adjacent soil samples were collected in numbers of 50 (Grofit), 47 (Arava 1) and 42 (Arava 2).

13.13.3 Sampling from Salt Cedar Windbreaks, Date Palms, and Weeds Within and Adjacent to the Yotvata Field

Within the cv. Milky Way bulb field section, at least 10 plants of each weed species were gently pulled from the soil with gloved hands and individually bagged. No disease symptoms were visible on any of the weeds.

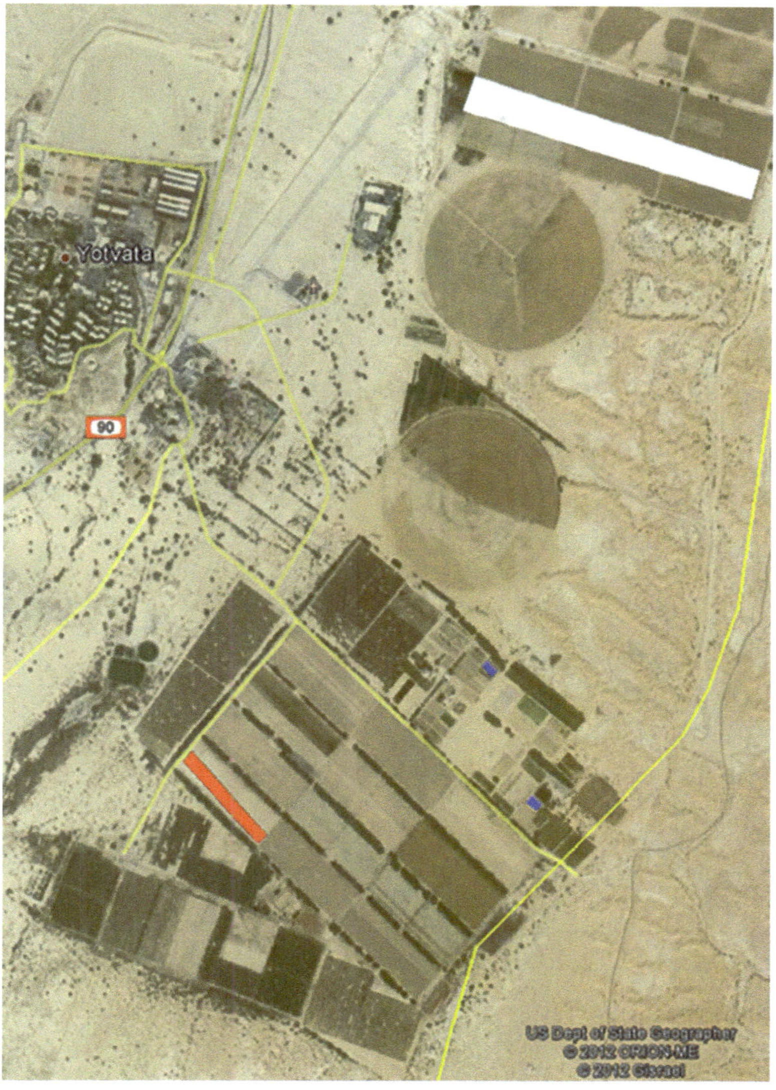

Fig. 13.2 Overhead map of the Yotvata area in southern Israel. The four sites used in this study included two commercial fields (*red and white large rectangles*) and two research fields belonging to the Volcani Institute's Arava Field Research Station (*blue rectangles*)

To assess the spatial distribution of *F. proliferatum* relative to the field location, three "perimeters" defined by their relative distance from the Yotvata field were sampled. The first and closest perimeter consisted of salt cedar trees planted as windbreaks along the north, west, and south borders of the Yotvata field; the second included date palm trees planted in blocks to the south and west of the Yotvata field; and the third consisted of natural vegetation and weeds growing along nearby Highway 90, which connects northern and southern Israel. For larger plants not easily pulled from the soil, a 6-inch branch cross-section was excised using shears sprayed with ethanol after each cut.

13.13.4 Fungal Isolations from Sampled Plants

Seeds Seeds of onion cv. Milky Way, from the same lot as those planted in the northern Israel set production fields, were washed to remove any fungicide and plated onto date agar (Isack et al. 2014).

Sets A single batch of apparently healthy onion sets, harvested in 2011 from a single northern Israel set production field, was split into two groups; one was shipped for planting in the southern Israel bulb production fields, while the other was sent directly from the set field to the Gamliel laboratory at the Volcani Institute.

Bulbs and Soil Mature onion bulbs, and soil samples collected immediately adjacent to each, were collected from the four southern Israel fields. For all of the above samples, tissue or soil was plated on date agar and fungal colonies resembling *Fusarium* were hyphal-tipped from aerial mycelium (Isack et al. 2014). Fungal colonies were examined for the presence of *F. proliferatum*-characteristic polyphialides and chains of microconidia (Leslie and Summerell 2006).

13.13.5 Fungal Isolation from Non-onion Vegetation

Weeds growing in the Yotvata field (*Malva nicaeensis* All., *Chenopodium murale* L., *Tamarix aphylla* (L.) Karsten, *Melilotus sulcatus* Desf., *astragalus* spp., *Citrullus colocynthis, Avena* spp., and *Phoenix dactylifera* L.), salt cedar and date palm seedlings adjacent to the field, and a variety of plants growing near a highway that served as the outermost perimeter of the Yotvata field, also were collected and *F. proliferatum* isolation was attempted as described above.

13.14 Results

13.14.1 Fungal Isolation from Plants and Soil

Onion seeds left over from batches planted, in 2011, in the northern set fields were devoid of *F. proliferatum;* no fungus was cultured from any of 1420 seeds. *F. proliferatum* was isolated from both bulb and soil samples collected from December to January 2012–2013 in southern production fields. Percentages of positive samples varied with the field, the sample type (bulb vs. soil), and the onion cultivar. The highest isolation frequency (84 %) was from soil samples collected in the "Milky Way" area of the Yotvata field, frequencies from collections of cvs. Ada, Gobi and Mata Hari were lower. Isolation frequencies from soils in the other three fields also were lower. Not surprisingly, *F. proliferatum* isolation frequency from each onion bulb was similar to that of the soil sample collected near that same bulb.

Most plant species tested, other than onion, were poor sources of the fungus. *F. proliferatum* was never isolated from windbreak salt cedars and only one isolate was obtained from 126 wild plants collected near the highway. In contrast, 13 % date palm samples from plantations east and south of the Yotvata field yielded *F. proliferatum*.

13.14.2 SSR PCR Amplification and Analysis

Of 309 genomic DNA samples, extracted from Israel isolates of *F. proliferatum* and PCR assayed using six SSR primers, 216 amplified consistently with all six primers. One, designated SSR primer 38, revealed the greatest number of alleles (8) based on differential amplicon sizes ranging from 372 to 402 bp, while the other primers yielded 3–6 alleles each.

AMOVA Analysis Analysis of molecular variance (AMOVA) describes the amount of genetic variation within and among populations using pairwise. Assessment of the 216 *F. proliferatum* isolates described above generated several population pairs having significant diversity (in order from most to least diverse):

- All isolates from cv. 'Milky Way' bulbs (south) vs. those from cv. 'Milky Way' sets (north).
- All isolates from sets (all cultivars) vs. those from bulbs (all cultivars; bulbs derived from a cohort of the same sets).
- All isolates from the north (sets only) vs. all isolates from the south (soils, onion bulbs, date palms, and weeds).
- All isolates from cv. 'Milky Way' soil vs. those from cv. Ada soil. Although cvs. 'Ada' and 'Milky Way' were separated in the Yotvata field by only a single furrow, there is significant diversity among the isolates from their soils.
- All isolates from bulbs vs. those from date palms (both in the south only; date palm samples consisted of mature trees in adjacent plantations as well as volunteer seedling weeds in the Yotvata field).

Comparisons having lower, but moderate, levels of genetic diversity include *F. proliferatum* isolates from all onions sets and bulbs vs. those from other plant hosts, salt cedar vs date palms, salt cedar vs. Yotvata field weeds, and 'Milky Way' bulbs vs date palms. The AMOVA comparisons of the soils from all four field sites show low but significant genetic diversity.

The fact that *F. proliferatum* isolates from onion sets, grown in the north, belong to a different population than all isolates collected in the south suggests that the sets were not the source of the fungus. Interestingly, AMOVA indicated that isolates from Yotvata soil collected at harvest time and those collected before the sets were planted are clonal. The fungus is known to survive in fields for several years (Cotten and Munkvold 1998) and the hypothesis that the Yotvata field soil was the source of the fungus responsible for the current outbreak of salmon blotch is consistent with the comparisons between soils of the Yotvata field and the other three fields.

We obtained isolates from volunteer salt cedar and date palm seedlings growing within the white onion bulb plots in the Yotvata field. AMOVA analysis indicated that these two populations are different, albeit of relatively low genetic diversity. Similarly, white onion bulb isolates comprised a different population than those from date palm seedlings. AMOVA results also indicated that isolates from white onion bulbs and salt cedar volunteers are clonal, suggesting that salt cedar can be an alternative host to salmon blotch strains of *F. proliferatum* in Israel. Interestingly, isolates from the 'Milky Way' portion of the soil are moderately different, genetically, from the isolates from the 'Ada' portion of the field, even though the two cultivars were separated by only one furrow, possibly reflecting multiple or uneven distribution of fungal populations in the soil. Furthermore, our failure to recover *F. proliferatum* from 'Ada' sets suggests that the 'Ada' bulbs became infected after their arrival in southern Israel.

13.14.3 BioNumerics Minimum Spanning Tree Analysis

BioNumerics analysis was done using data from multiple locus variable number tandem repeat analysis (MLVA), which compares isolates by the number of SSR repeats present in each. A minimum spanning tree of the 216 *F. proliferatum* isolates revealed four clusters, each containing isolates from a wide variety of locations (including both northern and southern sites) and substrates (multiple plant hosts and soils).

- The largest cluster (A; 147 isolates) included at least one each from date palms (1 isolate), 'Milky Way' sets (1), 'Mata Hari' sets (1), 'Gobi' sets (1), weeds (30), soils (63) and bulbs (50). Cluster A isolates are indistinguishable from one another by this analysis.
- Most bulb isolates are indistinguishable from one another and fall into cluster A, which also contains all but three of the isolates from weeds within the 'Milky Way' section of the Yotvata field.
- Five isolates from Yotvata soil (collected in the year before the sets were planted) fell into cluster A along with those isolated from the same field in 2012.
- Cluster B comprises 34 isolates from 'Mata Hari' sets (2), date palms (7), soils (10), weeds (3) and bulbs (10). Within cluster B, two circles contain a mixture of isolates from soil, weeds, and bulbs, and one circle contains two isolates from 'Mata Hari' sets and one from a 'Milky Way' onion bulb. Isolates from the date palm plantations constituted their own cluster, B, except for one isolate that fell into cluster A.
- Cluster C has 16 isolates from 'Milky Way' sets (7), 'Gobi' sets (2) and soil (2).
- Cluster D has 20 isolates from 'Milky Way' sets (16), bulbs (1) and soil (3).
- Thirteen bulb isolates (including some from each Yotvata, Grofit, Arava 1 and Arava 2) fall into clusters B and C.

- Overall, MLVA analysis groups the majority of the set isolates into clusters C and D. However, three cluster with the majority of the southern isolates and two others fall into cluster B.
- The isolates from the soils of the four fields are predominately grouped together in cluster A, but several are scattered in clusters B, C, and D.

BioNumerics results were consistent with those of AMOVA. Most weed isolates, including those from salt cedar volunteers within the Yotvata field (cluster A), match those of isolates from the Yotvata bulbs. *F. proliferatum* isolates from date palms form a separate cluster (B), mostly separate from the onion bulbs. Most soil isolates fell into cluster A, but the fact that a few are distributed elsewhere among the four clusters suggests possible movement of the fungus in the south.

13.14.4 STRUCTURE Analysis

STRUCTURE compares populations based on genetic similarities. For this analysis the 216 *F. proliferatum* isolates were first grouped into 14 sub-populations based on the field and substrate from which they were isolated and, for each sub-population, a probability of genetic similarity was calculated based on SSR data. Of the 14 sub-populations, only two were recognized in the analysis. Most cv. Milky Way (white) and Gobi (yellow) sets were separated from the southern isolates. However, two (one Milky Way and one Gobi) were >99 % similar to the southern populations. Unexpectedly, sets of cv. Mata Hari (red) have >95 % similarity to the isolates from southern Israel. Not surprisingly, *F. proliferatum* isolates from weeds within the Milky Way section of the Yotvata field are >99 % similar to those collected from bulbs and soil in the same field. The majority of the isolates from bulbs and soils in the Yotvata, Grofit, Arava 1, and Arava 2 fields grouped with the second population. The STRUCTURE data hint of possible hybridization between *F. proliferatum* isolates from northern and southern Israel, as indicated in the schematic by a mix of red- and green-designated populations. Isolates from the date palms near the Yotvata field were similar to the isolates in population 1, except for a few isolates that fall in either of the two populations.

This analysis provided evidence that some northern isolates could be hybridizing with some from the south. Hybridization by hyphal fusion could result if an *F. proliferatum* population was transported from north to south and persisted among the isolates already present there. Prevailing winds could further disseminate the fungus across southern Israel.

Surprisingly, a single isolate from cv. Milky Way onion sets showed >99 % similarity to the southern isolates. A possible explanation could be that the set yielding this isolate was contaminated with Yotvata field soil, but not planted, and was later transported to the lab. Alternatively, it could be explained by high diversity among *F. proliferatum* populations throughout Israel, brought about by fungal movement from the north to the south or vice versa.

13.14.5 Principal Component Analysis with GeneAlex

PCA, a multivariate analysis, identifies patterns in a diverse data set, such as one having multiple loci. PCA gives a spatial representation of the 216 *F. proliferatum* isolates (placed, again, in the same 14 sub-populations used in the STRUCTURE analysis) and their clusters.

- Overall, the isolates fell into two main clusters, one containing the set isolates from the north and two comprising the isolates from the south (weeds, bulbs, soil, date palms).
- Except for one set isolate, which clustered with *F. proliferatum* collected from southern Israel, all isolates from white onion (cv. Milky Way) sets clustered together.
- Three isolates from red onion bulbs (cv. Mata Hari) grouped with the isolates from the south.
- Isolates from yellow (cv. Gobi) sets were scattered in the PCA plot; one clustered with the southern population and two with the northern population.
- All isolates from southern field soils clustered with the bulbs, weeds, and date palms, except for one group that fell closer to the northern *F. proliferatum* population. Isolates from the date palms grouped with the southern *F. proliferatum* population, but fell within two clusters.

The date palm plantation west of the Yotvata field has been established for over 10 years, while the date palm isolates from the oldest plantation (20 years old), to the south of the Yotvata field group, are distinguishable from the other southern isolates. It is possible based, on this data, that multiple genotypes of *F. proliferatum* have been introduced to southern Israel over the years.

13.14.6 Unweighted Pair Group Method with Arithmetic Mean (UPGMA) Using NTSYS Software

UPGMA analysis yielded a dendrogram of the 216 *F. proliferatum* isolates based on pairwise comparisons of SSR data. Forty nine recognized genotypes fell within four main groupings. Multiple isolates are assessed to belong to the same genotype.

- The first group contains 12 genotypes, one of which contains soil isolates from the Yotvata field (collected both before and after the sets were planted), and the Arava 1 and Grofit fields. Another genotype contains isolates from salt cedar volunteers within the cv. Milky Way section of the Yotvata field and from cv. Milky Way bulbs of the same field.
- The second group contains the largest genotype, which includes some isolates from soils and bulbs of all four southern fields.

- The third major group contained all isolates from date palms near the Yotvata field, grouped into 5 distinct genotypes that differ genetically from one another, based on the pairwise analysis.
- Major group 4 contains all of the set isolates (collected in northern Israel), which were differentiated into three distinct genotypes.

As with the other analyses, UPGMA showed separation of set (northern) *F. proliferatum* isolates and the isolates collected from the south; however, notably, the *F. proliferatum* isolates from white onion sets form a clade separate from that of the other set isolates. These data support a conclusion that the sets are unlikely to be the source of the salmon blotch outbreak pathogen in Israel.

Southern soil isolates are distributed throughout the entire dendrogram, one small clade containing one isolate from each field, indicating that these isolates could be clonal and suggesting that the fungus can be spread, by wind or other means, to nearby fields. As with the other analyses, the date palm isolates form a unique UPMGA clade unrelated to those associated with the salmon blotch outbreak. Salt cedar windbreak isolates fell into the same clade as that of the white bulbs, indicating that these two groups are clonal and suggesting that salt cedar can be a host of the salmon blotch strain of the fungus and could have been a source for the recent outbreak. That we did not isolate *F. proliferatum* from the mature salt cedar windbreak is unexplained; perhaps we sampled tree tissues that were not colonized, or perhaps a physiological inhibitor in mature trees prevents the fungal colonization. A conclusion that the soil in the south could be a source of the fungus causing the salmon blotch outbreak is based on data supporting the grouping of the Yotvata soil isolates, collected before set planting, fall into the same clade as those collected early during the investigation of the 2012 outbreak.

13.15 Interpretations/Discussion

The work described represents a unique merger of technologies and strategies of traditional plant pathology, epidemiology and forensic sciences. The recent discovery and rapid severity increases of a new disease, salmon blotch of onions, in Israel served as a highly suitable framework for the field validation of several technologies previously developed and validated in the laboratory. From a plant pathology perspective, we hypothesized that the 2012 salmon blotch outbreak in southern Israel was caused by a strain or strains of *F. proliferatum* present in the onion sets grown in northern Israel and shipped for planting in commercial onion production fields in the south. An alternative hypothesis was that the pathogen was already endemic in the southern onion production areas and, for reasons that might relate to environmental or host factors, emerged as a serious pathogen only in recent years. To examine the case from a forensic perspective, the hypothesis concept is replaced by goals of determining whether an incident was the result of a criminal action, and if so deemed, identifying the source of a pathogen and its perpetrator for attribution purposes. The first question, whether or not the incident was the result of a crime, was addressed here by applying a decision tool designed to assist investigators in making such judgments in an agricultural setting (Rogers et al. 2012; Moncrief et al. 2014). To answer the second question, identifying the source of the pathogen, a fungal population biology analysis based on SSR strain typing was used to understand the diversity and relationships among and between populations of *F. proliferatum* found in a variety of host species or other substrates, locations in Israel, and times of collection. The data collected, analyzed in a variety of ways, provide substantive support for a specific conclusion to that question.

13.15.1 Disease Distribution in the Field

Disease distribution within a field can offer clues about pathogen behavior relating to the site(s) of initial entry into the field: whether the disease began at one focal point or several, whether pathogen entry was facilitated by prevailing winds or by insect vectors, and whether and in what directions within-field spread occurred. If a criminal action is suspected in a forensic investigation, the disease distribution also can suggest whether human-directed dissemination might have occurred. For example, spatial disease distribution was studied for wheat stripe rust epidemics by Cowger et al. (2005), who inoculated a 1.5 by 1.5-m area within a wheat field and monitored disease spread upwind and downwind from that area. No significant difference was found in disease severity based on wind direction.

In this work, salmon blotch distribution in the four southern Israel production fields was determined based on which onion bulbs or adjacent soil samples were positive for *F. proliferatum*. In the Yotvata field the high disease incidence (virtually 100 %) in cv. Milky Way made determination of a disease pattern challenging. Distribution of the fungus in cv. Ada bulbs, and in the soil samples adjacent to them, was less uniform, reflecting the lower disease incidence in that cultivar. In the Grofit and Arava I fields, disease patterns for cv. Milky Way were relatively uniform, but that in the Arava 2 field was less so. The information gathered about the disease distribution in these fields will be useful for future studies to examine the epidemiology of *F. proliferatum* not only in onions in Israel, but also for other disease models.

13.15.2 F. proliferatum *Isolation from Plant and Soil Substrates*

Onion Seeds If the outbreak strains of *F. proliferatum* reached the southern Israel onion production areas as passengers on the sets grown in the north, then the sets themselves must have become contaminated, either from the seeds or from the local environment. That *F. proliferatum* was never cultured from seeds of cv. Milky Way, left over from planting the northern set field, suggests that the fungus was not present in or on the seeds and that seeds were not the source of the *F. proliferatum* strains causing the 2012 salmon blotch outbreak. Although, in this work, the seeds were washed to remove any surface fungicide, it is possible that a physiological effect, such as the presence of a chemical inhibitor in the seeds, could prevent the fungus from growing out of the seeds.

Onion Sets Of the four onion set cultivars, those of cv. Milky Way had the greatest incidence of fungal contamination; 48 of these 50 set samples yielded *F. proliferatum* isolates. In contrast, the same number of sets from cvs. Ada, Gobi and Mata Hari resulted in only 0, 3 and 4 isolates, respectively, suggesting that *F. proliferatum* infestation of the onion sets is cultivar dependent, and that white cultivars could be more susceptible to the fungus than yellow and red cultivars. A similar cultivar-associated phenomenon was seen in the percentages of *F. proliferation* contamination in the southern production fields, as noted below.

Onion Bulbs The incidence of *F. proliferatum* in the Yotvata field onion bulbs was variable, the highest being 84 % in cv. Milky Way bulbs. Fewer or no isolates were cultured from cvs. Mata Hari, Gobi, and Ada. Interestingly, although no isolates were obtained from cv. Ada sets, 22 isolates were cultured from Ada bulbs, suggesting that the *F. proliferatum* isolates from bulbs were likely infested after being planted in the Yotvata field. *F. proliferatum* was cultured, but at low frequency, from bulbs of cvs. Gobi and Mata Hari. The percentages of *F. proliferatum* isolates cultured from cv. Milky Way bulbs grown in the other three fields in southern Israel varied from 45 to 60 %.

Soil In the Yotvata field, soil samples from the cv. Milky Way plot had the highest percentage (84 %) of positive fungal isolations. Overall, the numbers of *F. proliferatum* isolates obtained from Yotvata field soil were similar to those collected from the bulbs from the same field, a finding consistent with the hypothesis that the soil was the source of the fungal strains causing the salmon blotch outbreak, and also with the alternate hypothesis that fungus present in the bulbs contaminates the soil in its immediate vicinity. Neither hypothesis can be tested using the isolation data alone, since they do not reveal whether the *F. proliferatum* isolates from the bulbs match those from the soil.

Interestingly, *F. proliferatum* was cultured from Yotvata field soil samples collected in 2012 before any sets were planted. The fact that those isolates were similar, by phylogenetic analyses, to those collected during the 2012 outbreak is consistent with a conclusion that the soil could have been the source of the strains causing salmon blotch.

Weeds Within the Yotvata Field Weeds within the cv. 'Milky Way' area of the Yotvata field included salt cedar and date palm volunteers, both of which are known hosts of *F. proliferatum* in Israel (Gamliel, personal communication). The recovery of fungal isolates from field weeds, and their phylogenetic similarity to those of the onion bulbs and soil from the same field, presents the possibilities that (1) weeds can be a source of fungus that subsequently infects onion bulbs and soil, or (2) weeds acquired the fungus from the soil or from the infected onion crop.

Windbreaks, Date Palm Plantation, Highway Perimeters *F. proliferatum* was never cultured from any of the salt cedar trees planted as windbreaks along the Yotvata field borders, suggesting that either the fungus was not present in them or that the trees contained inhibitors of fungal growth. The fungus was, however, cultured from date palm trees in two plantations, both at least a decade old, located just east and south of the Yotvata field. Only one isolate was cultured from weedy vegetation near the Yotvata highway, specifically from *Acacia tortilis* (Forssk.) Hayne. It would not be surprising to find that the fungus is rare in that highway location, since this part of southern Israel is arid, irrigation used in agricultural production does not reach the roadsides, and vegetation along the highway is sparse.

13.15.3 SSR and Phylogenetic Analyses of **F. proliferatum** *in Israel*

While the disease incidence and in-field pattern data described above provide important insights into the history and evolution of the disease outbreak, conclusions about pathogen origins, host ranges and movements cannot be made without understanding the relationships among these populations. In this study, SSR analyses were used to determine relationships among *F. proliferatum* isolates from different populations, locations, hosts, and times of collection. Six previously described SSR loci (Moncrief 2014) were amplified from 216 out of 309 *F. proliferatum* isolates tested.

Overall, the results of the phylogenetic analyses are consistent with one another, and all point to the conclusion that the onion sets were not the source of the salmon blotch causal agent since they group separately from the rest of the isolates from southern Israel. Further, the *F. proliferatum* isolates from date palm plantations, which have been in the Yotvata area for over 20 years, are genetically different from the other southern isolates. *F. proliferatum* isolates from all four southern field sites are similar to one another and the isolates cultured from the soil in the Yotvata field, before the sets were planted, match those collected during our investigation. In this study, we were unable to collect samples in northern set fields to test for *F. proliferatum*, but previous work has shown that it is present there (Gamliel, personal communication). *F. proliferatum* isolates from salt cedar volunteers within the cv. Milky Way section of the Yotvata field match pathogen isolates from the soil and the bulbs collected in that section, based on the phylogenetic analyses. It is possible that *F. proliferatum* is endemic in various plants and soils in southern Israel.

In conclusion, SSRs are powerful molecular markers useful for identification, phylogenetic analysis and traceback of a fungus, and are useful for forensic applications. Their discriminatory power was demonstrated by their differentiation of *F. proliferatum* isolates from northern and southern Israel. Based on the SSR analyses, we conclude that the onion sets are not the source of the *F. proliferatum* causing the salmon blotch outbreak.

Acknowledgements The authors thank Ori Mishli (farmer, Yotvata, Israel) for permission to conduct this work on his property, and Daryl Gillette (Arava Research and Development Station, Yotvata, Israel) for his extensive field support and expert counsel. We are grateful to Drs. Carla Garzon and Steve Marek (Department of Entomology and Plant Pathology, Oklahoma State University) for their invaluable collaboration related to the development and analysis of the SSR assay. We thank Bianca Boehnke, Dimitri Warkentin and Jürgen Derpmann, University of Bonn, Germany, for their assistance in sample collection and processing. This research received funding from the European Union Seventh Framework Programme (FP7/2007–2013) under grant agreement n° 21752.

References

Alizadeh A, Javan-Nikkhah M, Fotouhifar KB, Motlagh ER, Rahjoo V (2010) Genetic diversity of *Fusarium proliferatum* populations from maize, onion, rice and sugarcane in Iran based on vegetative compatibility grouping. Plant Pathol J 26:216–222

Bayraktar H, Dolar FS (2011) Molecular identification and genetic diversity of Fusarium species associated with onion fields in Turkey. J Phytopathology 159:28–34

Bluhm BH, Flaherty JE, Cousin MA, Woloshuk CP (2002) Multiplex polymerase chain reaction assay for the differential detection of trichothecene- and fumonisin-producing species of *Fusarium* in cornmeal. J Food Prot 65:1955–1961

Bogale M, Wingfield BD, Wingfield MJ, Steenkamp ET (2005) Simple sequence repeat markers for species in the *Fusarium oxysporum* complex. Mol Ecol Notes 5:622–624

Breeze GR, Budowle B, Schutzer ES (2005) Microbial forensics, 1st edn. Elsevier Academic Press, Burlington

Chandra NS, Wulff EG, Udayashankar AC, Nandini BP, Niranjana SR, Mortensen CN, Prakash HS (2011) Prospects of molecular markers in Fusarium species diversity. Appl Microbiol Biotechnol 90:1625–1639

Cooke JCL (2005) Forensic genetic analysis of microorganisms: overview of some important technical concepts and selected genetic typing methods. In: Breeze RG, Budowle B, Schutzer ES (eds) Microbial forensics. Elsevier Academic Press, Burlington, pp 233–249

Cotten TK, Munkvold GP (1998) Survival of *Fusarium moniliforme*, *F. proliferatum*, and *F. subglutinans* in maize stalk residue. Phytopathology 88:550–555

Cowger C, Wallace LRD, Mundt CC (2005) Velocity of spread of wheat stripe rust epidemics. Phytopathology 95:972–982

Debourgogne A, Gueidan C, Hennequin C, Contet-Audonneau N, de Hoog S, Machouart M (2010) Development of a new MLST scheme for differentiation of *Fusarium solani* Species Complex (FSSC) isolates. J Microbiol Methods 82:319–323. doi:10.1016/j.mimet.2010.07.008

Dissanayake M, Kashima R, Tanaka S, Ito S (2009) Pathogenic variation and molecular characterization of *Fusarium* species isolated from wilted Welsh onion in Japan. J Gen Plant Pathol 75:37–45

Fletcher J, Bender C, Budowle B, Cobb WT, Gold SE, Ishimaru CA, Luster D, Melcher U, Murch R, Scherm H, Seem RC, Sherwood JL, Sobral BW, Tolin SA (2006) Plant pathogen forensics: capabilities, needs, and recommendations. Microbiol Mol Biol Rev 70:450–471

Gherbawy YAMH, Adler A, Prillinger H (2001) Genotypic identification of *Fusarium subglutinans*, *F. proliferatum* and *F. verticillioides* strains isolated from maize in Austria. Egypt J Biol 3:37–46

Grunow R, Finke EJ (2002) A procedure for differentiating between the intentional release of biological warfare agents and natural outbreaks of disease: its use in analyzing the tularemia outbreak in Kosovo in 1999 and 2000. Clin Microbiol Infect 8(8):510–521

Hopkins DL, Adlerz WC (1988) Natural hosts of *Xylella fastidiosa* in Florida. Plant Dis 72:429–431

Iqbal Z, Mehboob ur R, Dasti AA, Saleem A, Zafar Y (2006) RAPD analysis of *Fusarium* isolates causing "Mango malformation" disease in Pakistan. World J Microb Biotechnol 22:1161–1167

Isack Y, Benichis M, Gillet D, Gamliel A (2014) A selective agar medium for isolation, enumeration and morphological identification of *Fusarium proliferatum*. Phytoparasitica 1:1–7

Kuhlman EG (1982) Varieties of *Gibberella fujikuroi* with anamorphs in *Fusarium* section Liseola. Mycologia 74:759–768

Lee S, Moorman GW (2008) Identification and characterization of simple sequence repeat markers for *Pythium aphanidermatum*, *P. cryptoirregulare*, and *P. irregulare* and the potential use in *Pythium* population genetics. Curr Genet 53:81–93

Leslie JF (1995) *Gibberella fujikuroi*- available populations and variable traits. Can J Bot 73:S282–S291

Leslie JF, Summerell BA (2006) The *Fusarium* laboratory manual. Blackwell Publishing, Ames

Leslie JF, Pearson CAS, Nelson PE, Toussoun TA (1990) *Fusarium* spp. from corn, sorgum, and soybean fields in the central and eastern United States. Phytopathology 80:343–350

Leyva-Madrigal KY, Larralde-Corona CP, Calderón-Vázquez CL, Maldonado-Mendoza IE (2014) Genome distribution and validation of novel microsatellite markers of *Fusarium verticillioides* and their transferability to other *Fusarium* species. J Microbiol Methods 101:18–23

Logrieco A, Moretti A, Ritieni A, Bottalico A, Corda P (1995) Occurrence and toxigenicity if *Fusarium proliferatum* from preharvest maize ear rot, and associated mycotoxins, in Italy. Plant Dis 79:727–731

Moncrief I (2014) Validating forensic tools for crop biosecurity: case study investigation of salmon blotch of onions in Israel. Ph.D. thesis, Oklahoma State University

Moncrief I, Garzon C, Marek S, Gamliel A, Stack J, Issac Y, Fletcher J (2014) Adapting microbial forensics approaches for crop biosecurity: validating the use of simple sequence repeats (SSRs) for typing populations of Fusarium proliferatum associated with salmon blotch of onion in southern Israel. APS Annual Meeting, Minneapolis

Moncrief I, Garzon C, Marek S, Stack J, Gamliel A, Garrido P, Proano F, Gard M, Dehne H, Fletcher J (2016) Development of simple sequence repeat (SSR) markers for discrimination among isolates of Fusarium proliferatum. J Microbiol Methods 126:12–17

Moorman GW, Kang S, Geiser DM, Kim SH (2002) Identification and characterization of *Pythium* species associated with greenhouse floral crops in Pennsylvania. Plant Dis 86:1227–1231

Nelson PE, Toussoun TA, Marasas WFO (1983) *Fusarium* sepcies: an illustrated manual for identification. The Pennsylvania State University Press, University Park

Neumann MJ, Backhouse D, Carter DA, Summerell BA, Burgess LW (2004) Genetic structure of populations of *Fusarium proliferatum* in soils associated with *Livistona mariae* palms in Little Palm Creek, Northern Territory, Australia. Aust J Bot 52:543–550

Nirenberg H (1976) Investigations on the morphological and biological differentiation in the Fusarium section Liseola. Mitteilungen aus der 51 Biologischen Bundesanstalt fur Land- und Forstwirtschaft Berlin-Dahlem 169:117 pp

Nirenberg HI, O'Donnell K (1998) New *Fusarium* species and combinations within the *Gibberella fujikuroi* species complex. Mycologia 90:434–458

Palmero D, De Cara M, Iglesias C, Moreno MM, Gonzalez N, Tello JC (2010) First report of *Fusarium proliferatum* causing rot of garlic bulbs in Spain. Plant Dis 94:277–277

Proctor RH, Desjardins AE, Moretti A (2010) Biological and chemical complexity of *Fusarium proliferatum*. In: Strange RN, Gullino ML (eds) The role of plant pathology in food safety and food security. Springer, Dordrecht, pp 97–111

Radosavljevic V, Finke EJ, Belojevic G (2014) Escherichia coli O104:H4 outbreak in Germany – clarification of the origin of the epidemic. Eur J Public Health 25(1):1–5

Radosavljevic V, Finke EJ, Belojevic G (2015) *Escherichia coli* O104:H4 outbreak in Germany– clarification of the origin of the epidemic. Eur J Pub Health 25:125–129

Ren X, Zhu Z, Li H, Duan C, Wang X (2012) SSR marker development and analysis of genetic diversity of *Fusarium verticillioides* isolated from maize in China. Sci Agric Sin 45:52–66

Rocha O, Ansari K, Doohan FM (2005) Effects of trichothecene mycotoxins on eukaryotic cells: a review. Food Addit Contam 22:369–378

Rogers S (2011) Decision tool and molecular typing technique for a plant pathogen forensic application, using wheat streak mosaic virus as a model pathogen, entomology and plant pathology. Oklahoma State University, Edmon Low Library

Rogers S, Payton M, Allen RW, Melcher U, Carver J, Fletcher J (2012) Validation of a single nucleotide polymorphism genotyping method for Wheat streak mosaic virus. Investig Genet 3:10. HYPERLINK http://www.investigativegenetics.com/content/3/1/10

Sampietro DA, Marin P, Iglesias J, Presello DA, Vattuone MA, Catalan CAN, Jaen MTG (2010) A molecular based strategy for rapid diagnosis of toxigenic *Fusarium* species associated to cereal grains from Argentina. Fungal Biol 114:74–81

Singh R, Sheoran S, Sharma P, Chatrath R (2011) Analysis of simple sequence repeats (SSRs) dynamics in fungus *Fusarium graminearum*. Bioinformation 5:402–404

Stankovic S, Levic J, Petrovic T, Logrieco A, Moretti A (2007) Pathogenicity and mycotoxin production by *Fusarium proliferatum* isolated from onion and garlic in Serbia. Eur J Plant Pathol 118:165–172

Wolfe AD (2005) ISSR techniques for evolutionary biology. In: Zimmer EA, Roalson EH (eds) Molecular evolution: producing the biochemical data, part B. Elsevier Academic Press Inc, San Diego, pp 134–144

Xiang N, Xiao Y, Duan C, Wang X, Zhu Z (2012) Genetic diversity in *Fusarium solani* f. sp. *pisi* based on SSR markers. Biodivers Sci 20:693–702

Zietkiewicz E, Rafalski A, Labuda D (1994) Genome Fingerprinting by simple sequence repeat (SSR) – anchored polymerase chain reaction amplification. Genomics 20:176–183

Chapter 14
The Fragmentation of Plant and Food Biosecurity Research Networks: A Scientometric Analysis

Vincent Cardon and Marc Barbier

Abstract Scientometric analysis based on the mapping of complex networks performed with the Cortext manager platform reveals that biosecurity and bioterrorism have established research communities and literature, in which plant and crop biosecurity are by far less represented than human and animal issues. Biosecurity has not made plant health "disappear" and/or does not constitute a rival field of research. The conceptual apparatus of biosecurity is close to that of some other fields of research on plant health. Some specific consistent clusters of scientists and concepts related to biosecurity and agro-terrorism can be isolated however, in particular the members of the PlantFoodSec Network of Excellence. This result demonstrates the impact of European and international programs (such as CropBioTerror, PlantFoodSec, etc.) on the structure of research networks on agro-terrorism. The article opens with an analysis of qualitative material regarding the way this scientific production and agenda permeates (or not) through daily professional activities. Focusing on plant biosecurity and agro-terrorism, it targets some common issues in scientometrics and sociology of science about the boundaries of research domains and the emergence of new paradigms with specific concepts, methods, authors and cited references. It also echoes the range of questions and reflections at stake within the scientific communities related to biosecurity, and shows the effects of expertise-driven processes on the dynamics of knowledge.

Keywords Biosecurity • Plant health • Agro-terrorism • Bioterrorism • Scientific communities • Expertise • Risk analysis • Social studies of science • Sociology • Scientometric analysis • Network analysis • Lexical analysis

---

V. Cardon (✉)
University of Picardie Jules Vernes, CURAPP (UPJV-CNRS), CESPRA (EHESS-CNRS), Lisis (Inra-UMLV), Paris, France
e-mail: cardon@ehess.fr

M. Barbier
INRA, Lisis (Inra-UMLV), Marne-la-Vallée, France
e-mail: marc.barbier@grignon.inra.fr

M.L. Gullino et al. (eds.), *Practical Tools for Plant and Food Biosecurity*, Plant Pathology in the 21st Century 8, DOI 10.1007/978-3-319-46897-6_14

14.1 Introduction

In a context of globalization and standardization, food provision triggers an increase of human trade and exchanges of plants, seeds, and biological material, and hence insect pests and pathogens. It relies on large socio-technical systems (Hughes 1986) that produce, collect, assemble, transport, process, package and distribute bio-products or food resources. Corporate business, nations and scientists face new challenges regarding plant health and food quantity and quality, such as the biosecurity of those large food provision systems from lab to farm to fork (Evans and Waller 2010). After the terrorist event of 9/11, biosecurity has entered the array of European public problems and is a matter of politics (Dobson et al. 2013). It harnesses classical issues of tracking pervasive pathogens or insects, with specific human and technical capabilities for diagnosis (Miller et al. 2009). But one of the aftermaths of 9/11 and the anthrax episode at around the same time has been the extension of the domain of threat. Those events led to the hypothesis that terrorist attacks could target not only humans and animals but also crops. What if individuals or organizations wanted to spread fear over populations by intentionally introducing pathogens or pests into fields or the food chain? However, how can such a risk that has rarely become reality be assessed, and by who? How can biosecurity systems anticipate a proportionate reaction to what has not happened before? And more generally, how do scientists, countries and international institutions comprehend the eventuality of a dual use of knowledge and research? The *PlantFoodSec (PFS)* research Network of Excellence targets those issues. It aims at the constitution of a specific area of security research in relation to the prevailing European research framework but also more widely to the governance of this issue at the international level (Strange and Gullino 2010). The promoters of awareness and preparedness regarding that risk had to face the difficulty of its recognition as a national or international security challenge.

During the last decade, many scientific works have been published and conducted in the area of biosecurity and bioterrorism. Those contemporary biosecurity issues open up new sociological interrogations and perspectives of analysis, crossing the field of science studies, risk studies and governance studies (Collier et al. 2004; Rappert 2007; McLeish and Nightingale 2007; Rappert and Selgelid 2013; Rappert and McLeish 2014). Nevertheless those studies mainly focus on human and animal health, in part because for a long time there has not been any consensual definition of agro-terrorism (Suffert et al. 2009) and because the "one health" approach (Zinsstag et al. 2012) that bridges veterinary and human medicine does not find such an equivalent driver in the ways plant pathology or entomology and human health are related. However, it addresses relevant issues since the hybrid nature of the threat, involving both human beings and activities and non-human entities (pathogens, insects, etc.), challenges the cognitive framework of professionals and decision makers (Barbier and Prete 2010; Barbier and Boumrar 2011; Boumrar 2013).

Agro-terrorism can be defined "*sensu lato* (anticrop bioterrorism and use of bioweapons against crops) as the intentional use (as well as the threat or simulation of use) of plant pathogens (fungi, bacteria, viruses) or insects by any human individual or group in order to cause direct damage to crops or forests, or to indirectly affect the agricultural sector" (Latxague et al. 2007, p. 427). The purpose of the social studies involved in the PFS project from the beginning was to introduce knowledge, collaborative reflection and concrete results regarding the scientific expertise process at play. The project dedicated attention and resources to an analysis of the conditions and effects of this type of biosecurity research on the existing national systems and capacities and on scientific production. Our focal research interest regarded the specific impact – if any – of political agendas related to plant biosecurity on professional mandates and missions, on daily regular activities, and on the production of knowledge.

Our study is based on two main materials, qualitative and bibliometric. Firstly, it is based on an international comparison between France, Italy and the United Kingdom based on interviews with the main actors (scientific, governmental, etc.) involved in biosecurity issues at the national level. The composition of the PFS project and the active cooperation of its members have been decisive for this fieldwork. We could also benefit directly from stimulating discussions with scientists about the present status of the disciplines involved, such as plant pathology, entomology, crop protection, epidemiology, and biotechnology. Our initial focus on national action systems made it possible to take into account the main regulations involved, the organizational framework (administration, agency, public-private), the administrative and professional bodies for diagnosis and epidemio-surveillance capabilities, the knowledge centers and research capacities, etc., in our study of the constitution of knowledge systems for biosecurity.

Secondly, given that science can be considered as the conjunction of human activities, facilities and institutions, but also of concepts, publications and other information, the other part of the investigation uses new scientometric methods in order to explore how scientific communities dealing with biosecurity are structured and are evolving though time. A complex network analysis of the scientific literature on the subject reveals how concepts associated with biosecurity, but also authors, institutions and editors involved in the production of those concepts are tied together and contribute to structure this field of research. It relies on the constitution and analysis of various corpuses extracted from online scholarly multi-disciplinary research data platforms such as Web of Science and on the use of the specific algorithmic tools offered by the Cortext Manager Platform – developed by the French National Agronomic Research Institute for the Ifris Institute[1].

---

[1] For a detailed description of those tools, see the CorTexT Platform website (managerv2.cortext.net), They have been used by Tancoigne et al. (2014) in their work on the delineation and description of the field of researches about ecosystems, which can be considered as a methodological template for our own work on plant biosecurity and agro-bioterrorism.

In this chapter, we propose to focus on that last aspect of the research. One of our purposes was to determine the epistemological locus of plants and crops in the literature dedicated to biosecurity. Two main empirical questions guided the study. Are there identifiable sub-communities that build the dynamic of knowledge in the era of security research about plant protection (differing from the recognized institutions related to Plant Health)? Is there an effect of public or private funding programs in the dynamic of knowledge and instruments dedicated to security research about plant protection, or is the dynamic of security research influenced by patterns forged in "regular" plant pathology or entomological science?

In the first section, we describe the conceptual and methodological framework of scientometric analysis. Then we propose a statistical analysis of the evolution of the scientific fields targeted, before performing a lexical extraction and a complex network analysis to examine the structure of the scientific communities involved in the scientific production in biosecurity research. We propose in conclusion some remarks regarding the impact of agro-terrorism research on risk perception and professional daily activities.

14.2 Specifying the Plant Biosecurity Epistemological Community

Is there a new paradigm – plant biosecurity – emerging, with its specific concepts, methods, authors and cited references? The question cannot be answered without careful examination of how scientific communities dealing with biosecurity are structured. Scientific communities express themselves as interpersonal relationships between researchers and institutions but they can also be studied through the analysis of their production, such as articles in peer reviewed journals and books. This scientific material uses concepts, references, and more generally words, revealing how scientific objects are expressed and how those expressions and categories evolve.

Those scientific networks involving people, concepts and texts can be comprehended as scientific specialities, as defined by Chubin (1976). According to Morris and der Veer Martens "a research specialty is a self-organized network of researchers who tend to study the same research topics, attend the same conferences, read and cite each other's research papers and publish in the same research journals. A research specialty produces, over time, a cumulating corpus of knowledge, embodied in educational theses, books, conference papers, and a permanent journal literature." (Morris and Van der Veer Martens 2008, p. 274–275) This definition is very close to what Kuhn defines as a paradigm in science: "a paradigm is what the members of a scientific community share, and, conversely, a scientific community consists of men who share a paradigm." (Kuhn 1970, p. 176). Even though most social studies of science have focused on the social structure of scientific disciplines (evaluation systems, spaces of communication, accumulation of resources processes, etc.),

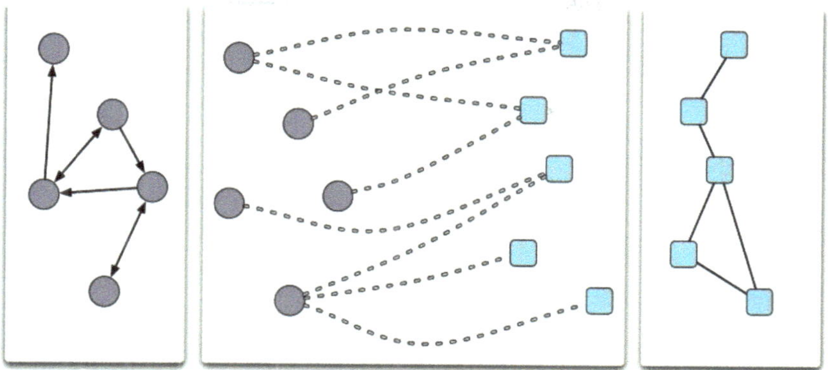

Fig. 14.1 Principle of heterogeneous network analysis (Source: Cointet 2009)

those fields of research can be revealed and detected though the cognitive properties that structure competing or evolving paradigms (Chen et al. 2002). Hence a complex network analysis of the scientific literature on biosecurity and agro-terrorism has been carried out, following the methodological principles and steps of scientometric analysis:

- Building corpus-based queries on bibliometric online platforms (database of records, author, DOI, Institutions, abstract, title, etc.)
- Lexical extraction (based on frequency and co-occurrence) of relevant terms
- Indexation/tagging of the corpus
- Mapping of homogeneous (lexical *or* social) and heterogeneous (lexical *and* social) networks (Fig. 14.1)

The analysis aims at revealing how concepts associated with biosecurity agro-terrorism, bioterrorism, but also the authors and institutions involved in the production of those concepts contribute to structure this field of research.

The first step of the analysis is to build a corpus to be extracted from the Thomson Reuters Platform Web of Science, (Scientific Citation Index database)[2], which requires us to choose concepts to define queries. Different queries (see Table 14.1) have been the basis of the constitution of a set of specific corpuses, starting from the most generic and broad ones (namely with the term "biosecurity") to the more specific ones (*e.g.* "agro-terrorism"). Each keyword has then been intersected with other focused ones to refine the understanding of the composition of each broad field of literature.

[2]Although a powerful bibliometric tool, WoS has some biases, and does not offer an exhaustive indexation of scientific journals. In particular it is very centred on international scientific outputs and thus minors areas of science that would be not included.

Table 14.1 Queries and corpuses build with Web of science

Basic search on Web of science (broad concepts):	Focus on:
Bioterrorism	Crop
Biosecurity	Human
Plant health	Animal
Crop protection	Plant
Agro-terrorism	Pathogen
	Plant pathogen

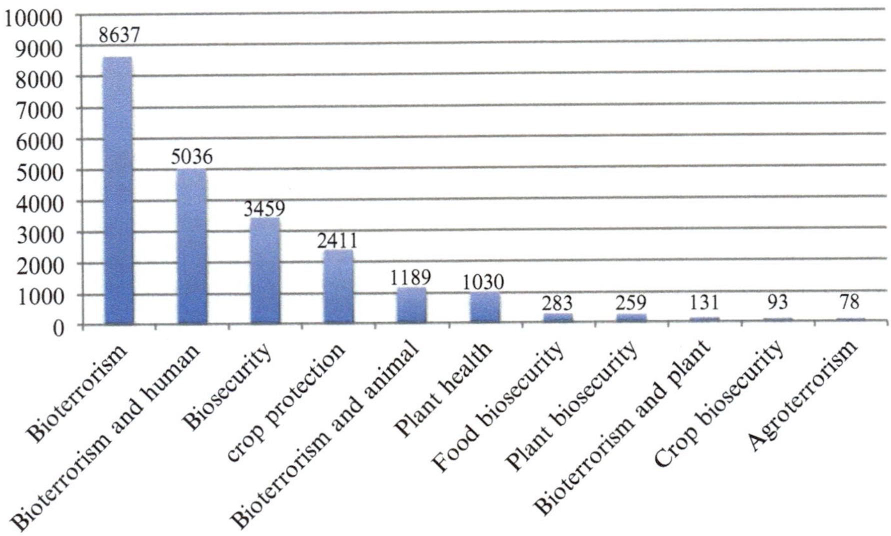

Fig. 14.2 Counting records for various queries (Source: Web of Science, analysis powered by CorTexT Manager)

The first element of description to consider is the "demography" of the population of articles associated with the various queries, meaning the distribution of records associated with each query (Fig. 14.2).

Biosecurity is a dynamic research area, with 3459 records. But plants, crops and food are not very central in that field of research: for instance, plant biosecurity counts only 259 records, 7.5 % of the total. Crop biosecurity scores only 93 records and food biosecurity, 283 records. Animal biosecurity is the center of that research area. The same phenomenon can be observed for bioterrorism. The number of articles dedicated to that topic is 2,5 times bigger than the number of articles dealing with biosecurity (8637 records) but the articles are event more centered on human health: 5036 articles on bioterrorism deal with human pathogens and health-related issues, 1189 with animal health. Plant is far less represented: ("Bioterrorism" AND "plant") only gives 131 records. Plant health and crop protection (3441 records) are also dynamic and rather superposed but when intersected with biosecurity, the volume of literature decreases dramatically. This is the first clue that those various concepts refer to rather separate literatures and thus scientific communities or research domains.

14.3 Temporal Evolution of the Corpuses

However, those aggregated results hide tremendous temporal evolutions and national variations. For instance, the temporal evolution showing the leading countries (top ten) in terms of contribution to the corpuses shows that the literature dealing with biosecurity has grown massively in the 2000s (Fig. 14.3).

This field of research has been spreading internationally. Until 1996, all the output was American. Since the mid 1990s, other countries have made deep inroads into this field of study, although the USA is still a major contributor. The national distribution of the production effort in biosecurity research reveals that island countries massively dominate this topic: during the 2008–2013 period, the accumulated contribution of the USA, Australia, UK and New Zealand has been close to 80 % of the yearly output in the field. The same result occurs when focusing on plant biosecurity. The comparison with the same statistical analysis on the plant health-related literature shows radical differences: while it is also a growing literature, it has a wider international distribution. While North America is also the leader in that field of research, Europe also contributes to a great extent, and in Europe, UK, Germany, Italy and France are the greatest contributors (Fig. 14.4).

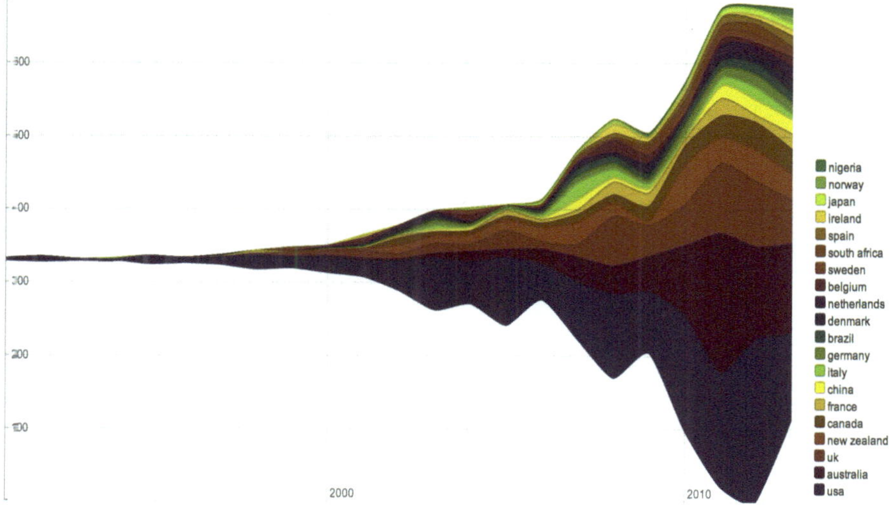

Fig. 14.3 Temporal evolution: Biosecurity and countries (1993–2013) (Source: Web of Science, analysis powered by CorTexT Manager)

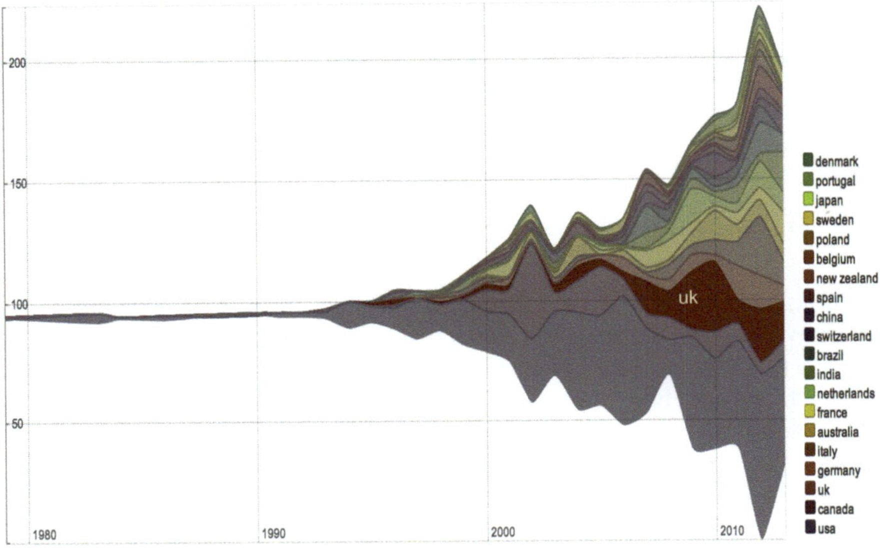

Fig. 14.4 Temporal evolution: Plant Health and countries (1990–2014) (Source: Web of Science, analysis powered by CorTexT Manager)

The emergence of agro-terrorist concerns arose a few years before the 9/11 attacks, however, and seems to be related to the context of globalization and the four "Ts" (trade, travel, transportation, tourism) (Waage and Mumford 2008). As shown in Fig. 14.5, mostly American scientists produced the scientific literature on bioterrorism after 2001.

Since the attacks of 9/11 and the episodes of envelopes containing anthrax, many American experts have taken up a response to the bioterrorism threat, and the scientific production on the topic expanded significantly in the early 2000s, with a peak in 2005, which shows in Fig. 14.5. Once again, the USA has been prominent in this field of research, and so they are on agro-terrorism (anti-crop and anti-livestock), a threat that has been tackled very seriously after 9/11 (Wheelis et al. 2002; Sprinkle 2003; Madden and Wheelis 2003). Two international symposia on agro-terrorism were organized in 2005 and 2006 by the US Federal Bureau of Investigation (FBI), under the patronage of the US Department of Homeland Security (DHS), with the active participation of the US Department of Agriculture (USDA) and the US Food and Drug Administration (FDA). In 2006, the second symposium brought together about 1000 participants (mostly American) from all professions concerned (justice, police, military, agricultural and food control authorities, research laboratories, universities), as well as private companies operating in the field of biosafety. The service of the USDA in charge of food safety (Food Safety and Inspection Service) asked its inspectors to integrate agro-terrorism concerns in all their inspections, now combining the concept of "food defense" with

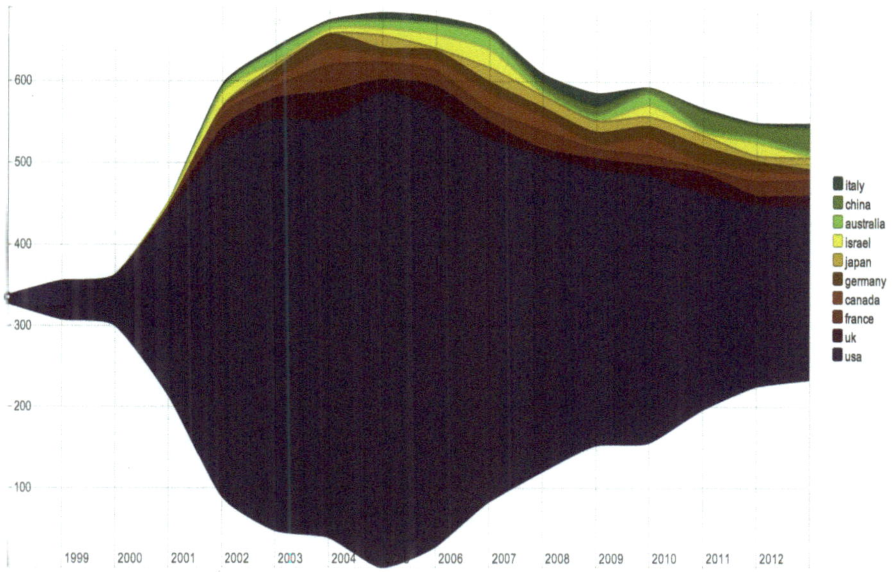

Fig. 14.5 Temporal evolution: Bioterrorism and countries (1999–2013) (Source: Web of Science, analysis powered by CorTexT Manager)

"security health." Part of the scientific community has been involved in this effort since 1999 by setting up an internal committee on bioterrorism and drafting a white paper. Although the regulatory concept of bioforensics (Fletcher et al. 2006) has been institutionalized only recently, an extensive network of diagnostic laboratories (the National Plant Diagnostic Network NPDN) has represented a huge effort to link distributed systems of plant diagnostic laboratories with major actors of research and universities, plant industries and state departments of agriculture (Stack and Baldwin 2008). Figure 14.6 shows the impact of this trend of concerns and action regarding biosecurity threats on scientific literature production. 9/11 clearly appears as a turning point.

In relation to this trend, an also because of international networking of scientific expertise, the European Union (EU) launched two successive programs funded by Framework Packages 6 and 7 (FP6 and 7) named CropBioterror (launched in 2004) and PlantFoodSec (launched in 2011) to build a knowledge capacity and develop awareness and preparedness concerning the risk of intentional threats against crops or the food chain and to assess possible socio-economic outcomes of such threats of opportunistic actions.

Those statistical analyses do not allow a full understanding of how those fields of research are structured. For that purpose concept analysis based on lexical extraction are required.

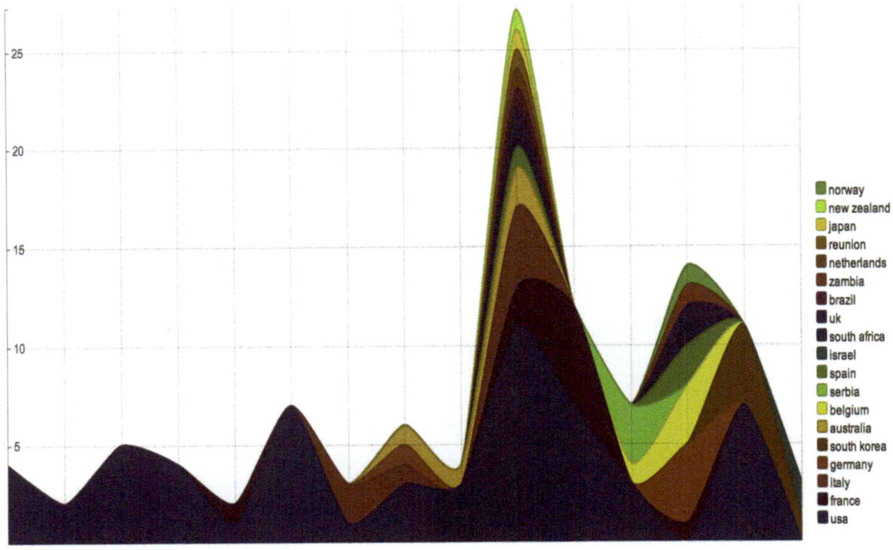

Fig. 14.6 Temporal evolution: Agro-terrorism and countries (1993–2013) (Source: Web of Science, analysis powered by CorTexT Manager)

14.4 Lexical Extraction and Network Analysis

A lexical extraction based on natural language processing (NLP) methods (1. parsing and tagging words, 2. tag chunking, 3. stemming and filtering, 4. ranking according to C-value measure) has been performed on the different corpuses. This NLP methodology enables in-depth analysis of the lexical composition of a scientific corpus by giving a precise view of the most frequent concepts used, and the most commonly addressed issues. It notably shows (Table 14.2) that articles dealing with biosecurity and bioterrorism are centered on human pathogens (in general or specific ones, such as *Bacillus anthracis*) and broad necessities or issues (health care, health system, preparedness, etc).

One of our research questions was whether or not it was possible to isolate some specificity of "plant biosecurity" literature, as compared to "plant health" related articles. A simple count of the most frequent concepts used in those two corpuses shows that the plant biosecurity conceptual framework closely resembles that of "plant health", but with some specificity (Table 14.3). It is more "risk" and "species" related, while productivity is a more important concept in the case of plant health. "Species" and "alien species" are more central in biosecurity-related literature.

Based on the selection of those multi-terms, a network analysis was performed in order to reveal the emergence of a structured conceptual apparatus for biosecurity research and the structuring effect of the various programs. In the following maps, a non-deterministic algorithm of clustering (Louvain) has been used, and performed using Chi^2 metrics to calculate the similarity between terms. The clusters are represented in large colour circles. The size of nodes is a measure of their degree within

Table 14.2 Top 25 concepts, query:" Biosecurity"/"bioterrorism"

Bioterrorism	Main form	Total number of occurrences
1	Bacillus anthracis	553
2	Bioterrorism agents	405
3	Public health	370
4	Health care	279
5	Biological weapons	270
6	Biological agents	183
7	Health emergency	180
8	Surveillance systems	175
9	Smallpox vaccine	173
10	Disease outbreaks	170
11	Smallpox vaccination	169
12	Bioterrorism preparedness	165
13	Emergency preparedness	161
14	Emergency department	142
15	Potential bioterrorism agent	131
16	Health departments	128
17	Anthracis spores	117
18	Emergency response	114
19	Vaccinia virus	112
20	Causative agent	111
21	Early detection	101
22	Health professionals	94
23	Inhalational anthrax	92
24	Health system	92
25	Public health emergencies	92
Biosecurity	**Main form**	**Total number of occurrences**
1	Risk factors	195
2	Influenza virus	126
3	Avian influenza	121
4	Risk assessment	90
5	Wild birds	44
6	Disease outbreaks	69
7	Biosecurity risk	78
8	Avian influenza viruses	43
9	Animal health	50
10	Animal diseases	50
11	Public health	57
12	Biosecurity programs	50
13	Dairy herds	46
14	Disease risk	41

(continued)

Table 14.3 (continued)

15	Herd size	35
16	Risk management	39
17	Risk analysis	44
18	Invasive species	55
19	Food safety	48
20	Climate change	44
21	Pig farms	34
22	Regression model	26
23	Food security	30
24	International trade	32
25	Polymerase chain reaction	47

Source: Web of Science, analysis powered by CorTexT Manager
Legend: In each corpus, a word extraction has been carried out and each table shows the ranking of those terms and their total number of occurrences

Table 14.3 Top 26 concepts, query: "Plant health" and "Plant biosecurity"

Plant biosecurity	**Main form**	**Total number of occurrences**
1	*Plant diseases*	22
2	*Plant species*	20
3	Plant biosecurity	19
4	Host plant	15
5	*Food security*	12
6	Plant material	11
7	Invasive alien species	11
8	*Risk assessment*	10
9	Disease outbreak	9
10	Host range	8
11	Plant virus	7
12	**Biosecurity measures**	6
13	Insect pests	6
14	Polymerase chain reaction	6
15	Suitable areas	6
16	Biosecurity concern	6
17	*Pest management*	6
18	Animal health	6
19	Plant pathology	6
20	**Invasion risk**	6
21	Processing plant	6
22	*Animals and plants*	6
23	Mosaic virus	5
24	Spider mites	5
25	**Future climate scenarios**	5
26	**Non-indigenous species**	5

(continued)

Table 14.4 (continued)

Plant health	Main form	Total number of occurrences
1	*Plant diseases*	49
2	*Risk assessment*	36
3	Pest management	30
4	*Plant species*	28
5	Biological control	27
6	**Animal health**	23
7	Soil fertility	23
8	Control agents	23
9	Health status	22
10	Tomato plants	22
11	Community structure	20
12	Plant health management	19
13	*Food safety*	19
14	Plant growth-promoting	19
15	*Insect pests*	16
16	Growth promotion	16
17	**Methyl bromide**	16
18	Plant health status	16
19	*Risk analysis*	16
20	**Plant stress**	16
21	**Plant health and productivity**	16
22	Water content	15
23	Dry weight	15
24	Root length	15
25	Disease severity	14
26	*Plant pests*	14

Source: WoS/Cortext
Legend: The terms that are common to both corpuses are represented in red, while the terms that are specific are written in green

the network; the size of a link is a measurement of the number of co-occurences; finally, the size of a circle is proportional to the number of articles that contribute to the cluster. The link between two nodes, for instance, such "an author cites such an author", is shown by a line. The geography of clusters has some meaning: clusters that are close share terms or are associated by brokering terms. It follows that those maps are a representation of the underlying structuration of concepts and meanings.

When exploring the words used in a wide corpus referring to bioterrorism, biosecurity and agro-terrorism, plant health shows to be embedded in the same cluster as animal health, and including risk assessment and food safety concerns. This cluster centered on the management of plant and animal diseases is close to another one referring more explicitly to concrete crisis, such as avian influenza (Fig. 14.7).

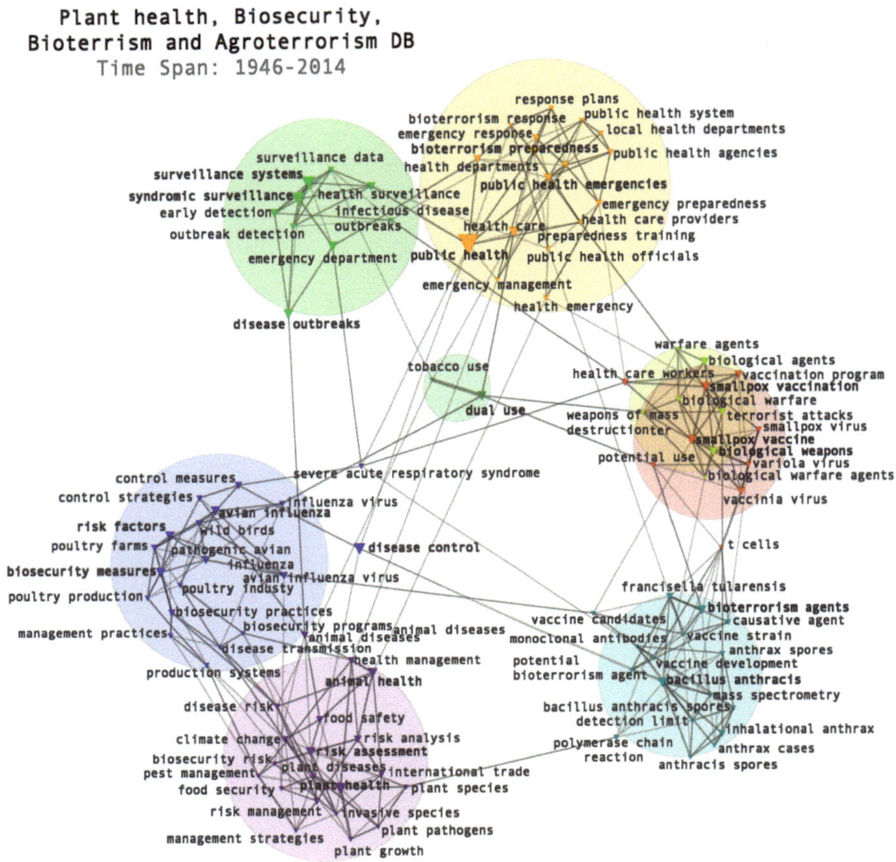

Fig. 14.7 Lexical networks, Query: "bioterrorism OR biosecurity OR plant health OR agro-terrorism" (10733 records) 1946–2014

When focusing on the bibliographic networks (examining who cites who) of the plant biosecurity corpus, the networking outcome of the various programs financed by the USA, NATO, Asialink or European funds, appears clearly (Fig. 14.8). Gullino, Fletcher, Stack, Suffert, Gamliel show in a dedicated cluster, which also involves Madden and Wheelis.

The same effect emerges from the analysis of collaborations (Fig. 14.9). Plant biosecurity is an atoll-shaped field in which a cluster clearly gathers around Gullino, Fletcher, Stack, Suffert, Gamliel, etc (Fig. 14.9).

When concentrating, finally, on agro-terrorism (Fig. 14.10), the literature shows some fragmentation but two cohesive clusters involving PlantFoodSec members clearly emerge.

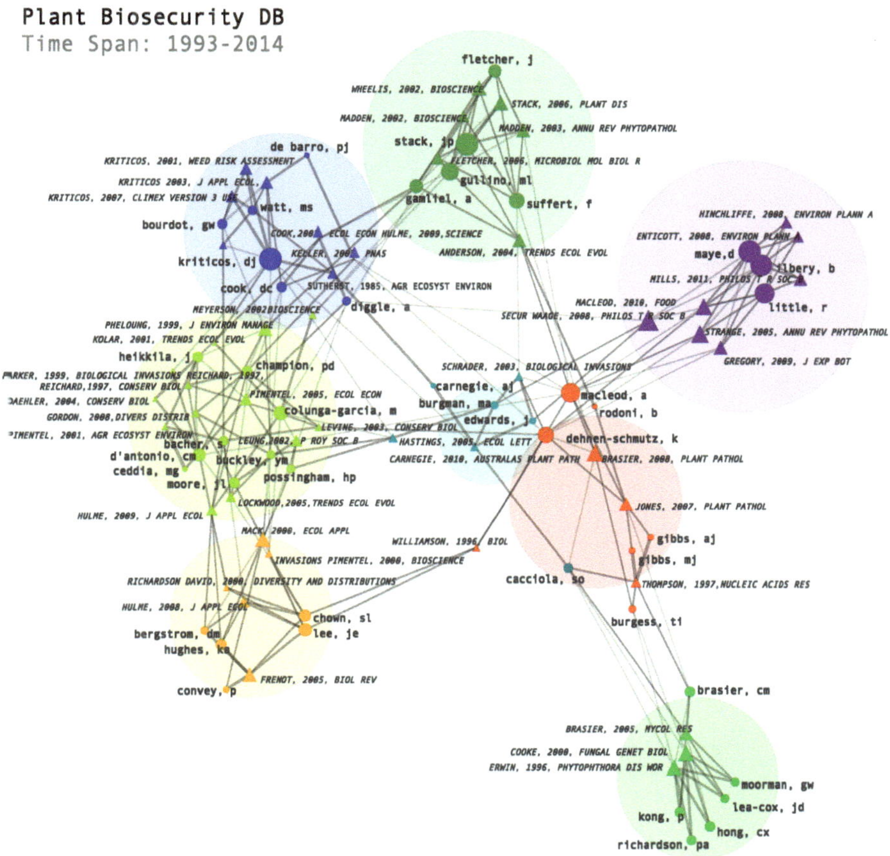

Fig. 14.8 Plant biosecurity (1993–2014): Authors and cited references, Chi², 50 nodes

To conclude this section, biosecurity and bioterrorism form constituted research networks and literature, in which plant (+ forests) and crop biosecurity are by far less represented than human and animal issues. Biosecurity has not made plant health disappear and/or does not constitute a rival field of research. The lexical/conceptual apparatus of biosecurity is close to that of some fields of research on plant health. Some specific consistent clusters of scientists and concepts related to biosecurity can be isolated though, in particular the members of the Network of Excellence (NoE) and the different cluster maps show the impact of the different European and international programs (CropBioTerror, PlantFoodSec, etc) in terms of structuring of research networks.

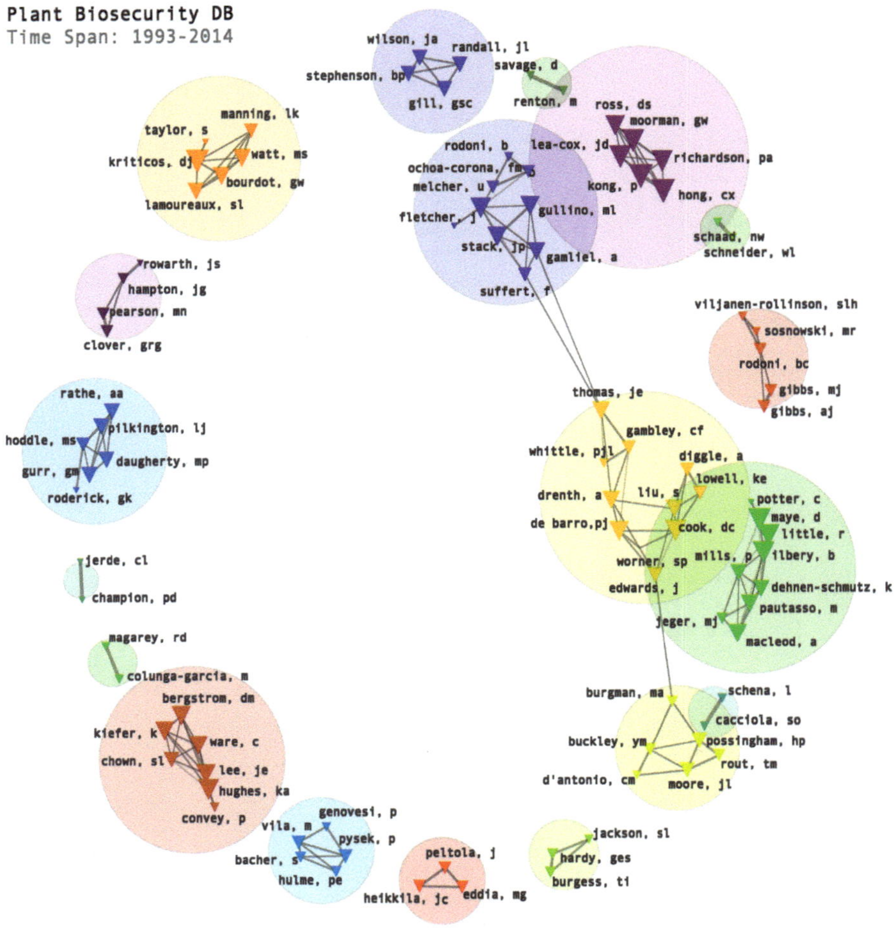

Fig. 14.9 Plant biosecurity (1993–2014): Authors-Authors, Chi², 100 nodes

14.5 Activity and Risk Perception of Professionals Involved in Plant Protection

We mainly focused on the analysis of scientific networks and showed that there was a growing literature on plant and food biosecurity and agro-terrorism and that this literature was organized in a way that shows the impact of political agendas and public funding on scientific production. The various programs implemented at an international level had effects on the volume and structure of those fields of research. But what about daily work in plant epidemio-surveillance actors? Does biosecurity and agro-terrorism-related research permeate and translate into professional practices? In other words, do biosecurity issues imply changes in daily plant and

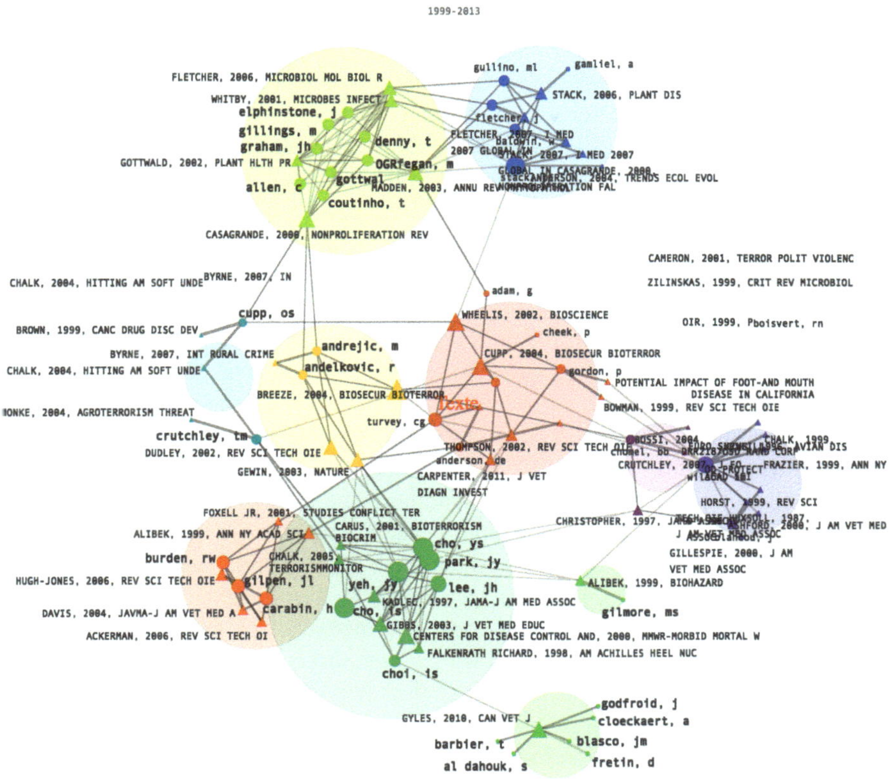

Fig. 14.10 Agro-terrorism (1999–2013): Authors and cited references, Chi², 50 nodes

crop protection? Do the professionals involved in crop protection consider the hypothesis of an agro-terrorism attack "credible" or "likely"? A qualitative international comparison between France, Italy and the UK was thus carried out. It aimed at describing the present co-evolutions (if any) of: research capacities, diagnosis and epidemio-surveillance capabilities, and public-private organizational settings. The fieldwork is mostly based on interviews (N = 38) with coordinators of PlantFoodSec (Unito, FERA, DEFRA, INRA members), risk assessors, plant health seeds inspectors, researchers (epidemiology and plant health), public officers, etc., but also with professionals involved in the qualification and assessment of the risk of bioterrorism, in the elaboration of responses to that risk, and more generally with professionals involved in biosecurity issues. Finally, we also interviewed some members of international organizations, such as EPPO and NATO. Our standard questionnaire has been adapted to each interviewee, depending on his function and role.

Some variables influence the way the threat is assessed and perceived, despite a growing effort to assemble national biosecurity policy-making, notably thanks to EPPO. The geographical variable is of some importance. For instance, Great Britain being an island shows differences compared to continental approaches. More generally, national frameworks of regulation, norms and procedures of biosecurity are not unified. The link with the private sector is more developed in the UK and Italy than in France and influences the assessment of the perceived risk. The link with the public (lay reporting and broadcasting) is more developed in the UK and Italy than in France and impacts the anticipation of the detection of deliberate introduction of plant pathogens and the reaction that should be triggered in that case. Professional boundaries are very different in the three countries: plant epidemiology is not a specialty as such in the UK and Italy (whereas in France it is). The administrative structures of the state vary greatly. France has more centralized political structures while there is more autonomy to Regions or States in the UK and Italy[3]. The type of crops grown, depending on whether they are strategic or not, influences risk perception. Finally, the role of the press and its importance in the issue vary greatly: possible media reaction is considered more of an issue in the UK than in France, for instance. But the three countries also share some strong similarities: they have very dense and efficient epidemio-surveillance networks. The professionals involved are integrated in strong national and international networks of experts. In the three countries, there is a growing importance given to molecular biology and real time PCR for diagnostics. Moreover, one can notice scientific and technological cooperation to build databases on pests, particularly in the frame of international organizations such as EPPO or through European projects.

The outputs of the fieldwork can be synthesized as follows. Few interviewees have an explicit and direct opinion on the *concept* of"biosecurity". They all share a concern regarding plant health and crop protection but when asked to define "biosecurity", very few are able to give a simple definition and this definition is very variable. However, they are more familiar with the concept in the UK, where the "biosecurity related" literature is the most developed. The general assessment of the risk of successful intentional outbreak is generally very low: few interviewees actually believe such an attack would happen. In Great Britain, this assessment is a little higher, and related mostly to food biosecurity. Two main reasons are evoked to justify this idea that the risk is not likely: the "motive" is too ill-defined, and they cannot identify a "likely pathogen candidate"; attacking food or animals would be more "effective" for a terrorist according to them. Moreover, those specialists of plants show a general difficulty to deal with "intentionality" other than with common sense: "one never knows".

In the three countries, professionals express a general "feeling" of preparedness, despite a rather ill-defined conception of the meaning and consequent extension of biosecurity. According to them, from a technical point of view, intentionality "would

[3] This implied a bias in our study in the Italian case study, mostly in the region of Turin. Specialized in leafy vegetables (*e.g.* "Roquette" and ornamental flowers, which can affect the risk assessment concerning intentional release threats formulated by interviewees).

not change much" the responsive approach that should be taken and the probable slight differences with "regular outbreaks" (such as the distributional patterns of the pest) would not be of great consequence. Besides, interviewees show few insights, in general, on the changes in the "chain of command" that would be implied by intentional use of a pest or pathogens against crops. According to them, the normal chain of reaction would prevail and lead "naturally" to the appropriate response, with a nuance: there is a general feeling that Ministries responsible for internal security, rather than for agriculture, health or environment, would take the lead. But interviewees have no precise idea of how this would happen in a governance perspective.

Finally, preparedness still appears as an ongoing process. In the cases we studied, scientific regime changes can be considered mediated by "normal/regular science" progressively integrating diagnostic tools promoted by and developed thanks to biosecurity and security related agendas. Scientific domains contemporarily put under the umbrella of biosecurity (Plant pathology, entomology, plant epidemiology for instance) have gained or adopted a regulatory science dimension long before the present contemporaneous biosecurity turn: they have long been involved in quarantine, bio-invasion and food chain safety oriented policy. They do not seem to have become "more" regulatory for that matter. Nevertheless, existing networks and professional specialties "translate" biosecurity agendas into their own categories of understanding, constraints, professional systems and methods (building lists, detection methods, etc.). As shown by our scientometric account, there is no radical shift in regulatory science – a "scientific revolution" – associated with biosecurity and concerns about agro-terrorist threats. But those agendas, promoted by various programs funded by national and international research agencies and institutions, will likely influence on the long run the general ecology of scientific domains involved in plant protection.

References

Barbier M, Prete G (2010) Un regard sociologique sur la biopolitique des maladies émergentes et réémergentes. In: Barnouin J, Sache I (eds) Les maladies émergentes- Epidémiologie chez le végétal, l'animal et l'homme. Quae, Paris

Barbier M, Boumrar J (2011) Entre risque hybride et légitimation d'une politique de sécurisation. Eléments pour une compréhension du risque agro-terroriste et de sa gestion. In: Valoriser l'intégration du risque. L'Harmattan, Paris

Boumrar J (2013) Les spécificités des coopérations et de leurs modalités de mise en oeuvre dans le cas de la construction et de la gestion simultanées d'un risque hybride et potentiel : une étude empirique exploratoire sur le terrorisme biologique. UVSQ

Chen C, Cribbin T, Macredie R, Morar S (2002) Visualizing and tracking the growth of competing paradigms: two case studies. J Am Soc Inf Sci Technol 53:678–689

Chubin DE (1976) The conceptualization of scientific specialties. Sociol Q 17:448–476

Cointet J-P (2009) Dynamiques sociales et sémantiques dans les communautés de savoirs – Morphogenèse et diffusion. Ecole polytechnique

Collier SJ, Lakoff A, Rabinow P (2004) Biosecurity: towards an anthropology of the contemporary. Anthropol Today 20:3–7

Dobson A, Barker K, Taylor SL (eds) (2013) Biosecurity: the socio-politics of invasive species and infectious diseases. Routledge/Taylor & Francis Group, London/New York

Evans HC, Waller JM (2010) Globalisation and the threat to biosecurity. In: Strange RN, Gullino ML (eds) Role of plant pathology in food safety and food security. Springer, Dordrecht/New York

Fletcher J, Bender C, Budowle B et al (2006) Plant pathogen forensics: capabilities, needs, and recommendations. Microbiol Mol Biol Rev 70:450–471

Hughes TP (1986) The seamless web: technology, science, etcetera, etcetera. Soc Stud Sci 16:192–281

Kuhn T (1970) The structure of scientific revolutions. University of Chicago Press, Chicago

Latxague E, Sache I, Pinon J et al (2007) A methodology for assessing the risk posed by the deliberate and harmful use of plant pathogens in Europe. Bull OEPP/EPPO 37:427–435

Madden LV, Wheelis M (2003) the threat of plant pathogens as weapons against us crops. Annu Rev Phytopathol 41:155–176

McLeish C, Nightingale P (2007) Biosecurity, bioterrorism and the governance of science: the increasing convergence of science and security policy. Res Policy 36:1635–1654

Miller SA, Beed FD, Harmon CL (2009) Plant disease diagnostic capabilities and networks. Annu Rev Phytopathol 47:15–38

Morris SA, Van der Veer Martens B (2008) Mapping research specialties. Annual Rev Inf Sci Technol 42:213–295

Rappert B (ed) (2007) Technology and security: governing threats in the new millennium. Palgrave Macmillan, Basingstoke/New York

Rappert B, Selgelid MJ (2013) On the dual uses of science and ethics: principles, practices, and prospects. ANU E Press, Acton

Rappert B, McLeish C (2014) Web of prevention: biological weapons, life sciences and the governance of research. Earthscan, London/Sterling

Sprinkle RH (2003) The biosecurity trust. Bioscience 53:270–277

Stack JP, Baldwin W (2008) The need for secure communications networks and global connectivity. In: Gullino ML, Fletcher J, Gamliel A, Stack JP (eds) Crop biosecurity. Springer, Dordrecht, pp 103–109

Strange RN, Gullino ML (eds) (2010) The role of plant pathology in food safety and food security. Springer, Dordrecht/New York

Suffert F, Latxague É, Sache I (2009) Plant pathogens as agroterrorist weapons: assessment of the threat for European agriculture and forestry. Food Security 1:221–232

Tancoigne E, Barbier M, Cointet J-P, Richard G (2014) The place of agricultural sciences in the literature on ecosystem services. Ecosystem Serv 10:35–48

Waage J, Mumford J (2008) Agricultural biosecurity. Philos Trans R Soc B: Biol Sci 363:863–876

Wheelis M, Casagrande R, Madden LV (2002) Biological attack on agriculture: low-tech, high-impact bioterrorism. Bioscience 52:569

Zinsstag J, Mackenzie JS, Jeggo M et al (2012) Mainstreaming one health. EcoHealth 9:107–110

Chapter 15
Developing an International Communications Network: The PlantFoodSec Model

Anita Kocic, Marta Bonifert, Alessandro Bertin, Paola Colla, and Maria Lodovica Gullino

Abstract Dissemination, awareness and communication on plant and food biosecurity are important issues among stakeholders and the general public. The development of a communication plan was an essential tool to achieve the goal of awareness. It took into account specific communication aspects and was defined by the general objectives, a specific approach for internal and external correspondence, distinct tools and techniques as the main means of communication, an individual visibility path, a definition of target groups which eventually determine the communication network, a timeline, budget considerations and the evaluation of the performed communication activities. All these aspects were significant to create the basis for establishing an international communication network. These elements of the communication plan are illustrated in this chapter in a practical way by examples of concrete actions of the PlantFoodSec model.

Keywords Biosecurity • Dissemination • Awareness • Exploitation public • Relations • Communication plan • Communication network • Communication goals • Communication tools • Project deliverables • Promotional material • Target groups • Visibility • Visual identity

---

A. Kocic (✉) • M. Bonifert
Regional Environmental Center for Central and Eastern Europe (REC),
Ady Endre ut 9-11, 2000 Szentendre, Hungary
e-mail: AKocic@rec.org; EDOFFICE@rec.org

A. Bertin
Spin-To, Comunicare per innovare, Via Roma 366, 10122 Torino, Italy
e-mail: bertin@spin-to.it

P. Colla • M.L. Gullino
Center of Competence for the innovation in the agro-environmental field (AGROINNOVA),
University of Torino, Largo Paolo Braccini 2, Grugliasco, Italy
e-mail: paola.colla@unito.it; marialodovica.gullino@unito.it

M.L. Gullino et al. (eds.), *Practical Tools for Plant and Food Biosecurity*, Plant Pathology in the 21st Century 8, DOI 10.1007/978-3-319-46897-6_15

15.1 Introduction

What shall we communicate, within which timeframe, to whom and in particular how can we communicate that in an appropriate way? To have these questions in mind is indispensable while developing a communication plan within EU funded projects and while establishing a professional network in a globalized world in the international sphere. An international communication network is essential to improve and maintain a good form of cooperation and to show the value of a specific research project.

The PlantFoodSec project had the goal to improve awareness of stakeholders and the general public in biosecurity issues, taking into account that information sharing is a critical issue as well as the balance between confidentiality and public access (dual use issues). It was part of a comprehensive strategy to enhance knowledge of target groups and to inform relevant stakeholders about expected results and outcomes. The results were split between the creation of a database of relevant stakeholders, interested in the themes of biosecurity, the use of an online registration for the newsletter and a press area. Another result was the creation of a virtually restricted online area with the purpose to support the network members to work together and to share information and documents in real time. Furthermore, it aimed to improve the communication within the research community and to enhance effective flows of information about goals and results achieved during the project. The improved knowledge on scientific networks for European citizens, the general public, targeted groups, end users and decision makers was considered as priority in this regard. Eventually, another result was the advanced collaboration of networks on a Europe-wide scale.

The specific objectives were split into the enhancement of tools for spreading excellence, exploiting results and disseminating knowledge through the engagement with stakeholders and the public at large. Improved communication was set as another objective through the development of better reciprocal knowledge and through the promotion of trust-building – both among consortium members themselves but also with similar scientific and technological networks. Furthermore, it was aimed to promote a mutual learning process and the systematic exchange of information and good practices for effectively using research findings. Another specific objective was to advance tools for promoting and sharing success stories and to encourage their submission by beneficiaries. Finally, it was aimed to highlight and promote achievements and take into account a variety of communication means and target groups as appropriate, thereby specifically referring to policy-makers, governmental bodies, interest groups, the industrial sector, the media and the public.

In the following chapter the authors describe the main ambitions of the international communication network on the basis of the PlantFoodSec model. It will be also described which communication tools were considered useful and necessary in this process to reach the main purpose of the project – namely the goal to improve awareness in stakeholders and general public in biosecurity issues. Furthermore, concrete actions to maximize the impact of communication efforts in view of increased visibility efforts will be touched upon. With the aim to visualize communication strategies in a more practical way, the authors present a comparison of common communication tools which are widely applied in research projects of the European Union (EU) Framework Program 7 (FP7), with the PlantFoodSec communication model.

The chapter will highlight the need for a communication plan in general as well as its objectives, the internal and external correspondence, specific tools and techniques, the visibility and visual identity, the timeline, budget considerations and the evaluation of performed communication activities.

15.2 The PlantFoodSec Communication Plan

The term communication defines the effort to disseminate a specific message to a specific target group, while using the most suitable communication tools (European Neighbourhood Info Centre 2013: 4). For communication in EU FP7 projects, a communication plan is the core element in developing an international communication network and supposed to be used as a guideline for all foreseen project communication activities. Communication within European research projects has the intention to expose instruments in which research is contributing to a European 'Innovation Union' (Communicating EU Research & Innovation 2012: 7). Its overall goal is to predefine the intention as precisely as possible, while all communication actions should be the supporting elements in reaching that intention (European Neighbourhood Info Centre 2013: 4). The objectives of the communication plan are indispensable and underpin the way of reaching the main goal. Whilst defining measurable and feasible objectives, the communication plan has the purpose to indicate a timely framework and available resources. However, every communication plan does have a flexible form and changes might be useful in the course of the project (European Neighbourhood Info Centre 2013: 11). It is mandatory for all projects, which are supported by the EU, to regularly communicate processes and activity flows and to appropriately and timely inform about possible discrepancies (EU Commission 2011: 8).[1]

The PlantFoodSec project in general contributed to a new awareness of the threat of biological weapons to agriculture, forestry, livestock and poultry and the preventive role that bioscientists can play. The project was implemented against the background that scientists have an important role to play in assessing the risk and in informing the public and policymakers concerning actions that can be taken to prevent biological warfare and bioterrorism and to minimize the impact of any use of biological weapons. The expected impact of "Network of Excellence" funding scheme within the European 7th Framework Programme is to increase the quality and impact of relevant training and research in Europe by bringing together the top

[1] A template for a communication plan from the EU commission is available at the following website: https://ec.europa.eu/europeaid/sites/devco/files/communication_and_visibility_manual_en.pdf

specialists and encourage the exchange of knowledge, development of new ideas and new trends in the respective area thus enhancing the innovation process. Therefore, the PlantFoodSec project communication plan was an imperative. It was designed and implemented in accordance with national and EU rules and guidelines with the intent to improve awareness of stakeholders at the global and EU level and the general public on the importance of agricultural biosecurity and its related issues. It had specific dissemination requirements which were considered throughout the implementation period. Besides, it had the final purpose to ensure the availability of the results achieved, whenever possible, to a wide audience of end users. It had moreover the principle to safeguard that policy makers are informed in such a way that non-proliferation agreements on bioweapons can be continually updated. It is important to bear in mind that a consistent message should be given to the media and the public. Therefore, the project utilized a database of experts to assure that both media and the public have access to the best available information. Eventually, it was implemented to strategically address end users and expand the network, with the aim of reaching the main project target – the creation of an enduring Network of Excellence.

This Network should move away from being 'virtual' to become 'real', to create a European Centre of Competence. The external factors that determine whether project impacts will be achieved, are mainly related to the interest of the end users. In the PlantFoodSec model, the role of end users was determined by strategic guidance. This guidance was guaranteed by the PlantFoodSec Security Panel members within the project Advisory Committee. The committee involved high profile scientists dealing with plant and food biosecurity and representatives of end users. Its main function was to strategically direct the work with discussions, proposals and comments.

In order to develop the communication plan, it was necessary to answer basic questions about its characteristics and the enhancement of communication with other scientific networks within the research community at the national and international level. Besides defining the target group, it was relevant to identify the ways of communication and to decide who from the consortium will establish the individual contacts. The latter is valid for the use of multiple strategies, different tools for different audiences, different approaches and structural set ups in different countries. Further on, it was necessary to define which concrete information should be shared with the target groups, how to present the shared information, how often to inform the target groups and how to receive their feedback. The plan considered earlier experiences and examples of best practices from partner countries and from other related projects in view of sudden outbreaks, such as *E. coli.*[2] All project bodies, namely the Steering Committee, the publication and dissemination board within the Steering Committee and the Project Security Panel, gave additional support to the communication plan, evaluated it whenever necessary during project meetings and suggested changes for improving its effectiveness.

---

[2] *Escherichia coli* – abbreviated as E. coli – are bacteria, which are harmless for humans. However, some bacteria strains can cause disease in humans. They can multiply in foods, the environment, and intestines of people and animals.

Suggested changes for improving the project effectiveness also came through the feedback from external reviewers who participated intermittently in consortium meetings. During the PlantFoodSec mid-term phase, the feedback on outputs was specifically directed to appropriate end users and for ensuring the Network of Excellence sustainability.

In a follow-up, the communication plan was adapted with a comprehensive communication strategy with particulars of relevant stakeholders. In addition, the planning of an extraordinary workshop in Brussels, Belgium, was initiated with the purpose to target EU representatives. Any other supplementary tasks, such as the development of additional and unplanned deliverables,[3] have been handled without difficulties, in line with the available budget and in agreement with the EC through increased efforts of the project staff.

Due to the nature of certain issues which were object of this project, a balance between confidentiality and public access to information had to be found and shared among partners. No classified outputs of potential dual use were disseminated within the WP7. It was indicated which of the data provided was subject to unrestricted accessibility and which was not. Sensitive information was not released inadvertently and the information contained on available websites was carefully validated. Data, that was deemed to be sensitive and whose accessibility should be limited to European Commission (EC) officials only, was clearly indicated.

This refers to the consortium procedures for disseminating information: partners had the chance to check whether the information is correct or perhaps confidential. The collected data was internally controlled and updated by all partners. The Security Panel members had a supervising role related to knowledge accessibility to third parties and classified and dual use issues, especially in relation with the production of deliverables and in terms of training activities. It considered suggestions from the coordinator and the group of experts of the Steering Committee. The publication and dissemination board[4] was the decision-making body regarding the level of confidentiality of information.

With regard to human resources of the PlantFoodSec communication component, it involved all consortium partners, including their respective staff and researchers. Professional communicators and communication experts have been engaged by the main actors of WP7. Above all, they helped to suitably frame all disseminated messages which drew the aimed attention and attracted the interest of the audience, while connecting them to what the audience wanted to know. The variety of communication methods and means put an emphasis on purposeful liaison with the target audience.

[3] What fell under additional deliverables is the production of a video, the publication of a novel in which PlantFoodSec was mentioned, a promotional activity on Italian Television, the publication of numerous press clippings and posts on Social Media.

[4] The project Steering Committee established a publication and dissemination board composed of scientists and led by the project coordinator. It was responsible for the authorization of publications in international journals and presentations at congresses.

Good management was ensured by having developed a separate dissemination and communication timeline with clearly defined intermediate stages and allocated financial resources right from the beginning. These aspects could guarantee a measurable evaluation of results. The timeline stated the beginning and the end of the project activities. Moreover, it included all relevant tasks and external as well as internal deadlines for all project related activities on the way and clearly stated the responsibilities for the realization of the tasks. The PlantFoodSec communication budget was defined accurately in the application phase of the project and it was defined realistically to reach the agreed communication aim and goals. The costs of the communication were not on one level during the whole project period. They were above average in the beginning for attracting possible stakeholders and also in the end for communicating the outcome and the benefits. The costs were as well above average during the organization of the three project workshops with external invitees and travels abroad.

One major part of the communication plan, namely the visibility and visual identity, is required for all projects which are funded by the EU (EU Commission 2011: 8).[5] Measurements for visibility are obviously aimed at emphasizing that the project is supported by EU funds. One concrete tool for this is the placement of the EU flag on a dominant spot in all project materials.[6] The EU is providing basic obligations and suggested materials in this respect (European Neighbourhood Info Centre 2013: 9 f.) and diverse EU websites provide various templates for this purpose.

When it comes to the creation and maintenance of the visual identity of a project, different tools exist whereby the logo has to be mentioned first. A logo is a powerful tool to increase the recognition value of the project. The PlantFoodSec logo was build up on the identity of a previous project, implemented between 2007 and 2010: the CropBioterror European Project.[7] Its image and standards assured a harmonized presentation of all project related activities and supported the development of a strong brand and identity of PlantFoodSec. The logo was conveyed through a number of channels: advertising, print collateral, website, HTML emails and PowerPoint templates. All these products displayed the required European emblem and the FP7 programme logo to give appropriate prominence. In addition, they specified at all time that the foreground was generated with the assistance of financial support from the European Union, hence, that PlantFoodSec received community research funding from the EU. It used the following corresponding sentence: "The research project receives funding from the European Union Seventh Framework Programme (FP7/2007–2013) under Grant Agreement n° 261752".

---

[5] There are some exceptions when the EU renounces the right of visibility. These apply for instance, if the visibility of the EU would hazard the objectives of the project.

[6] There are strict directives from the EU how to use their logo in any project. More detailed information is available at the following link: http://www.enpi-info.eu/enpi/userfiles/PB_Visibility%20EN.pdf

[7] "Crop and Food Bio-security, and Provision of the Means to Anticipate and Tackle Crop Bioterrorism" (Crop BioTerror) has been funded by the European Union under the 6th Framework programme.

15.3 Internal and External Communication – A Two-Way Approach

As it is particularly required to develop an effective, efficient and high-grade communication plan both within the project team as well as visibly with interested third parties, the communication plan is envisaged to be developed and seamlessly implemented during the lifetime of the project, ensuring a two-way communication – internally and externally. For the internal communication of research projects, it is necessary to keep the relevant persons informed, to respond to their concerns and to actively involve them in the project implementation. The external communication has the scope to serve the target audience with regular, meaningful and solid materials about the project, its progress and outputs.

For ensuring effective communication, the PlantFoodSec plan dealt with internal and external communication separately and it was regularly updated in the light of gained experience. On the basis of the communication plan implemented during the PlantFoodSec lifetime, specialized activities have been considered for implementation in the context of the project.

15.3.1 Communication Among PlantFoodSec Partners and Their Roles and Responsibilities

The internal communication objectives included the effective interchange of information between all project partners on general communication issues. At any time, the project partners were well informed about the actual status and project activities and communicated actively with other project partners in order to coordinate their communication related activities successfully. It was recognized that the success of PlantFoodSec depended on the commitment of every project participant.

The work of the researchers has broadened the boundaries of communication, giving the communications team new inputs about their achievements and expanding the overall newsfeed. Accelerated research results and increased levels of activity have meant more opportunities to keep in touch. The dissemination, awareness and communication team was keen to maintain constant progress on all communication levels.

The general activities of the internal communication focused on the update of project developments and the development of newsletters, leaflets and posters. Comprehensive communication guidelines and internal progress reports have been produced on a regular basis. Communication sessions have been held during the project meeting events to summarize the advancement and status of the work in a 6-monthly period between two project meetings. The most effective means of communication for PlantFoodSec were continuously suggested, as the consortium was provided with updated information at each project meeting. The progress of any dissemination activity from any partner, such as publications or participation in events in which the project was mentioned, was reported by the task leaders and the respective partners. This dissemination activity reporting was particularly important for the daily management and regular update of the EU Participant Portal.

The Regional Environmental Center for Central and Eastern Europe – REC (www.rec.org) acted as communication leader and ensured the compilation and coordination of provided inputs, materials and information, such as research results, lessons learnt and perspectives from consortium members. The REC furthermore coordinated the communication on project deliverables within the consortium and ensured the scientific editing of texts provided by the project partners. Additionally, REC hosted the launch workshop in its Head Office premises in Szentendre, Hungary, and provided logistic assistance to other meetings and events. SPIN-TO, a private communication and PR agency based in Torino and Milan, Italy, (http://www.spin-to.it/eng/) ensured the technical part, such as the graphic design of the dissemination material and was responsible for the website development and management as well as media relations. Therefore, REC and SPIN-TO, with a strong background in training and communication, were main actors in planning and implementing the communication component but also on spreading project information.

Subcontractors worked on psychological aspects of communication and provided structured and strategic approaches to dissemination, especially taking into account consumers and the general public. When it comes to the direct communication with the EC, UNITO as project lead partner had an overall responsibility for the project and monitored all communications activities. Its representatives were responsible for the direct and sole communication with the EC.

All project partners have signed a Consortium Agreement before the start of the project, setting the principles of the consortium management and placing the relationship between the partners and their roles and responsibilities on a legal basis for the duration of the work. The Consortium agreement among the partners covered aspects related to patenting as well as all other matters regarding the exploitation and dissemination of the results of this project. The EC was the governmental instrument for decisions on these subjects. In order for partners and third parties to be able to exploit the outcomes of the project, the results deriving from it were clearly described by measurable elements and completely documented by the coordinator on a regular basis in form of progress reports which allowed for comprehensive ascertainment of their applicability.

15.3.2 Targeting the External Audience

Having described the PlantFoodSec internal communication which led to the production of project deliverables and to reaching the milestones, the external communication needs special attention, as it attracted the interest of the outside audience and potential partners for future collaborations.

Regarding the external target group in EU funded research projects, it is fundamental that the target group is provided with professional, newsworthy and accurate information about the project in a way that explains the project in an easily digestible way for the general public. It is strategically important to initially analyze an accurate and deliberate priority audience which should be addressed with key messages. While defining the key message for the targeting group, it is crucial to reinsure that all relevant parties are receiving the required information at a particular time. In addition, their current awareness level has to be clear, namely their knowledge and attitudes towards the thematic issue. The messages should thereby vary according to the audience and its awareness level. It should be defined whether the specifically adapted and correctly addressed messages are trying to educate or to change the audience or whether it is aimed to mainly inform about the current developments.

The PlantFoodSec external communication objectives focused on the promotion of effective dissemination of news and information relative to the development of the project. It was aimed to encourage active participation of researchers, policy makers and all other interested third parties in the project initiatives, such as workshops and dissemination events, in order to guarantee an exchange of experience and knowledge, enhance the quality of project's activities and disseminate the project results. Further, it was intended to support and enlarge the project network of end users. The end users have been continuously contacted and well informed about the ongoing project activities as well as the results and benefits of the project with the goal to assure continued interest in the project.

Another essential feature of activities, aimed at spreading outputs externally, was the educational program for researchers and other key staff. This function had a special connection with the communication activity and built on synergies, as the consortium members had recognized the parallels between scientific communication and academic training, and that the steady supply of skilled staff was indispensable for the sustainability of European excellence in biosecurity. While the exchange activities were directed to the consortium personnel and aimed at harmonizing expertise across the network partners, other training activities in the frame of educational programs, academic courses and summer schools were open to external participants. They ensured a transfer of the existing knowledge within the network and communicated it outwardly. In view of the existing lack of capabilities in crop and food biosecurity field, the training component was an essential feature of the network and the increase of awareness and knowledge.

15.4 Means of Communication – An Overview of Specific PlantFoodSec Techniques and Tools

Whichever purpose the disseminated key messages might fulfill, the communication techniques and tools represent the way how to promote the topic externally. Tools can be distinguished between those which focus on creating and maintaining the visual identity and visibility, others which are useful to inform and to raise awareness, and tools that are eventually needed to establish and maintain the media relationship.

As far as PlantFoodSec techniques are concerned, all tools were designed to formulate and to preserve the project identity while raising awareness with all identified stakeholders. Enhancing channels of communication within the PlantFoodSec network was one of the main goals right from the early stages of the project development. Hence, the team developed various ways to address the different target groups on the subject of plant and food pathology under the WP7.

The setup of a database with tangible food security contacts and more than 1600 entries has served as a platform for identifying stakeholders and developing a communication network. The stakeholders were well defined with a relatively homogenous group of people which was further specified. Besides, the database with direct audience and intermediaries helped to extend the network significantly.

During the organized workshops for end users, these contacts were crucial. Moreover, they were given the change to receive regular project updates via the project mailing list. The database was constantly updated as the project proceeded and the number of contacts has risen, the more connections were part of the network.

The database distinguished several national and regional target groups and their representatives for PlantFoodSec as shown in the list below:

- The scientific community in general
- Epidemiologists
- Farmers
- Plant industries
- Food industries
- Agroindustries
- Private entities/institutes/diagnostic labs
- Retailers
- Manufacturers and distributors
- Health experts
- Chief plant health officers in Europe
- Policy makers
- Regulators
- Public health sector (Ministry of Health)
- Ministry of Environment

- Specialized state agencies which are in charge of emergencies, local and national laboratories
- National and European level authorities responsible for plant health and for security (end users)
- Ministry of Agriculture
- Ministry of Interior
- Extension educators
- National Plant Protection Organizations (NPPOs)
- Non-Governmental Organizations (NGOs)
- Public (citizens)
- University and PhD students
- Other researchers

15.4.1 The Project Website, Newsletters and Leaflets

PlantFoodSec tools that fell among the traditional dissemination of project deliverables were agreed and defined as tasks in the submitted proposal and comprised the preparation of various dissemination materials exclusively in English language. In the 5 years project duration, these included the development of an internet website, three flyers, ten editions of the digital 6-monthly newsletter, five press releases and the preparation of up to fifteen publications in the form of PhD thesis and scientific papers, as well as articles, posters and technical communications.

To start the very first PlantFoodSec dissemination activity with a tangible communication gesture, a press conference was organized right at the project kick-off meeting. It was attended by EC representatives and the local and national press. Focused actions with a selected number of European media took place regularly in view of major events during the project progression. All coming press releases, which were issued during the project implementation, lead to a considerable number of articles in national and international media and websites. The press releases served mainly to attract media attention to significant events and publications. A full press review is available on the project website (www.plantfoodsec.eu).

The project website was developed and constantly updated with a restricted area for project participants and provided information about the project itself, significant updates and the progress of the results. It played a major role and was created in order to respond to two specific objectives: firstly, to be a clear, user-friendly instrument to communicate and disseminate to a general public and stakeholders the results and the activities related to the project and secondly, to be a useful working instrument for the Partners. It has evolved into an important platform for disseminating general information such as activities, the project progress and all produced deliverables. Likewise, it was an online repository for the project documentation. It is worth mentioning that the website was put online simultaneously with the kick-off meeting in February 2010 which was earlier than initially planned in the proposal. Hence, the project had right from the beginning a platform to announce news and events. In addition, the website included a public area containing news, publications, events, press reviews and a description of the project's main contents, aims and involved partners. It entailed a 'partners only' reserved area, which was accessible with a username and a password, divided in one shared area for general documents and specific folders for each WP to improve a better organization of the documents and to ensure intra-partnership communication. Finally, the website included a press area, which was dedicated to journalists.

As per the technical features, it built on an owned and 'tailor made' Content Management System (CMS) which used specific technologies. The digital newsletters and web portal news helped the audience to keep regularly up-to-date with the general progress of the project and its results. The leaflets offered a concise and visually attractive way to disseminate PlantFoodSec to a wider audience, mainly during promotional activities. Relevant technical outcomes deriving from the project action were published in international scientific journals.

15.4.2 Workshops and Dissemination Events

The organization of workshops and conferences was another technique of the communication plan to engage stakeholders into a dialogue. The three workshops were organized at regular intervals and at key stages of the project, subsequently followed by the workshop proceedings. Whenever appropriate, external experts and stakeholders have been invited to participate. The launch workshop was scheduled along with the second consortium meeting in 2011. Besides discussing relevant scientific and technical aspects, the workshop offered a podium to deal with initial set-ups, such as setting the rules and procedures or introducing the project bodies.

The next and extraordinary "Roundtable on Plant and Food Biosecurity" in Brussels in 2013 was a chance to gather and present mid-term results to stakeholders. It aimed to underline the importance of research in plant and food biosecurity at the European level and was an outcome of the external reviewer's suggestion to adapt the communication path. High-level participants, such as national policy makers or MEPs, speakers coming from European Union institutions (Directorate-General Enterprise and Industry, Joint Research Center, Directorate-General for Health and Consumers, Directorate-General for International Cooperation and Development, Directorate-General for Research and Innovation) and other organizations (United Nations Industrial Development Organization UNIDO, European CBRN Centre[8]) actively promoted the status and purpose of the research activities.

The mid-term workshop took place in the scope of the 10th International Congress of Plant Pathology (ICPP) in Beijing, China, in 2013. The ICPP (http://www.icppbj2013.org/) is organized every 5 years and provides a forum for the

[8] The Center initiates and coordinates research, training and education within the field of safety and security, more specifically dealing with chemical (C), biological (B), radioactive (R), nuclear (N) and explosive (E) materials.

presentation and dissemination of the most recent advances and developments in plant pathology, with the aim of promoting international collaboration among researchers from different countries and regions. It represented a unique opportunity for researchers from all over the world to meet and discuss the latest research results. The theme for the 10th congress, was "Bio-security, Food Safety and Plant Pathology: The Role of Plant Pathology in a Globalized Economy". Selected PlantFoodSec partners contributed to the concurrent session "Plant Food Security: A Network of Excellence on Biosecurity", by presenting six lectures and addressing various topics ranging from international cooperation and communication, to the identification of deliberate contamination, platforms for collaborative diagnostics, and the misuse of scientific research. A presentation was held on "Plant and Food Biosecurity and Communication – Fostering Collaboration from Virtual to 'Real' Networks". The congress provided a global forum for introducing the PlantFoodSec model as international communication network. The project was additionally featured at daily poster sessions. The closing workshop was organized in the light of the final consortium meeting in Brussels in 2016 and provided the opportunity to present ultimate results and outcomes. The choice of Brussels as the venue for the concluding event was a strategic move, with the main goal to present to high-level EU stakeholders what has been achieved during 60 months of PlantFoodSec implementation in the FP7 framework.

The establishment and fostering of professional networks goes hand-in-hand with the participation in other international dissemination events on biosecurity. Therefore, the active attendance in topic related meetings played an important role in the approach to exchange scientific and technical project experiences. Many PlantFoodSec partners were involved in other ongoing (EU funded) research projects and active in various research networks and user groups that deal with agriculture, food and biosecurity. The character of these events can be classified in different groups: major international events, meetings of other thematically relevant projects or restricted EC meetings. The project representatives participated either with the purpose to disseminate promotional material or in the capacity as speakers. The identification of synergies and possible collaboration with these projects and networks was continuously realized and enabled wide-ranging dissemination throughout different communities of scientists and users. Hence, PlantFoodSec partners followed, whenever considered useful, meetings, workshops and conferences and ensured a direct contact with the external audience. These networking activities have taken place both in and outside of Europe and project findings were disseminated to the scientific community with papers which have been published in peer-reviewed journals. Any dissemination activity of this kind was reported on the Participant Portal of the EC and included sufficient details and references to enable the Commission to trace the activity in due time following the publishing date.

In view of measuring the communication efforts and success with regards to dissemination of deliverables, the register below shows the number of developed products and performed activities:

- 10 consortium meetings with internal, face-to-face dialogues and group discussions
- 20 peer-reviewed publication quoting the project, 33 papers in proceedings of scientific conferences, 3 articles in edited books in scientific press at the project end
- More than 30 oral communications about the project in several scientific congresses
- More than 150 press clippings (web and print and articles) published in the general press
- 7 Press releases
- 6 external dissemination events and exhibitions on biosecurity and research
- 10 Newsletters
- 3 Leaflets
- 1 roundtable with high-level participants from the EU
- 1 promotional activity on Italian Television
- 1 Video developed by the DG ENTR
- 3 Workshops (launch, mid-term, closing)
- 2 Posters
- 1 Project website, including a public area, "partners only" reserved area, press area

15.4.3 Additional Communication Tools

Additional communication tools have been initiated, mostly to strengthen the visibility through other media channels. The fact that PlantFoodSec was targeted by the European Commission's DG Enterprise and Industry for the video "A Safe Future for All – 2011"[9] proves that the project's importance and added value in the field of biosecurity research was well recognized by the EU.

The project was also introduced during an interview broadcast on Italian television. "The Invisible War", a novel about bioterrorism by E. Accati and M.P. Simonetti, Lineadaria Editore, Biella (IT), 2012, is another example of additional tools. The innovative book, a rare thriller based on the theme of agriculture, is part of the dissemination activities of the PlantFoodSec project and available in English and Italian language. The book can be seen as a valuable tool for raising awareness of bioterrorism while important information is conveyed in an accessible and exciting way. Besides, PlantFoodSec events were mentioned on the institutional websites of the consortium partners and thereby disseminated through channels of their network.

In the past few years, Social Media (SM) has become an increasingly vital communication engine. The original PlantFoodSec communication plan, as submitted in the proposal in 2010, did not foresee the set-up of SM accounts. However, it was

[9] The video is available online at: http://ec.europa.eu/enterprise/policies/security/videos/index_en.htm

possible to catch up with the trend and to ensure that PlantFoodSec was mentioned in Twitter accounts of a number of other current projects as well as in major EU, institutional and personal accounts (e.g. @EU_H2020, @AgriNewTech, @spinto_it,@GullinoL, @FoodSecurityUK, @FeraScience, @unito, @LIFE_Programme, @EUEnvironment, @EU_Commission, @agroinnova, @EmphasisProject).

15.5 Final Remarks

The quantitative and qualitative evaluation of the communication activities has various benefits and is required for all projects in the FP7 project framework of the EU. Besides justifying the activities externally, the evaluation helps to measure the internal process of the project. It gives an overview by analyzing weaknesses and discovering possibilities for improvement. Identified indicators and signs give possibilities to evaluate the entire communication plan and particular communication efforts.

In the PlantFoodSec model, the impact of the communication was measured, among others, based on the identified and established contacts and their integration into the project implementation. The performance indicators concentrated on the number of organized workshops and participations in scientific external events on biosecurity, hence, the number of direct contacts with the external audience which eventually lead to increased networking activities. Further on, the indicators were based on the knowledge management and the communication inside the network in view of the methodology for data classification and the developed tools and databases. Finally, the overall PlantFoodSec success was evaluated by measuring its impact in the biosecurity field.

Various activities brought the research to the attention of as many relevant people as possible. The examples of interpersonal and mass media communication affected numerous dialogues and the project offered a wealth of opportunities for productive discussions at individual level. They enhanced channels of communication within the project network and were spreading excellence, exploiting results and disseminating knowledge. In terms of communication this means that partners reached out the project documentation as far as possible. The characteristic of the network allowed an easy and quick transfer of results both to end users and to policy makers. Especially the fact that the partners had long-lasting professional links with relevant institutions helped to endeavor and carry out their part of the dissemination activities in a viable way. The PlantFoodSec voice was represented in a widespread way and had an overall impact on three continents worldwide. Due to the selection and integration of non-European countries, PlantFoodSec could outreach in Europe, Asia (with the partners from Turkey and Israel) and North America (with partners from the United States). The close and smooth relationships through clarified roles among the partners was a 'hidden', yet major aspect for the success of PlantFoodSec.

The appeal of the website design as promotional tool played a role in captivating the website visitors. It helped to measure the communication efforts and impact from a statistical point of view. According to the visitor statistics, the website has been visited by an international audience from more than 30 countries worldwide. Altogether, it was followed by 943.847 accesses, 19.777 different visitors and with 673.562 pages visualized (update January 8th 2016).

Much new energy was injected by strategic planning of the workshops which added value to the status and purpose of the project research activities. Especially the final activities strived to ensure that the PlantFoodSec voice is even more confident. A tangible gesture was, for instance, to organize the closing consortium meeting vis-a-vis the final workshop for presenting "Tools for plant and food biosecurity" in Europe's capital Brussels in the last month of the project implementation. Targeting directly EU representatives highlights the fact that a proper validation and reputation was given to the project.

The overall impact of the PlantFoodSec model is the creation of a new international communications network. The early development of the PlantFoodSec communication plan contributed to the establishment of focused relationships with the scientific community in its broader sense and identified proper interactions. The plan raised awareness on risks that plant and food systems were facing in Europe during the project implementation and beyond. It showed that the partnership had the competences to name and suggest solutions for handling these risks and that the consortium was aware of specific research needs and topics which required further development in the future. The overriding requirement for an effective response to agro terrorism emergencies predicates links and effective communication among all players, and their linkage to the existing national preparedness and emergency response system. In this sense, PlantFoodSec closed some of the research gaps while having created associations between the different sectors of biosecurity and by having showed the need for the continued building on increased joint research efforts.

Since the final outcomes included a decision-making tool to be used by law enforcement offices, risk assessment tools and a database of international expertise, the project identified priorities for research and regulatory policy and reached stakeholders for raising the awareness of policy makers. The identified research needs moved the project forward and the momentum of PlantFoodSec was kept to concretize a European Centre of Competence on Plant and Food Biosecurity. By that, it manifested the awareness that communication is a continuous process even once the project has ended. PlantFoodSec provided timely scientific inputs and its products are available for a long-term use to any European and or even global stakeholder that deals with agro terrorism. The sustainable communication has therefore been guaranteed.

The positive impact of the PlantFoodSec communication component, as part of European research projects, demonstrated ways in which research is contributing to a European 'Innovation Union'. It provided tangible proof that collaborative research adds value by showing how joined international efforts under the European guidance have achieved more than would have otherwise been possible. This notably applies in scientific excellence, the contribution to competitiveness and a better use of the results. It was made sure that these efforts are taken up by decision-makers to influence policy-making. They are taken up by industry and the scientific community to ensure continuity through follow-up. The coherent implementation of research activities has to be seen in the light of linking science (researchers) and politics (government representatives) under a common technical reference system, such as a stakeholder platform. An ultimate goal would be the inclusion of agricultural biosecurity as a priority in future EU work plans.

References

Communicating EU Research and Innovation: A guide for project participants (2012) Europe. Available online at: https://ec.europa.eu/jrc/sites/default/files/minisites/euradiss-2012/downloads/20121025-euradiss-eu-guide.pdf

EU Commission (2011) Financing agreement: annex I – general conditions: project/programme financing. Available online at: https://ec.europa.eu/europeaid/sites/devco/files/general-conditions-financing-agreement-2012-edf_en.pdf

European Neighbourhood Info Centre (2013) Visibility and communications for EU-funded Projects: frequently asked questions

FP7 in Brief How to get involved in the EU 7th Framework Programme for Research (2007) Available online at: https://ec.europa.eu/research/fp7/pdf/fp7-inbrief_en.pdf

Chapter 16
Training and Knowledge Transfer to Support Strong, Comprehensive National and International Plant Biosecurity Preparedness

Christine M. Henry, Charles R. Lane, Paola Colla, and Maria Lodovica Gullino

Abstract **Plant biosecurity** has been defined as the protection of natural and managed plant systems from the emergence/introduction of pests that would negatively affect the productivity, sustainability or diversity of plant systems. Threats to plants may be as a result of accidental or deliberate introduction of a plant disease/pest with the goal of reducing productivity or as a result of an introduction of contaminants that would render the resultant produce unusable. Experience from previous outbreaks and incidents demonstrated that it is essential that a high state of readiness is maintained to enable countries to deal with outbreaks in crops. Training and routine exercises are the key to maintaining overall preparedness. The chapter describes some experiences and hints for future training programmes based on the experience of the authors particularly in the EU PLANTFOODSEC Network of Excellence.

Keywords Training • Student • Academic • Skills • Plant pests • Plant disease • Human pathogen • Mycotoxin • Food security • Agriculture • Biosecurity

---

C.M. Henry (✉) • C.R. Lane
Fera Science Ltd, York, UK
e-mail: christine.henry@fera.co.uk

P. Colla • M.L. Gullino
Universita Degli Studi di Torino, Torino, Italy
e-mail: paola.colla@unito.it; marialodovica.gullino@unito.it

M.L. Gullino et al. (eds.), *Practical Tools for Plant and Food Biosecurity*, Plant Pathology in the 21st Century 8, DOI 10.1007/978-3-319-46897-6_16

16.1 General Introduction

Biosecurity was originally defined as a strategic and integrated approach that encompasses the policy and regulatory frameworks (including instruments and activities) that analyse and manage risks in the sectors of food safety, animal life and health, and plant life and health, including associated environmental risk. Biosecurity covers the introduction of plant pests, animal pests and diseases, and zoonoses, the introduction and release of genetically modified organisms (GMOs) and their products, and the introduction and management of invasive alien species and genotypes (FAO 2003).

However, this definition has become modified over recent years to encompass issues surrounding bioterrorism, or its sub-category agroterrorism, and is being used in common parlance to include bioterrorism e.g. Wikipedia adds that – 'These preventative measures are a combination of systems and practices put into its place at bioscience laboratories to prevent the use of dangerous pathogens and toxins for malicious use, as well as by customs agents and agricultural and natural resource managers to prevent the spread of these biological agents'.

The concept of bioterrorism is not new but in 2001, a wake – up call was sounded when *Bacillus anthracis* was disseminated through the U.S. postal system in a deliberate attempt to harm humans. (Jernigan et al. 2001). In the future, terrorists may resort to non-conventional means such as biological materials, which have the capacity to infect thousands of people, destroy agriculture, infect animals and crops and poison food. The deliberate release of a biological agent or chemical either against livestock/crops or into the food chain for the purpose of undermining political/economic stability and/or generating fear has been defined as agroterrorism. As scientific techniques have become more sophisticated, including the advent of genetic modification, the scope for producing novel biological weapons has increased. In addition, there is now heightened awareness of a possible agroterrorism attack and an onus on governments to implement measures to protect public safety and national food supplies.

The risk of "bioterrorist" attacks is low, but the consequences of such acts could be devastating. An accidental *E. coli* O104 outbreak in May 2011 claimed 49 lives and infected 4180 people in France and Germany, as well as causing economic losses in the horticultural produce industry of £54 million in the UK, 200 m euros(€) a week in Spain, 80 m€ a week in the Netherlands, 30 m€ a week in Germany and France, 4 m€ a week in Belgium and 3 m€ a week in Portugal. The intentional or unintentional release of a biological agent need not target humans to disrupt our societies and harm our economies. In May 2005 a letter was sent to the New Zealand Prime Minister's Office claiming that Foot and Mouth Disease had been deliberately released on the island of Waiheke. Subsequent investigations, and a second letter, confirmed that the claimed release was a hoax but not before a bill had been accrued of at least $1.5 million excluding staff time and compensation costs (MAF Media Release May 2005). Perhaps not equally damaging, but nevertheless able to cause severe economic disruption, would be deliberate releases of a plant pest/

pathogen into agriculture or the environment could cause the destruction of crops, which are food sources, but could also cause interference in open trade across the EU and with non-EU trade partners. Oerke (2006) reported that losses in crops such as wheat and potatoes can reach 70 % where no crop protection control is applied. Attacks on iconic environments can cost countries both economic and social distress.

Thus, we need think of biosecurity as encompassing both the accidental and deliberate release of naturally occurring or genetically modified biological materials that cause damage to humans by affecting food safety, cause damage to animal life and health, and cause damage to plant life and health. The role of training and exercises is to support prevention, management and control of such releases that may damage the food production chain, agriculture and the environment.

16.2 Why Biosecurity Training Programs and Exercises Are Needed in Strong, Comprehensive National and International Plant Biosecurity Preparedness Programs

Experience from previous large scale outbreaks has shown that preparation is essential to the success of containing these. The recent EU report on the Ebola outbreaks in West Africa (2015/C 421/04) reiterated the need for the strengthening of expertise as regards the prevention of the spread, as well as control, management of serious cross-border threats and treatment in the fields of CBRN-E (Chemical, Biological, Radiological, Nuclear and Explosive) and natural and man-made disaster management, e.g. by expert networks as well as Europe-wide simulation exercises to test cross-sectoral coordination, including strengthening of risk-assessment and risk-management, of preparedness research and notably also with regard to diagnostic methods.

The management of disaster risks and crises of different kinds in the CBRN-E field is ruled by a number of international, EU and national policies covering various sectors (e.g. civil protection, security, health) and operational features such as preparedness, prevention, detection, surveillance, response, and recovery. A range of research and technological developments, as well as capacity-building and training projects, are striving to support the implementation of these policies. However, the complexity of the policy framework and the wide variety of project initiatives often leads to a lack of awareness about project outputs by the wide range of "end-users" (CoU 2015).

One of the main objectives of the EU funded PLANTFOODSEC project, as all other Networks of Excellence funded by the European Commission, was to tackle the problem of research fragmentation and to have a long-term integration effect on research into improving methods in the field of plant and food biosecurity.

16.3 What Key Issues Should Be Addressed and Why

It is essential that a high state of readiness is maintained to enable countries to deal with outbreaks in crops. Training and routine exercises are the key to maintaining overall preparedness.

It is important to train all of the actors who would be in the front line in the case of outbreaks. Ashford et al. (2003) emphasized that the most critical component for bioterrorism outbreak detection and reporting is training the frontline healthcare profession and the local health departments to recognize the possibility that the outbreak was not caused accidentally. This needs to be applied to the health of crops as well as people.

There is also a need for exercises, in order to plan and prepare for responses to outbreaks making sure that all participants understand their roles in the process and there is a joined up fast response.

Raising the overall appreciation of the potential threats in the minds of the general scientific and wider communities is also important so that they can act as a sentinel community, spotting any unusual occurrences that may be related to bioterrorism incidents.

Key areas to concentrate on are:

- Increasing awareness of biosecurity, dual use (potential use for both civilian and military applications and/or can contribute to the proliferation of Weapons of Mass Destruction (WMD)), etc. within the scientific, policy and inspection areas and also targeting agronomists and food producers who may be the front line in an outbreak.
- Training new entrants to the profession and established staff in new tools particularly with a view to their use in biosecurity, e.g. diagnostics, epidemiology, risk assessment, etc.
- Strengthening the Network (in its widest sense) by exchanges of staff and students between partners to increase the mutual understanding of working practices across disciplines and countries.
- Training in contingency planning for control of outbreaks of plant pests and diseases and human pathogens on plants.

To increase the level of awareness across a wide range of players in the EU PLANTFOODSEC project we used a range of types of media and events:

- Academic courses on dual-use consequences of bioresearch and on ethics of bioresearch (lectures in partner universities). The courses covered issues such as the implications of misuse of research results in relation to biological terrorism and warfare and professional responsibility as well as liability (professional code of conduct). These were given by experts as part of current academic courses so participants received course credits at the end of their courses. These had good audiences and had the merit of spreading knowledge to new entrants to the profession.

- Cross-border training/workshops (audience: diagnosticians, university classes) building on the pre-existing knowledge of the network participants. These were generally given as part of planned workshops or university courses held in a range of countries in Europe but also in Israel, USA and Turkey. They attracted audiences of interested participants in the chosen sector.
- Regular trans-national, multi-sector training courses on preventing, preparing for, containing, and responding to bioterrorism and/or naturally occurring disease outbreaks. These were held as lectures or workshops attached to regular meetings and often attracted a good audience as people were already committed to attending the core meetings.
- Training on diagnostic (laboratory and field level). These were a mixture of lectures on diagnostics as part of university courses, or as part of conferences/workshops already planned.
- Training on legislation and contained use licensing. Seminars were given to plant health inspectors. However, to reach a wider audience a video demonstrating the measures required to contain pests and pathogens was developed and will be freely available on the project website.

16.4 General Biosecurity Principles, Agreements and Legislation

Biosecurity requires a strategic and integrated approach to analysing and managing relevant risks to human, animal and plant life and health, and associated risks to the environment (FAO 2007). The original approach to biosecurity was planned to deal with accidental introductions of hazards such as biological, chemical or physical agents in food with the potential to cause an adverse health effect, plant, animal or pathogenic agents injurious to plants or plant products, plant pests of potential economic importance, unauthorised living modified organisms (LMOs or GMOs) or alien species likely to threaten biodiversity. This concept has now been extended to cover agroterrorism threats.

Various international agreements are in place to enhance biosecurity:

- The International Plant Protection Convention (IPPC) is an international plant health agreement, established in 1952, that aims to protect cultivated and wild plants by preventing the introduction and spread of pests. (https://www.ippc.int/). It reviews standards, develops international standards and resolves disputes etc. It liaises with the National Plant Protection Organisations.
- International Health Regulations (IHR 2005): Provides a framework for the coordination of the management of events that may constitute a public health emergency of international concern, and improve the capacity of all countries to detect, assess, notify and respond to public health threats.
- Cartagena Protocol on Biosafety (2000) under the Convention on Biodiversity: Ensures the safe handling, transport and use of living modified organisms

(LMOs- GMOs) resulting from modern biotechnology that may have adverse effects on biological diversity.

- The Agreement on the Application of Sanitary and Phytosanitary Measures (SPS Agreement) of the World Trade Organization: Protects against risks from additives, contaminants, toxins or disease-causing organisms in food beverages, feedstuffs, entry of plant or animal carried diseases, and so on.
- The Convention on Biological Diversity: covers the conservation of biological diversity, the sustainable use of its components, and the fair and equitable sharing of benefits arising from the use of genetic resources.

16.5 International Agreements and Initiatives Specific to Bioterrorism

In terms of weapons of mass destruction there is an overall international agreement covering weapons directed against humans:

The Convention on the Prohibition of the Development, Production and Stockpiling of Bacteriological (Biological) and Toxin Weapons and on Their Destruction (1972): covers the prohibition and elimination of all types of weapons of mass destruction, and convinced that the prohibition of the development, production and stockpiling of chemical and bacteriological (biological) weapons and their elimination.

Several international groups have been established to consider vulnerability to agroterrorism and measures that could be implemented to prevent an agroterrorism attack or mitigate the effects if such an attack occurred. One of the earliest groups to be established, the Australia group, was formed in 1985 and aimed to harmonise export controls so that certain precursor chemicals could not be traded and used to develop chemical weapons (Australia group website). Later their remit was extended to cover biological weapons and lists of controlled organisms and material were drawn up.

Following the destruction of the Twin Towers in 2001 in the USA, the US government expended a huge amount of effort on the development of its terrorism response (Ryan and Glarum 2008). In 2002, the USA approved Public Law 107–188: the Public Health Security and Bioterrorism Preparedness and Response Act. This Act directs the Secretary of Health & Human Services (HHS) to establish and maintain a list of biological agents and toxins that have the potential to pose a severe threat to public health and safety, so called 'select agents'. The Act also requires that all facilities and individuals in possession of select agents register with HHS. A similar programme was created to cover animals and plants, and was implemented through the Animal & Plant Health Inspection Service (APHIS). At that time the draconian terms of PL 107–188 had an unfortunate side-effect on research and other activities using the select agents, with some research projects being abandoned and stocks destroyed because facilities could not comply with the rigorous controls imposed. This illustrates the need for a balanced approach to legislation to prevent agroterrorism and the need for better training in use of biological materials in contained use.

In the US the Homeland Security Presidential Directive 5 (HSPD-5) issued February 28, 2003 requires that all federal departments and agencies adopt the National Incident Management System (NIMS) in their domestic emergency management. The NIMS is designed to provide a consistent nationwide approach to federal, state, and local governments to work effectively together to prepare for, respond to, and recover from domestic incidents, regardless of cause, size, or complexity, and this will include plant pest incidents if necessary.

As a response to the growing bioterrorism threat, in 2007 the EU initiated a debate on how to deal with the bioterrorism threat without duplicating the current frameworks already in place to deal with accidental releases of biological agents and toxins. This resulted in the publication of a "Green Paper on Biopreparedness". The European concept of biopreparedness covers a broad scope of activities relating to the protection of public health and wealth against all potential risks whether they arise from a terrorist attack, other intentional release, accident or naturally occurring disease affecting humans, animals or the food chain. This biological all-hazards approach encompasses all aspects of the prevention, protection, first response capacity, prosecution of criminals/terrorists, surveillance, research capacity, response and recovery. It also covers the steps necessary to minimise the threat of deliberate contamination of the food supply through biological agents and to protect against biological warfare.

The Green Paper on Biopreparedness acknowledged that the existing EU Directives were largely set up prior to 2001, and were aimed at prevention of natural infection processes or accidental contaminations/releases – not the deliberate release of pests and pathogens. It recommended that it is important that existing measures across the EU and its Member States (MS) are coordinated. To this end, the EU has carried out Europe-wide exercises under the Community Mechanism for civil protection assistance (Council Decision 2001/792/EC, Euratom).

Disaster risk and crisis management issues are covered in a direct or indirect way, either providing legally-binding frameworks of actions by EU Member States in the form of Directives, general frameworks in the form of Communications or technical specifications in the form of e.g. Decisions.

Additional legislation may be available under MS' counter-terrorism legislation, such as in the UK where the Counter-Terrorism Act 2008 refers to earlier terrorism acts, and the Biological Weapons Act 1974 makes it illegal to "use biological agents or toxins for hostile purposes". This gives a definition of "biological agents" as "any microbial or other biological agent", which would appear to cover plant pathogens if used for hostile intentions, i.e. deliberate introductions for the purpose of establishing and spreading serious plant diseases. An EU Council Framework Decision (2002/475/JHA) of 13 June 2002 on combating terrorism (EC 2002) includes the manufacture, possession, acquisition, transport, supply or use of "biological weapons", as well as research into, and development of, biological weapons, to be deemed to be terrorist offences, and therefore come under the powers of the Framework Decision.

Disaster risk and crisis management issues for the management of natural and man-made hazards are covered in a direct or indirect way at EU level, either in the form of Directives, of Communications or Decisions (CoU 2015). The policy is represented by the EU Civil Protection Mechanism (Decision No 1313/2013/EU of the European Parliament and of the Council), and the operational dimension is coordinated by the Emergency Response Coordination Centre. Disaster risk management is also addressed through the European Agenda on Security (The European Agenda on Security, COM (2015) 185 final) and the Decision No 1082/2013 on serious cross-border threats to health.

Finally, intergovernmental agencies are also involved in security policies, namely the European External Action Service (EEAS) implementing the EU Common Foreign and Security Policy, and Europol, which is the EU Law Enforcement Agency, both of them assisting EU Member States (CoU 2015).

16.6 Political and Regulatory Frameworks

Traditional systems where biosecurity is managed on a sector basis through separate policy and legislative frameworks (e.g. for animal and plant life and health, food safety and environmental protection) are still common (FAO biosecurity toolkit report 2007). A comparison of the EU, UK, USA, Israeli and Turkish plant health approaches suggested that Plant Health regulations are still separate from Food Safety regulations. Also separate criminal legislation is likely to be enacted if terrorism was suspected.

16.7 Issues Specific to Plants/Crops/Food Safety

Risk analysis is key to biosecurity- this should include assessment of deliberate release as well as accidental release.

Diagnostic methods and traceback methods are needed to confirm outbreaks and trace sources.

Eradication and control methods are needed to clean up after outbreaks.

16.7.1 The Main Current Issues for Plant Health Are

Increasing risk of spread of new plant pests (includes diseases) – globalisation of trade, climate change, new risk pathways.

Huge worldwide losses caused by pests -Food demand is expected to increase by 50 % but currently, a quarter of the world's crops are lost to pests.

- Changes in importance of existing pests – climate change, pesticide usage-increased pesticide resistance, reduction in pesticide availability, develop use of alternative IPM strategies, plant resistance breakdown
- Challenges in improving early detection and diagnosis
- Effective and rapid response
- Pressures on resources in government and industry
- EU Implementation of new legislation

16.7.2 The Main Current Issues for Food Safety Are

16.7.2.1 Mycotoxins

Mycotoxins are a particular contaminant of biological origin found in food and animal feeds. They are produced by a range of fungi, for example *Fusarium, Aspergillus, Penicillium, Alternaria,* some of which are plant pathogens. The toxins can cause significant human and animal disease particularly in developing countries. They are highly regulated, in the EU in particular, and this may cause distortion of trade. Conditions favouring mycotoxin formation may be more common in climate change scenarios -drought stress, insect damage, cold/wet harvests, etc. A lot is unknown about the potential consequences of mycotoxins e.g. – synergistic effects, toxicity of mixtures and long term exposure effects.

If mycotoxins are present in food products, their complete elimination is impossible. Despite the potential damage that can result from consumption of aflatoxins, fumonisins, and other dangerous mycotoxins, our knowledge of these compounds in food products is limited, and there are few regulations addressing minimal allowable limits of such toxins in food.

Immunological and HPLC-based methods are used for identification of mycotoxins but masked mycotoxins may elude analysis leading to underestimates of exposure. A lack of validated methods and occurrence data, e.g. for T2 and HT2 toxin and lack of maximum levels being set e.g. Fusarium toxin- nivalenol may limit enforcement detection.

16.7.2.2 Human Pathogens on Plants (HPOPs)

Healthy diets are increasingly featuring fresh fruits and vegetables that have not been cooked and the use of pre-prepared food (see Fletcher et al. 2016, Chap. 4 of this book). There is also an increase in global trade of these. This has increased the risk that outbreaks in food will occur and focussed attention on prevention of contamination. The most common HPOPs found in food outbreaks associated with plants/crops are pathogenic *Echerichia coli* strains (O157:H7, O26, O104 and O145), *Salmonella enterica* subp. *enterica* serotypes and *Listeria monocytogenes*. Also, complicating this picture, in the *E. coli* O104:H4 outbreak in the EU area, the unusual bacterial pathogens that were implicated had undergone horizontal gene transfer, resulting in unusually high virulence and evading detection.

As the food supply chain for Europe and associated nations is complex, often crossing national boundaries, strengthened channels for international cooperation and coordination will facilitate effective action before, during and after outbreak incidents. Traceback and containment measures cannot occur without rapid, reliable, easily interpreted and standardized analytical methods to detect, identify and discriminate among similar strains of these pathogens. Rapid molecular, methods and culture-enriched conventional analytical methods are commonly used but new methods using fast and whole genome sequencing (WGS) are being introduced.

Before WGS can be adopted globally for foodborne illness investigation, international standards (Standard Operating Procedures and/or quality thresholds below which data is not acceptable) must be implemented. The Global Microbial Identifier is currently working on proficiency testing for both data production and analysis. Additional needs include bioinformatics training for researchers and government employees, and improved analytical tools.

16.8 Current Training Initiatives and Programmes

Little of the training that happens worldwide is actually paid for by the beneficiary of the training. It is much more common for training to be funded by government or charitable bodies. Some examples of the current initiatives in training for biosecurity and related areas are given below:

16.8.1 EU Funded Initiatives

16.8.1.1 Training Programmes Directed at Improving Harmonization of Regulatory Programmes

<u>Better Training Safer Food initiative</u> (http://ec.europa.eu/chafea/food/about.html)

"Better Training for Safer Food" (BTSF) is a Commission initiative aimed at organising a Community (EU) training strategy in the areas of food law, feed law, animal health and animal welfare rules, as well as plant health rules. It supports the implementation of the EU regulations to ensure the verification of compliance with feed and food law, animal health and animal welfare rules (under Article 51 of Regulation (EC) No 882/2004 Regulation (EC) No 882/2004). It provides funding for training of e.g. inspectors. The priorities for funding are set at an annual meeting and tenders are then put out for suppliers of training. Priorities can be set by the EU Food & Veterinary Office (FVO) as a result of audits carried out in specific areas.

A separate initiative called BTSF World funds training for non-EU Member countries through tenders, specifically a call was issued in 2012- Organisation and implementation of training activities on improving and strengthening the sanitary and phytosanitary framework in non-EU Member countries ("BTSF World") under the "Better Training for Safer Food" initiative and covered training in plant and animal health and food safety in Africa, South and Central America, the Caribbean, Asia, the Pacific and Eastern European countries.

TAIEX (Technical Assistance and Information Exchange instrument of the European Commission) programme (http://ec.europa.eu/enlargement/tenders/taiex/index_en.htm)

The programme is largely directed at enlargement of the EU and provides training for countries that are hoping to join the EU. TAIEX supports public administrations with regard to the approximation, application and enforcement of EU legislation as well as facilitating the sharing of EU best practices. It is largely needs-driven and delivers appropriate tailor-made expertise to address issues at short notice.

Currently the TAIEX mandate to provide assistance covers: Croatia (new Member State – still benefitting from assistance programmed); Turkey, the former Yugoslav Republic of Macedonia; Montenegro, Serbia, Albania, Bosnia and Herzegovina and Kosovo; Turkish Cypriot community in the northern part of Cyprus; Algeria, Armenia, Azerbaijan, Belarus, Egypt, Georgia, Israel, Jordan, Lebanon, Libya, Moldova, Morocco, Palestine, Syria, Tunisia and Ukraine.

Twinning (http://ec.europa.eu/enlargement/tenders/twinning/index_en.htm)

Twinning is a European Union instrument for institutional cooperation between Public Administrations of EU Member States and of beneficiary or partner countries. Twinning aims to provide support for the transposition, implementation and enforcement of the EU legislation (the Union *acquis*). It builds up capacities of beneficiary countries' public administrations throughout the accession process, resulting in progressive, positive developments in the region. Twinning strives to share good practices developed within the EU with beneficiary public administrations and to foster long-term relationships between administrations of existing and future EU countries.

Twinning projects bring together public sector expertise from EU Member States and beneficiary countries with the aim of achieving concrete mandatory operational results through peer to peer activities. Current potential beneficiaries of this programme are: Albania, Bosnia and Herzegovina, Croatia, the former Yugoslav Republic of Macedonia, Kosovo, Montenegro, Serbia and Turkey, Algeria, Egypt, Israel, Jordan, Lebanon, Tunisia, Armenia, Azerbaijan, Georgia, Moldova, Ukraine, Libya, Morocco, Palestine, Syria, Belarus, and Russia.

EDES COLECAP project (http://edes.coleacp.org/en/edes/page/20334-methodological-approach)

The overall objective of the EDES project is to increase the contribution made by the food trade to alleviate poverty in African, Caribbean, and Pacific Group of States including to strengthen their national (or in some cases regional) policies in the area of food safety, in order to improve their food quality and adapt it to the international standards in force in importing countries and to consumer demands, with a view to encouraging significant growth in traceable and certified exports.

Bilateral training

Various types of specific bilateral training is carried out usually funded by EU or possibly other governments but these generally only fund travel and subsistence.

16.8.1.2 EU Exchange Programmes Aimed at Personal Development Covering Biosecurity Areas

Marie Skłodowska-Curie actions – Research Fellowship Programme (http://ec.europa.eu/research/mariecurieactions/index_en.htm).

The EU funded Marie Skłodowska-Curie actions support researchers at all stages of their careers, irrespective of nationality. The MSCA also support industrial doctorates, combining academic research study with work in companies, and other innovative training that enhances employability and career development. In addition to generous research funding, scientists have the possibility to gain experience abroad and in the private sector, and to complete their training with competences or disciplines useful for their careers. These are competitive and not targeted at biosecurity so provide mostly opportunities for academic progress.

The Erasmus Programme (http://www.erasmusprogramme.com/)

Erasmus is an EU exchange student programme that has been in existence since the late 1980s. Its purpose is to provide foreign exchange options for students from within the European Union and it involves many of the best universities and seats of learning on the continent.

16.8.1.3 Research Funding Covering Biosecurity Training

Horizon 2020 (https://ec.europa.eu/programmes/horizon2020/en/what-horizon-2020)

Horizon 2020 is the current financial instrument implementing the Innovation Union, a Europe 2020 flagship initiative aimed at securing Europe's global competitiveness. Its main focus is on driving economic growth and creating jobs. The main goal is to ensure Europe produces world-class science, removes barriers to innovation and makes it easier for the public and private sectors to work together in delivering innovation and improving the economy. As part of the projects funded under H2020 and the previous programmes (FP7, etc) training is considered to be an important function but is largely closely related to the techniques developed in the specific project. PLANTFOODSEC is an example of a Network of Excellence funded under the FP7 programme with the main aim of strengthening excellence by tackling the fragmentation of European research and where the main deliverable is a durable structuring and shaping of the way that research is carried out on the topic of the network, in this case biosecurity.

16.8.2 International Training Funding Sources

Food and agriculture Organisation of the United Nations (FAO) http://www.fao.org/elearning/#/elc/en/home

The FAO's mission on food security means that it has an interest in biosecurity. Its focus is less on biosecurity in terms of training although it offers courses on food safety (concentrating on Codex) and crop improvement (plant breeding) through its e-learning centre.

Gates Foundation (http://www.gatesfoundation.org/What-We-Do/Global-Development/Agricultural-Development)

The Gate foundation funds agricultural development efforts, primarily in Sub-Saharan Africa and South Asia and concentrates mainly on increasing farm productivity through better sustainable practices but concentrate less on biosecurity.

United Nations (UN)

The UN has funded training in specific area of biosecurity e.g. UNEP GEF fund supported training on compliance with the Cartagena Protocol for GMOs (http://www.unep.org/biosafety/Projects.aspx).

16.9 What Training Approaches Work Well, and Why

The development of a biosecurity education program needs careful thought from the outset to understand the desired outcomes of any piece of work commissioned. Broadly speaking, is the program to raise awareness or to develop skills? Although in practice it may be a combination of the two with the former leading to the latter. With clarity at the start, it makes measuring impact and success of any implementation more effective. The following step-by-step approach has been developed over several years of experience in running plant biosecurity education programs.

1. Decide if you are planning an engagement programme to raise awareness or to develop skills or combination of the two
2. Identify the audience – decide if the program is aimed at non-specialists (e.g. members of the public) or specialists who are looking for continuous professional development (CPD)
3. Carry out a training needs assessment-to identify skill needs gaps
4. Decide on the education level of the participants e.g. further or higher education, graduate, postgraduate etc,
5. Identify potential partners as-'co-design is the best strategy'
6. Think about how to deliver the training- decide on possible formats/platforms – face-to-face training, instruction manuals, e-learning such as webinars, or combination.
7. Decide if you need a competency assessment (formal examination, online assessment etc) or a formal qualification
8. Decide how are you going to measure effectiveness of the training-and if you are going to follow up on this.

From experience, there are also a number of 'Guiding Principles' that will help any education program be more successful.

16.9.1 Partnership Working and Motivations for Participation

Success of any education program can be measured, in part, by the level of uptake of the target audience and other interested parties. Therefore, the involvement of these groups at an early stage in the project is desirable. By working in partnership with these groups it ensures there is a good understanding from both parties about the motivation for participation. Why people participate is a fundamental question, most commonly a National Plant Protection Organisation (NPPO) or equivalent statutory body will have clear goals about strengthening plant biosecurity; whereas, the participants may have different reasons to participate. For example, in the UK, the NPPO organisation has worked with OPAL (Open Air Laboratories; http://www.opalexplorenature.org/treesurvey) to develop a tree health survey for citizens. OPAL wants to encourage members of the public to get more involved with nature and working outdoors. It is motivated in enabling people to connect with trees and understand why tree health is important – their value to ecosystem services and our own health and well-being. Whereas, the NPPO that is asking for help from citizens to survey for the presence of specific tree pests and diseases of current concern, many of which are not known to occur in this country. Therefore, if the activity was purely based looking for quarantine pests and diseases, members of the public could soon become disengaged as although they have spent time looking for these pests and diseases there would be no 'reward' for their efforts. Whereas, by working in partnership the survey has been designed to encompass a general overall assessment of tree health condition, survey for pests and diseases where there is a reasonable chance of finding them, and finally, a short list of quarantine pests and diseases. This particular plant biosecurity education program has been successful in not only raising awareness about plant biosecurity but also in gathering evidence about the distribution of plant pests and diseases. This example clearly demonstrates the value of working in partnership and understanding the motivations of both participants.

16.9.2 Mandatory or Voluntary Participation

The greatest uptake in any training program will always be if there is a mandatory/legal requirement in order to carry out a specific job. For example, an arboricultural contractor will have to demonstrate competency to use core machinery such as a chainsaw, woodchipper etc in order to stay in business. In the UK, this will be by attending and successfully completing a training course for the specific skill which will include an assessment and issuing of a certificate of competency. Agricultural consultants, in the UK, in order to provide advice concerning pesticides or fertilisers will have to demonstrate their competency by professional qualifications such as BASIS and FACTS (https://www.basis-reg.co.uk/). However, in order to practise as a plant health professional or working in plant biosecurity there is no such similar officially recognised demonstration of competency. Therefore, Defra recently launched a new voluntary Register for Plant Health Professionals with the Royal Society Biology (https://www.rsb.org.uk/careers-and-cpd/registers/plant-health-register) to help meet this need.

16.9.3 Resources Must Be 'RARE'

RARE is useful mnemonic when designing teaching resources to ensure they are:

Relevant – meet the needs of the audience
Accessible/affordable-resources are easily available and at a price affordable to the participants
Robust-materials can be used time and time again and without constant support
Engaging – must be enjoyable and participants find the experience rewarding

By ensuring any resources produced meet these 'RARE' criteria they have a much greater chance of being used on a regular basis. Although you may have produced the most comprehensive and technically challenging teaching resources, if they are not fit for purpose (because they are pitched at the wrong audience or require regular replenishment of expensive training materials that the host organisation cannot afford) they will not be deployed as desired.

Fig. 16.1 Images from the course on Fusarium and mycotoxins held at Fera in 2015, illustrating its diversity

16.9.4 Teaching Platform

The manner in which any teaching is delivered is an important consideration. Although face-to-face training has been the traditional mainstay of most education, participants are now familiar with the concept of e-learning as accessibility to computers and the Internet increases rapidly. E-learning has many benefits that include allowing people to fit training around existing commitments but also to learn at a pace appropriate to them. For example, within the EU Life + funded Observatree project (http://www.observatree.org.uk/), developing a network of trained volunteers to help protect the health of UK trees, eLearning has be used increasingly through the life of the project. In the first years, tree pest and disease recognition training was delivered face-to-face with good outcomes. However, as the project has developed and experience of the volunteer surveyors improved, this previous training has moved to webinars allowing face-to-face training to be used for more interactive skills-based workshops. The volunteers have welcomed this change as it allows them to review these complex training materials repeatedly, thus tailoring their learning to meet their own needs.

Practical training courses, especially where different disciplines can be brought together are valuable for building bridges between skilled personnel across the crop-food chain. An example of this was the course on 'Diagnosis of plant pathogens implicated in mycotoxins and chemical/immunological methods for mycotoxin detection' held at the Fera Science Ltd in York during the summer of 2015.

The course provided training on the identification and prevention of problems in winter wheat and maize crops especially concentrating on Fusarium ear blight of wheat. It also covered a day on regulations for mycotoxins, sampling and rapid test methods – especially those linked to Fusarium. The practical sessions were combined with lectures and discussion sessions providing background for the practical sessions and other relevant topics. One of the merits of the course was that it brought together a wide range of skills from agronomic to immunological and chemical detection. However, there was also sufficient time allocated to each topic to allow an in depth exploration of the topic (Fig. 16.1).

16.9.5 Staff Exchanges and Other Specific Training Approaches (e.g. Interns)

There is also a role for longer term training opportunities such as staff exchanges between country labs/institutes where a more in depth understanding of the approach to biosecurity in different countries including diagnostic methods, control methods, etc. These also help cement relationships and increase trust between groups, allowing the harmonisation of methods, etc.

During an example of the exchanges carried out during the project, Filiz Yeni from METU in Turkey visited Oklahoma State University/USA – National Institute for Microbial Forensics & Food and Agricultural Biosecurity (NIMFFAB) from May 6, 2013 – September 6, 2013. While she was there she worked with Prof. Dr. Jacqueline Fletcher (NIMFFAB Director, OSU) and Assist. Prof. Li Maria Ma (OSU) to learn how to develop a Multiple-Locus Variable number tandem repeat Analysis (MLVA) procedure for strain discrimination. She worked with Non-O157 STEC serogroups including O26, O111, O103, O121, O45, and O145 and also tried to develop a MLVA method for strain discrimination of *Salmonella enterica* subs. *Enterica*. She also worked with Ian Montcrief to learn the ISSR technique for strain discrimination of *Fusarium proliferatum* which causes salmon-colored blotches on yellow, red and white onions, using DNA samples collected from Germany, North America and Israel in the framework of the PLANTFOODSEC Project.

This type of training provides the expertise to trace pathogens during outbreaks to determine the origin of the disease. These exchanges are essential to the success of any Network as a coherent working entity as they enhance the mutual understanding of methods being used in partner laboratories.

16.9.6 Academic Courses

The inclusion of undergraduate degree courses and PhD studentships as part of the long-term training plan will promote the training of future experts in this area.

16.9.7 Simulation Exercises

As pointed out by Stack (2010), the management of a plant disease outbreak requires the co-operation of many individuals and organisations, covering all the activities associated with reporting, response and recovery.

It is thus important to transfer research outputs to "users": in this case the simulation exercises may bring together the people who are likely to be closely involved in incidents and there is nothing quite as successful in terms of training as having experienced the 'real thing'.

Under the PLANTFOODSEC project a workshop was organised in Brussels in January 2016 to test the tools developed by the Network in a simulation of a plant disease outbreak in Europe: the tools listed in Table 16.1 were tested for performance along the course of the outbreak, from risk assessment to management, based on a simulation of a plant disease outbreak (brown rot of potatoes, *Ralstonia solanacearum*): demonstrating how the project tools may be used to manage it, from detection to containment.

A **storyboard** was developed as follows to show how the operational tools developed in the project can be used in a real outbreak situation to counteract it. The workshop scenario covered a deliberate contamination of potato fields in Seine Maritime, France, with the quarantine pest *Ralstonia solanacearum*.and it comprised:

1. Introduction on the threats resulting from plant pathogen misuse as anti-crop bioweapons followed by the general presentation of the case-study (description of the pest/target crop, i.e. the pathosystem *Ralstonia solanacearum*/potato and its Pest Risk Assessment).
2. Detection and diagnosis of the pathogen and demonstration of the virtual diagnostic network
3. Generalization of the containment measures by the French Agricultural Ministry, and the presentation of the tools for management and containment
4. Application of the decision tool to determine whether the pathogen was intentionally or accidentally introduced
5. Demonstration of modelled surveillance programs

16.9.8 Measuring the Impact

Evaluating the impact of any intervention is a difficult process although essential for any successful program. Approaches may include both gathering qualitative and quantitative data usually immediately at the end of any training but may also include follow-up surveys. The assessment will come back to the original question what was the purpose of the training-raising awareness or developing skills? There are many approaches of measuring impact, however, from experience asking participants to evaluate their current knowledge or awareness prior to the training and then

Table 16.1 Tools developed during PLANTFOODSEC

Target crops and target pathogens	A list of 451 target plants and crop products relevant to Europe was drawn up. In addition, 522 pests were identified. Criteria for prioritisation were also identified.
Tool for the prioritisation of target human pathogens on plants	Repeated outbreaks of human illnesses attributable to the contamination of fresh produce and other plant-derived foods by human pathogens on plants were analysed in order to provide guidance on the prioritisation of the risks involved in any future HPOP incident.
Analytical methods for the identification of microbial or toxin contamination	Methods available for the assessment of foodborne contaminants from exemplar food matrices, as well as analytical methods available for the identification of microbial or toxin contamination, were critically reviewed. In addition, food microbiology laboratories in the EU and non-EU project countries that are able to respond in the event of an outbreak of foodborne illness were identified.
Decision tool to determine whether a foodborne illness was introduced intentionally	Primary factors that make it possible to discriminate between a deliberately initiated outbreak and an accidental outbreak were assessed and used to develop a decision tool for determining the likelihood that an outbreak was criminally induced.
Forensically discrimination technology for a foodborne pathogen (E. coli STEC)	A forensically valid microbial strain discrimination technology, based on multilocus variable tandem repeat assessment (MLVA), was developed for non-O157:H7 *Escherichia coli*, a foodborne pathogen of increasing EU concern.
Risk assessment tool	A tool was developed to enable rapid assessments of agro-terrorism scenarios. It has been demonstrated on almost 100 scenarios covering a wide range of potential motivations, biological agents, pathways and receptor systems in order to provide a comparative measure of risk. The tool makes it possible to assess the effects of potential prevention and mitigation measures.
The PLANTFOODSEC web-based virtual diagnostic network	The virtual diagnostic network allows information to be gathered, searched and reported, and also makes possible information flow between experts and field workers to access summary information on disease outbreaks in Europe.
Management programmes	The measures to be taken in order to prevent the establishment and spread of harmful crop pathogens have been established by identifying activities and responsibilities following pathogen introduction.

after the training is useful. You may well also want to consider questions such as 'as a result of this training what will you do differently?' or 'what was the most important piece of information you wish to share with a colleague? Are you'. However, it may be difficult to measure the impact immediately after the training where much of it is about long-term behavioural change with respect to plant biosecurity.

16.10 Conclusions and Recommendations

Our experiences within the PLANTFOODSEC project and from National training in biosecurity suggest that a range of types of training/knowledge transfer is the best approach to maintaining the biosecurity readiness and spreading awareness of the topic areas. Short presentations can be very effective if given as part of other courses, e.g. University courses, or to audiences that have already ‘bought into’ the subject area, e.g. specialist groups. However, spreading the word to the general public requires a different strategy, based on social media, etc.

These will never answer the need for specialist courses for people who will be in the frontline as technical experts and these will require much more in depth training as provided by the summer schools and other longer training courses. An example of this approach was suggested by European Academies’ Science Advisory Council (EASAC 2014), new grant schemes should be considered to ensure that relevant work in universities and public research institutions is appropriately coordinated with the activities of the plant health authorities.

In cases where training is of vital importance to safeguard agriculture, for example where dangerous organisms are being used in contained use for research purposes, obligatory training in their management is vital for the assurance of biosecurity.

Exchanges of staff are very important for establishing Networks as they allow for cementing of the relationships and sharing of expertise.

It is important to make available undergraduate and postgraduate studentships/courses in these topics to ensure the training of future experts in this area.

In all cases it was important to collect feedback from the participants and to provide them with some form of certificate to prove that they had received the training, whether this was as part of a recognised scheme (e.g. a degree) or as a one-off certificate for the course. This provides an extra incentive for people to attend the courses.

Most important of all may be the simulation exercises as these bring together the people who are likely to be closely involved in incidents and there is nothing quite as successful as practising the ‘real thing’.

With budgets becoming more restricted everywhere it is important not to lose sight of the need to develop and maintain a skilled workforce in the plant and food biosecurity area. This is both important to develop in-depth skills in a particular subject area but also to enhance cross sector understanding so that in an emergency response we are starting from a high level of mutual understanding.

References

Ashford DA, Kaiser RM, Bales ME, Shutt K, Patrawalla A, McShan A, Tappero JW, Perkins BA, Dannenberg AL (2003) Planning against biological terrorism: lessons from outbreak investigations. Emerg Infect Dis 9(5):515–519

Community of Users on Disaster Risk and Crisis Management (CoU) (2015) Building on a mapping of EU policies and their links to technical and scientific challenges – Draft working paper 10 December 2015

EASAC (2014) EASAC report 24 – Risks to plant health: EU priorities for tackling emerging plant pests and diseases. http://www.easac.eu/home/reports-and-statements/detail-view/article/risks-to-pla.html

EC (2002). EU Council Framework Decision (2002/475/JHA) of 13 June 2002

EU Green Paper on Biopreparedness. http://eur-lex.europa.eu/

European Council conclusions on 'Lessons learned for Public Health from the Ebola outbreak in West Africa – Health Security in the European Union' (2015/C 421/04). http://eur-lex.europa.eu/legal-content/EN/TXT/?uri=uriserv:OJ.C_.2015.421.01.0006.01.ENG&toc=OJ:C:2015:421:TOC

Food and agriculture organisation of the United Nations (FAO). Biosecurity for agriculture and food production. Committee on agriculture, 17th Session, Rome, 31 Mar–4 Apr 2003. http://www.fao.org/biosecurity/

Food and Agriculture Organisation of the United Nations (FAO). Biosecurity Toolkit(2007). ftp://ftp.fao.org/docrep/fao/010/a1140e/a1140e.pdf.

IHR (2005). International Health Regulations http://www.who.int/topics/international_health_regulations/en/

Jernigan JA, Stephens DS, Ashford DA, Omenaca C, Topiel MS, Galbraith M et al (2001) Bioterrorism-related inhalational anthrax: the first 10 cases reported in the United States. Emerg Infect Dis 7:933–944

MAF Media Release May (2005) http://maxa.maf.govt.nz/mafnet/press/250505fmd.htm

Oerke (2006) J Agric Sci 144:31–43. doi:10.1017/S0021859605005708)

Ryan J, Glarum J (2008) Biosecurity and bioterrorism. Published by Butterworth-Heinemann. Print Book ISBN: 9780750684897; eBook ISBN : 9780080569185

Stack JP (2010) Diagnostic networks for plant biosecurity. In: Knowledge and technology transfer for plant pathology. Springer, Dordrecht, pp 59–74

Wikipedia entry on biosecurity. https://en.wikipedia.org/wiki/Biosecurity

Chapter 17
Making the Most of International Opportunities and Experiences for Researchers' Training Within a Large, Multinational EU Project: The Students' Perspective

Tania Llera, Ian Moncrief, Yochai Isack, Filiz Yeni, Vincent Cardon, Giovanna Gilardi, James Woodhall, Giuseppe Ortu, and Maria Lodovica Gullino

Abstract The European Network of Excellence PLANTFOODSEC, has developed a series of courses and student exchanges among the partners, offering a unique opportunity for training and growth for the researchers involved. In this chapter, the impact of research periods abroad for both the researcher and the

---

T. Llera (✉) • G. Gilardi • G. Ortu • M.L. Gullino
Agroinnova, Università degli Studi di Torino, Torino, Italy
e-mail: tania.lleraduran@unito.it; giovanna.gilardi@unito.it; giuseppe.ortu@unito.it; marialodovica.gullino@unito.it

I. Moncrief
Oklahoma State University, Stillwater, OK, USA
e-mail: ian.moncrief@okstate.edu

Y. Isack
Department of Plant Pathology and Microbiology, the Robert H. Smith Faculty of Agriculture, Food and Environment, The Hebrew University of Jerusalem, Jerusalem, Israel

Laboratory for Pest Management Research, Institute of Agricultural Engineering ARO, The Volcani Center, Bet Dagan, 50250 Rishon LeZion, Israel
e-mail: yochai.isack@gmail.com

F. Yeni
Food Engineering Department, Middle East Technical University, Ankara, Turkey
e-mail: filizyeni@gmail.com

V. Cardon
University of Picardie Jules Vernes, CURAPP (UPJV-CNRS), CESPRA (EHESS-CNRS), Lisis (Inra-UMLV), Paris, France
e-mail: cardon@ehess.fr

J. Woodhall
Plant Protection Programme, Fera, Sand Hutton, York, UK
e-mail: james.woodhall@fera.gsi.gov.uk

M.L. Gullino et al. (eds.), *Practical Tools for Plant and Food Biosecurity*, Plant Pathology in the 21st Century 8, DOI 10.1007/978-3-319-46897-6_17

sending and hosting institutions are analysed from the point of view of the students. It is widely known that the international experiences and networks of scientists increase the quality and impact of the research conducted, and they can also impact the willingness to work abroad for long periods of time or help to define more clearly the researchers' career. The results reached by the students interviewed is in line with these findings.

Keywords Higher education • Research internationalization • International training • Young researchers • Interdisciplinary training • Multicultural research environments • Scientific career development

17.1 Introduction

The Network of Excellence project on biosecurity, Plantfoodsec, has allowed the participating institutions to exchange students and researchers that have travelled to other countries to increase their knowledge on the research issues, methodologies and approaches that each member of the international team has. These exchanges have further contributed to the strengthening of the Network, as the participants have carried back to their countries valuable experiences, knowledge and contacts.

The importance of the participation in international projects is almost unnecessary to mention. In particular, the ERA (European Research Area) is an important instrument in the exchange of personnel, skills, knowledge and technologies inside the EU. The participation in projects with many different partners in the EU has become fundamental for some research centres and universities, shaping their approach to work and funding hunt. Not only the international projects in the ERA have a great part of the economic resources dedicated to research in the EU, they also are managing to put together the best scientific teams on each field, being able to develop world class research.

The internationalisation of higher education is a known trend in the last decades. The knowledge economy pushes for the competition for the best scientists, PhD students and post-doctoral staff that find interesting offers to work in the best research centres in the world. The most urgent scientific issues that are being faced by large groups of researchers working together in many countries, this situation requires that the education of the students allows them to work in international and interdisciplinary settings (Nerad 2010). The main destination of the participants in the Plantfoodsec exchanges was the USA, followed by the UK.

17.2 Mobility Has Become a Necessary Part of the Curriculum of a Scientist

Mobility for young scientists (and young professionals in general) has become a rite of passage that many persons overcome during their undergraduate studies, as exchange students in programmes as Erasmus, the exchange programme devised by the EU to foster student exchange in Europe. It was introduced in 1987 and it offers the possibility of studying in another European country for up to 12 months with a small grant. The researchers in this group did not participate in undergraduate exchange programmes, and for some of them, this was their first and only experience working or studying abroad. This experience has had a big impact for many of them, as it has helped them to focus their careers and to become a part of an international network. Nowadays, research conducted on this topic shows that mobility is a necessary but not sufficient condition in the curriculum of scientists: "*Thus, researchers are encouraged to incorporate a stint abroad as part of their career development portfolio. Collaboration with researchers from other countries is increasingly also being seen as a merit in itself in much the same way as collaboration with actors outside the academe.*" (Jacob and Meek 2013)

Researchers with different levels of mobility have different levels of success and contribute in different ways to the institutions that host them. There is a direct relationship between mobility and impact of the research, caused by the internationalisation of the work teams and the improvement of the scientific network.

Abramo et al., showed in a work published in 2011 that the scientists that collaborate more at the international level were, in general, more successful than those who do not. A recent article in Nature affirmed that "*Internationally co-authored papers are more highly cited because the authors are more likely to be doing excellent research*" (Adams 2013).

The impact that the participation in the programme has had on the researchers work has been evident, as it has helped many of them into defining more clearly the next steps in their careers. For example, Mr. James Woodhall says: "*Yes it has made me think about working in a university environment more, people seemed eager to learn at a university.*"

Ms. Filiz Yeni affirms: "*Yes, this experience cleared doubts about the way I should direct my career to. The more different fields about food science I gained experience, the more I have become sure about the field I need to work on.*" (Fig. 17.1)

Fig. 17.1 Researcher Filiz Yeni, from Turkey, worked for 4 months at the Oklahoma State University, National Institute for Microbial Forensics and Food and Agricultural Biosecurity (NIMFFAB), U.S.A

17.3 Mobility Benefits Not Only the Scientists, But Also the Projects and the Institutions

Diversity in work groups is considered an important contributor to innovation, a fundamental part of the scientific process (De Dreu and West 2001). For this reason, international teams are highly valued, as they are able to combine different background experiences and skills that allow the development of new methodologies and approaches to scientific and social issues. The European Union and the impulse for the creation of a European Research Area look to foster the cooperation among the Union partners, that, with their diversity and different organisational and methodological approaches can enhance the level of the research conducted. Inside the EU the discussion about brain drain/brain gain has sense only in the countries that cannot offer opportunities to their own scientists to return with the knowledge and the networks acquired in their period abroad, but not to every country whose scientists travel abroad. As stated in recent research: "*…science is an expensive activity even for the wealthy; thus all countries are dependent for their economic development, not only on their own research capacity but on being able to access and absorb knowledge produced elsewhere.*" (Jacob and Meek 2013)

Nowadays, the discussion is more centred in the "brain circulation" (Ackers 2005) needed to participate in big international projects, as scientists that have worked abroad have usually wider networks that include institutions and people in other countries that can become a part of a competitive project team and whose methodologies and routines are already known, also at the administrative level. For the researchers participating in this exchange, when asked if this experience had prompted new collaborations with their host organisation, the main answer was "Not yet". Mr. Giuseppe Ortu points out "*My organization has a long story of*

Fig. 17.2 Mr. Ian Moncrief, from the USA, worked for 5 weeks at the Institute of Agricultural Engineering, ARO, The Volcani Center, Bet Dagan, Israel

collaborations with both host organizations and, of course, each activity has a positive effect to reinforce these connections." while Ms. Giovanna Gilardi, a colleague of Mr. Ortu indicates that a new project has already been planned with her host organisation. Mr. Ian Moncrief affirms "*Yes, just before my graduation, there were collaborations with my committee members with Mr. Gamliel. There is still ongoing research between the two labs*". So it seems that the researchers exchange served more to the reinforcement of ongoing collaborations than to the development of new ones. Nevertheless, the use of "not yet" by most of the researchers suggests that they hope these collaborations will take place in the future (Fig. 17.2).

However, even if there are not ongoing collaborations yet among most of the researchers that have participated in the exchanges, almost all of them declare that they have kept in touch with some of their colleagues in their host organisation. For example, Mr. Ortu asserts: "*Yes, I have met very competent researchers and I am in contact with them. Also, if some problem happens during my lab experiments I can discuss it with them in order to have important suggestions. In my opinion this is very important.*" And Ms. Filiz Yeni has the same perception: "*Yes, within the ongoing project activities, we still keep in touch. Some of the colleagues from the host institution also came to visit our institution a few months ago in order to exchange knowledge and discuss about new collaboration opportunities.*"

And also Mr. Isack feels his participation in the programme's exchanges has open new communication possibilities with colleagues abroad: "*The connection is not on a daily basis, but I feel I gained friends with whom I can consult freely (even I did it several times). In addition, I was happy to advise and assist the members of the program as I could.*"

Fig. 17.3 Mr. James Woodhall from the UK, spent 7 weeks working at NIMFAB, Oklahoma State University, Stillwater

Mr. Ian Moncrief has kept in contact with the head of the laboratory where he worked in Israel: *"Yes, I do keep in touch with the head of the laboratory, Abraham Gamliel. I do keep up with a little bit some his current and former students on Facebook. I have attended conferences with him after my internship."*

Mr. James Woodhall has had the opportunity to keep in touch and meet again some of his former colleagues at Oklahoma State University: *"Yes I was fortunate enough to attend the APS meeting in the summer after my placement and OSU was well represented there and caught up with lots of faculty and students present at that meeting. Its only been a 6 months since my time at OSU so maybe I will collaborate with people in the future."*

Scientists tend to work better with the people they have worked before in other projects (Cummings and Kiesler 2008). In a moment in which technology allows researchers to contact each other and share knowledge, methodologies and results in real time, or to perform analysis in the distance, the development of the network may relie in contacts that have been established in person before. The development of information technologies means that the possibility and eventual necessity to develop research abroad depend mainly on the study field and the location of the necessary tools to conduct the research (Ackers 2013) (Fig. 17.3).

17.4 Short-Term Exchanges Contributes to the Development of a Mobility Culture and Works for Internationally Established Research Centres

Short-term exchanges contribute to the development of a mobility culture inside the institutions that participate in international projects that helps to increase the production and impact of the research carried on by them. The harm caused by "academic inbreeding" or lack of mobility of the scientists can be challenged by the participation of the institution on international projects that allow the scientists, even the more "non mobile" to work in short periods abroad, participating in courses and networking events, encouraging them to adopt a more collaborative mind set with colleagues in other countries. As established by recent research, this kind of mobility has a positive impact and thus, should be promoted: "*However, this research shows that it is possible to diminish academic inbreeding detrimental effects by creating opportunities and incentives for inbreds to be more mobile. The analysis indicated that, unlike pure inbreds, mobile inbreds tend to have scientific output trends that are closer to that of more mobile faculty*" (Horta 2013). Moreover, when the preferences of Agricultural students for their experieces abroad have been studied, both graduates and undergraduates expressed their interest in short-term, faculty-led exchanges (Chang et al. 2013).

At the same time, research conducted on participants on the Erasmus project in Europe, shows that students with international experiences are more likely to work and live abroad later in time: "*Our results indicate that the effects of educational mobility programmes go far beyond affecting the decision to study abroad for some time period but rather reach far into the labour market an it will be interesting to follow the sample graduates as their careers unfold. But already at this early stage our results indicate that even short-term mobility investments can lead to significant further mobility investments later on.*" (Waldinger and Parey 2007)

The impact of the exchange experience on the mobility of the participants is clear, some of them have declared that their experience abroad has made them to think about the possibility of working or living abroad in the long term.

Mr. James Woodhall affirms: "*Yes, it's the longest I have lived away from the UK and I am definitely more comfortable with the idea of living and working abroad now.*" and Ms. Filiz Yeni says: "*I think yes. I may think of working permanently abroad after I complete my PhD studies.*" The experience had impact also on Mr. Ian Moncrief: "*I am considering working abroad in the next couple of years. I enjoyed my interactions in Israel and I like sharing different views on science with others.*" And for Ms. Giovanna Gilardi, this period abroad has changed her view about working permanently abroad.

While others have valued the short-term experience and intend to keep on participating on this kind of international experience in the next years. Mr. Giuseppe Ortu declares: "*No, I think that is very important to work abroad also for medium-lenght periods.*" And Mr. Yochai Isack: "*These days I'm wondering about the short study period abroad, but in any case want to go back to Israel.*"

When asked about their biggest achievements during their exchange periods, the researchers pointed out the development of the capacity to communicate with scientists from other countries and other fields. Intercultural communication competence "*which is most often viewed as a set of cognitive, affective, and behavioral skills and characteristics that support effective and appropriate interaction in a variety of cultural contexts*" (Bennett 2008) is an important asset for researchers that are required to work abroad, in international settings or as a part of a large group of international organisations. The participation in an international exchange programme is one of the available methods to enhance this set of skills that can have a determinant role in the career of a scientist. Ms Filiz Yeni affirms: "*I think the most significant contribution of this period to my professional life is that I learned the importance of always being open to learn more on food science even it is outside my research field.*" Mr. Ian Moncrief makes a similar reflection "*I think professionally, this experience has taught me how to interact and communicate with international scientists. I have learned to listen how other countries think about the challenges that face their countries when it comes to biosecurity. Some of my accomplishments include, being able to travel and live in another country on my own and develop relationships with people. As far a research goes, we were able to complete our field work, which was the first field work I experienced in my studies.*" And also Mr. Yochai Isack thinks the same, the new connections and ways to communicate were a fundamental outcome of his short period abroad "*The professional side, it was the most teaching experience; I learned to work in BSL3 laboratory. The different approaches and different opinions made me look at things differently. The personal side, the ability to communicate regularly with colleagues from different countries gave me confidence in my own abilities to express ideas clearly.*" PhD student Ms. Giovana Gilardi, shares the same point of view: "*…the opportunity to learn about other research areas different from mine. Being in contact with students from different parts of the world.*"

The researchers point out the importance of communicating with people from other countries in a constructive and productive way, something that is particularly important in a field like Food and Agricultural sciences, as it works with issues and situations that are not necessarily limited inside the borders of one country. At the same time, the importance of learning about other areas of research, of working with an interdisciplinary team emerges. In international projects that involve not only scientists from different parts of the world, but also from different academic backgrounds and working in different kinds of organisations, the experience of collaboration not only is an opportunity to improve the hard skills on the research field, but also the soft skills that allow researchers the collaboration with stakeholders, target population, other researchers from different disciplines and other organisations with diverse objectives and goals. It's necessary that researchers acquire the skills necessary to overcome the difficulties and unexpected situations that emerge in international, interorganisational and interdisciplinary projects (Schmidt et al. 2012; Perz et al. 2010) (Fig. 17.4).

The inclusion of young researchers and PhD students is one of the effects that EUFPs (European Union Framework Programmes) have had in the research centres

Fig. 17.4 Mr. Yochai Isack from Israel has participated in a 5 day course at Kansas State University Department of Plant Pathology Manhattan, Kansas, and has hosted international students in his lab as well

and universities. The EU Framework programmes were introduced in 1984 to put all the EU activities concerning research under one coherent framework. The participation in this kind of projects needs institutional support for the administrative and bureaucratic exigences, but has a positive impact in the organisation and funding of many university departments, specially in the applied science areas.

The researchers interviewed affirm that the administrative issues related with their period abroad (transport, accommodation, etc.) were easily managed by the host institutions, allowing them to focus on the research and the experience in their new teams. The preparation of the participant institutions is key to the development of exchanges practices that ease the participation of scientists, regardless of their origin and destination. The kind of project developed, *Network of Excellence*, requires international experience and skills connected to the management of long projects. That includes, of course, the management of the mobility of the researchers between the participating institutions. In a recent study carried on at La Sapienza University in Rome about the institutional changes prompted by the participation on European international projects, the researchers noted: "*Those [financial instruments] introduced by EUFP6, such as Integrated Projects and the Network of Excellence, are suited to more experienced groups, already included in wide international networks and able to carry out long-term, complex and mainly applied collaboration projects.*" (Primeri and Reale 2012).

The researchers from the EU interviewed, as well as the participant from Turkey, said that they had broaden their knowledge about the internal workings of the EU

Fig. 17.5 Mr. Vincent Cardon, from France, sociologist studying agroterrorism, during his post-doc at INRA

funded projects, from the organisational and administrative point of view, while the US and Israel participants were more unaware of these issues. Ms. Filiz Yeni comments: "*During my participation in the project I learned much about project management activities. As METU we organized two technical trainings, one summer school and one project meeting during last three years. These efforts have contributed a lot to my professional skills.*" Mr. Vincent Cardon also points out he learned about EU project's indicators and requirements, and Ms. Giovanna Gilardi gained "*a little more*" knowledge about them (Fig. 17.5).

17.5 International Exchanges Go Beyond the Professional Sphere

The impact of a period of study abroad is not limited to the career of the participants and their research sending and host organisations, but leaves a mark on the lives and views of the students that have an inmediate effect and last for a lifetime (Paige et al. 2009; Potts 2015). The participants answers show that they valued their experience abroad as a way to know other cultures and broaden personal horizons, not only in the professional sphere. None of the participants recalled any cultural or linguistic barrier, as the main destinations of the exchanges were English speaking countries. Those who travelled to non English speaking countries found colleagues that were fluent in English or had knowledge of it, and did not learn the local languages (Hebrew and Hungarian). Even culture was not regarded as a problem, some of the anecdotes shared by the researchers are in fact connected to cultural misunderstandings. For example, Mr. Ian Moncrief recalls: "*Mr. Gamliel's lab group generally eats lunch together and most of the time it was a type of salad. My first week there, they prepared lunch for me and they served salad as well as fruits and some pita breads. I do not eat salad at home because I do not like it really. I did not want to say anything at first so I did my best to eat as much salad as I could, but I never finished it.*

I guess they caught on and asked if I was full because I never finished the salad. Finally, I told them that I am really not a salad eater and they sort of felt bad I think. They asked me what foods do I eat, and I told them that I love pasta, just noodles and some butter and I am good. So the next day at lunch they called me to the table and there was a bowl of pasta. I had no idea they went and got pasta for me and just about every lunch for the next 4 weeks or so was pasta for me.

Mr. Gamliel's lab group were so generous to me during my visit and they did everything to make sure I was comfortable. I was part of their family for a while and I am forever grateful to them."

Also, even being from the UK did not guarantee for Mr. James Woodhall the absence of linguistic misunderstandings in the US: "*Well lamb is my favourite meat in the UK. You don't see much lamb on menus in the US so I was very happy and almost ordered lamb fries when I saw it on the menu in one restaurant. It's a delicacy there. However, when someone told me what lamb fries actually were (sheep's testicles) I quickly changed my order!*"

Researcher Filiz Yeni recalls this anecdote of her experience with her different environment at Oklahoma: "*Yes I remember a funny one. It was the first morning that I was in Oklahoma. I suddenly heard sirens and tried to understand what was happening but there seemed nothing strange to me. A few minutes later a saw someone on the street, however, they did not seem worried. I felt quiet strange. Hours later someone informed me that it was just a tornado drill and nothing dangerous had happened. So I realized that tornadoes were a real threat in Oklahoma.*"

All the researchers considered their experience as a successful one on the professional and personal levels. When asked if they would repeat the experience and what kind of advice would they give to a student that thinks about spend some time abroad, they all agree that this experience has been beneficial for their lives and careers. Mr. Ian Moncrief says: "*I would definitely do this experience again! I would tell an International student to not only do this type of study, but to embrace a new culture while he is there.*"

Ms. Filiz Yeni asserts: "*I definitely would repeat the experience. Besides from the professional contributions, I made good friends there. I strongly advice to an international student to participate in such a project. This experience may help them reshape their plans about their careers.*"

Mr. James Woodhall gives important advice on how to make the most of the period abroad: "*I would certainly repeat the experience. I would tell an international student to not have too many pre-conceptions and go with the flow. Let the host organisation show how they work first before you plan your experiments. Don't spend hours in the lab doing experiments, you can do that anywhere, go and speak to people there, find out there views and ideas and also get experiences you could not do at your home institute. Also, look at how they work there, how is the lab organised, what pipelines they have for generating data and producing papers.*

Finally, bring some food from home – in my case it was tea and HP sauce!" And the same advice is given also by Ms. Giovanna Gilardi: *"I would tell them to seize the positive but also the negative aspects of the educational experience. Make the most from the opportunity to be involved in a different research group. Evaluate innovations as suggestions for improvement."*

Mr. Yochai Isack encourages students to meet other students from around the globe: *"I would love to repeat this experience, I wholeheartedly recommend students to strive to get to know students from around the world."*

Mr. Vicent Cardon's advice to a possible international student is simple, but effective: *"Go!"*

17.6 Conclusions

The exchange experience has been regarded has positive by all the participants. It has already had an impact in their careers and lifes: they would participate in other similar experiences and some are now open to work abroad for a long term or even permanently. Also, the experience has given more focus or renewed their interest in their research field.

The participants have enlarged their networks and their intercultural communication compentences, points considered particularly important for a successful scientific career. Moreover, with their presence and collaboration, they have contributed to the internationalisation of their host structures and research groups, carrying with them the richness of their knowledge, skills and cultural background.

The agriculture and food research field works with pressing and complex issues connected with the life of people in the whole world. The nature of the studies necessary to understand and manage these issues allows the researchers to work in large international and interdisciplinary research networks, that collaborate in the production of scientific knowledge and its transferring to the stakeholders and the general public. This work can be made more fluid and engaging thanks to the exchanges of the students and researchers that are part of the network, with the aim of including the different methodologies, priorities and points of view that come with their cultural, educational and institutional differences.

References

Abramo G, D'Angelo CA, Solazzi M (2011) Are researchers that collaborate more at the international level top performers? An investigation on the Italian university system. J Informet 5(1):204–213. doi:10.1016/j.joi.2010.11.002

Ackers L (2005) Moving people and knowledge: scientific mobility in the European Union1. Int Migr 43(5):99–131

Ackers L (2013) Internet mobility, co-presence and purpose: contextualising internationalisation in research careers. Sociología y tecnociencia: Revista digital de sociología del sistema tecnocientífico 3(3):117–141. http://uvadoc.uva.es/handle/10324/9304

Adams J (2013) The fourth age of research. Nature 497(30 May 2013):557–560. doi:10.1038/497557a

Bennett JM (2008) Transformative training: designing programs for culture learning. In: Contemporary leadership and intercultural competence: understanding and utilizing cultural diversity to build successful organizations. Sage, Thousand oaks, pp 95–110

Chang CW, Pratt O, Bielecki C, Balinas M, McGucken A, Rutherford T, Wingenbach G (2013) Agriculture Students' interests, preferences, barriers and perceived benefits of international educational experiences. NACTA J 57(3a):97

Cummings JN, Kiesler S (2008) Who collaborates successfully? Prior experience reduces collaboration barriers in distributed interdisciplinary research. In: Proceedings of the 2008 ACM conference on computer supported cooperative work. ACM, pp 437–446

De Dreu CKW, West MA (2001) Minority dissent and team innovation: the importance of participation in decision making. J Appl Psychol 86(6):1191–1201. doi:10.1037/0021-9010.86.6.1191

Horta H (2013) Deepening our understanding of academic inbreeding effects on research information exchange and scientific output: new insights for academic based research. High Educ 65(4):487–510. doi:10.1007/s10734-012-9559-7

Jacob M, Meek VL (2013) Scientific mobility and international research networks: trends and policy tools for promoting research excellence and capacity building. Stud High Educ 38(3):331–344. doi:10.1080/03075079.2013.773789

Nerad M (2010) Globalization and the internationalization of graduate education: a macro and micro view. Can J High Educ 40(1):1

Paige RM, Stallman EM, Jon JE, LaBrack B (2009) Study abroad for global engagement: the long-term impact of international experiences. In: NAFSA: association of international educators 2009 annual conference, Los Angeles. Retrieved September, vol 10, p 2010

Perz SG, Brilhante S, Brown IF, Michaelsen AC, Mendoza E, Passos V, ... Selaya G (2010) Crossing boundaries for environmental science and management: combining interdisciplinary, interorganizational and international collaboration. Environ conserv 37(4):419–431

Potts D (2015) Understanding the early career benefits of learning abroad programs. J Stud Int Educ 19:441, 1028315315579241

Primeri E, Reale E (2012) How Europe shapes academic research: insights from participation in European union framework programmes. Eur J Educ 47(1):104–121. doi:10.1111/j.1465-3435.2011.01511.x

Schmidt AH, Robbins AS, Combs JK, Freeburg A, Jesperson RG, Rogers HS, Sheldon KS, Wheat E (2012) A new model for training graduate students to conduct interdisciplinary, interorganizational, and international research. Bioscience 62(3):296–304

Waldinger F, Parey M (2007) Studying Abroad and the effect on international labor market mobility. Econ J 121:194–222. doi:10.1111/j.1468-0297.2010.02369.x

Chapter 18
The Need for International Perspectives to Solve Global Biosecurity Challenges

John D. Mumford, Maria Lodovica Gullino, James P. Stack, Jacqueline Fletcher, and M. Megan Quinlan

Abstract Global biosecurity presents international challenges because the majority of instances of novel organism introductions are due to international movements of goods, food and people and the likelihood of introduced agents crossing political boundaries. The inherent vulnerability of environments to introductions of alien, or non-indigenous, biological agents is due to the greater ecological vulnerability to these exotic entrants in the receiving environment. Agencies and individuals responsible for approving intentional introductions of beneficial organisms recognize this relationship and consider potential impacts in risk assessments prior to release of the organisms. However, some of those responsible for detection and control of novel pathogens and pests, introduced either inadvertently or intentionally, lack extensive training in ecology, environmental biology, and pathology, and may therefore underestimate the risk from such events. The latter is a key factor in the case of food safety. Europe is particularly vulnerable to cross-border movement of introduced agents, and one response to this has been the recent revision of plant health regimes throughout the European Union. Other responses include project-based initiatives, such as PLANTFOODSEC.

J.D. Mumford (✉) • M.M. Quinlan
Centre for Environmental Policy, Imperial College London, London, UK
e-mail: j.mumford@imperial.ac.uk; m.quinlan@imperial.ac.uk

M.L. Gullino
Centro di competenza per l'Innovazione in campo agro-ambientale (AGROINNOVA), Università degli Studi di Torino, Grugliasco, Italy

DISAFA – Dipartimento di Scienze Agrarie, Forestali ed Alimentari, Università degli Studi di Torino, Grugliasco, Italy
e-mail: marialodovica.gullino@unito.it

J.P. Stack
Department of Plant Pathology, Kansas State University, Manhattan, KS, USA
e-mail: jstack@ksu.edu

J. Fletcher
National Institute for Microbial Forensics & Agricultural Biosecurity, Department of Entomology and Plant Pathology, Oklahoma State University, Stillwater, OK, USA
e-mail: jacqueline.fletcher@okstate.edu

M.L. Gullino et al. (eds.), *Practical Tools for Plant and Food Biosecurity*, Plant Pathology in the 21st Century 8, DOI 10.1007/978-3-319-46897-6_18

Much of the existing framework for biosecurity has evolved over decades due to the need for States to protect the public from unsafe food, and from economic and sociocultural losses to biodiversity and agricultural resources. While malicious intentional releases are rare compared to conventional trade related unintentional introductions of agents, the security paradigm (the possibility of intentional introductions) should be added to more traditional biosecurity approaches that focus on inadvertent and accidental incursions. While there is a need to distinguish the unusual from the ordinary, in both source and receiving areas, security-related risks should be set within that context, in terms of risk assessment (for appropriate scaling) and for management of factors common to conventional plant health risks. This chapter considers the existing international risk frameworks and how to adapt them to include the security paradigm by moving from the traditional concepts (agent-pathway-receptor systems) to also consider motivation.. Motivation for harm may arise from experiences at home or abroad, and the pathway for access, transport and delivery of harmful agents would link a foreign source to the receptor environment in a global system. The adapted processes provide a general framework for analysing malicious biosecurity risks in a consistent and proportionate manner. For food safety in particular, novel agents introduced to the food supply maliciously may not be anticipated or identified initially through the traditional risk assessment. For this and other cases, the formation of networks of experience and technical excellence, such as that accomplished by PLANTFOODSEC, will help to fill the gaps and address the weaknesses of individual national programs. A call is made to create a mechanism and assign a coordination role for a sustainable international cooperation in addressing the full spectrum of global biosecurity concerns.

18.1 Introduction

Biosecurity has been considered within the PLANTFOODSEC (Plant and Food Biosecurity – Network of Excellence: https://www.plantfoodsec.eu/aboutbiosecurity_scenario.php) project as the protection from harm caused by biological agents or, more specific to plant biosecurity, the protection of all plant resources and the food supply from the natural or intentional introduction, establishment and spread of plant pests, pathogens and noxious weeds (Preface). Security is one paradigm of a set of related biosecurity risk approaches that share many common characteristics, analytical processes and outcomes (Fig. 18.1) (Mumford et al. 2013a). The PLANTFOODSEC project has been carried out within a security perspective, consistent with the European Security Research Advisory Board (ESRAB) definition of security research as "*…research activities that aim at identifying, preventing, deterring, preparing and protecting against unlawful or intentional malicious acts harming European societies; human beings, organisations or structures, material and immaterial goods and infrastructures, including mitigation and operational continuity after such an attack…*" (EC -European Commission 2006).

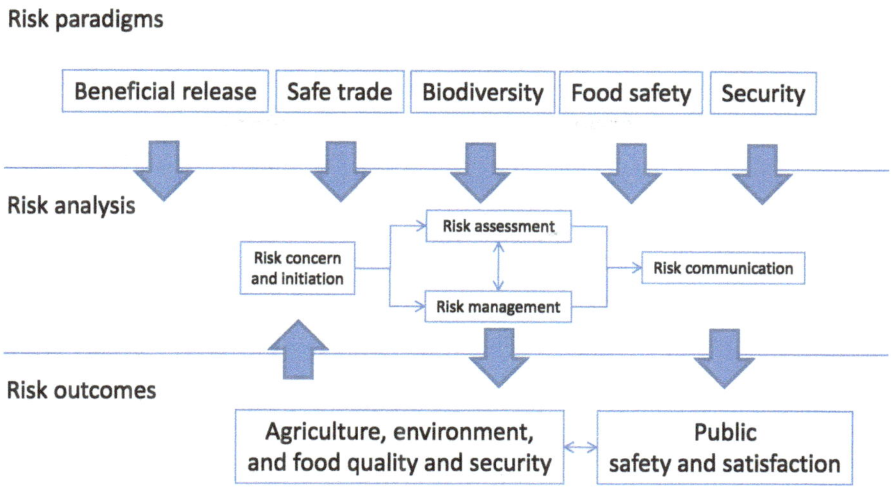

Fig. 18.1 Risk paradigms for global biosecurity (Mumford et al. 2013a)

A common concern of these biosecurity risks is the potential introduction of any agent or biological contaminant[1] into a cropping system, natural environment or food supply chain that could cause harm to the public.

An introduced agent would generally have the greatest impact if it is alien or non-indigenous[2] to the affected, or targeted, environment or food. The origin of the agent is important because, in addition to greater biological vulnerability to alien agents, detection and management of these agents would likely be outside the experience of those who must prevent or control them. As novel agents, they may also be more alarming to the public (Suffert et al. 2009; see also Chap. 2). In the case of conventional (not novel) biosecurity risks, the organisms and pathways may be relatively well understood, even if challenging to prevent and control.

The international perspective on the other biosecurity challenges (Fig. 18.1) is essential when considering the security paradigm. Those involved in the trade, importation or production of possible target crops or food are already cooperative partners in the assessment and management of biosecurity risks. The legal and

[1] The Codex Alimentarius Commission (CAC) defines a contaminant as "any substance not intentionally added to food or feed, which is present in such food or feed as a result of the production (including operations carried out in crop husbandry, animal husbandry and veterinary medicine), manufacture, processing, preparation, treatment, packing, packaging, transport or holding of such food or feed, or as a result of environmental contamination" (CAC 2015). While this definition does not require that the contaminant be harmful, nor does it refer to intentional introduction of an agent, this chapter includes those possibilities. (Food additives are substances intentionally added to foods.) The project has focussed on living, biological agents, rather than other forms of contaminants such as chemical substances.

[2] A discussion of the meaning of the terms *introduction* and *alien species* appears in the Appendix 1: Terminology of the Convention on Biological Diversity, ISPM 5 Glossary of phytosanitary terms (IPPC 2015), comparing details of how they are used under the CBD and the IPPC. The IPPC would prefer non-indigenous to describe such a population.

institutional frameworks are well tested and also embedded in each region and country. This makes the international perspective on biosecurity an important starting point, although novel agents may require additional or different partners to respond to and address these new threats.

18.1.1 International Legal Structure

Protecting domestic agriculture from non-indigenous pests and diseases is generally understood to be a public good, because it promotes both economic and food security. It is a role undertaken by governments, with the cooperation of importers, shippers and travellers (Mumford 2002). At the international level, the two main instruments that deal with prevention and management of organisms harmful to plants are the International Plant Protection Convention (IPPC) and the Convention on Biological Diversity (CBD).

The earlier instrument is the IPPC (FAO 1997; IPPC: https://www.ippc.int/en/), which came into force in 1952. The IPPC created an international regime "to secure common and effective action to prevent the spread and introduction of pests of plants and plant products, and to promote appropriate measures for their control". Its authority regards transboundary movement and the introduction of pests not already established into new areas, rather than general pest control of native or cosmopolitan pests. The IPPC has traditionally focused on safe trade and food security, although its mandate extends to environmental objectives related to natural flora and fauna and aquatic plants. The IPPC became more explicit about how an introduced agent's predicted impact on ecosystems justifies regulation of incoming trade when its international standards for phytosanitary measures (ISPM) 11 on Pest Risk Analysis was expanded to discuss invasive species (IPPC 2013; IPPC Secretariat 2005). This international plant health agreement has 182 signatories and is referenced by the World Trade Organization (WTO: https://www.wto.org/) in relation to phytosanitary rule making under the WTO Agreement on the Application of Sanitary and Phytosanitary Measures (WTO 1994), commonly referred to as SPS.

The CBD (UNEP 1992, CBD: https://www.cbd.int/) has protection and sustainable use of biodiversity as one of its primary objectives and supports implementation and provides guidance on its principles as part of its ongoing work program. The CBD calls on its parties to "prevent the introduction of, control or eradicate those alien species which threaten ecosystems, habitats, or species" (Article 8 h). It advocates a "three-stage hierarchical approach" of prevention, eradication and containment, with support measures based on identification of challenges and prioritization of objectives, prevention and early detection and control and long-term containment. The CBD is not a standard setting organization, however, and it has collaborated closely with the IPPC on plant biosecurity issues. The CBD has been ratified by 196 of the signatory countries and territories. It also holds a supplementary agreement on biotechnology, the Cartagena Protocol on Biosafety, which has been signed by many (although not all) of the signatories to the CBD.

Both agreements – the IPPC and CBD – consider the intentional release of beneficial organisms into agricultural or unmanaged natural environments. The Food and Agriculture Organization (FAO: http://www.fao.org/home/en/) '*Code of Conduct for the Import and Release of Exotic Biological Control Agents*' (IPPC 1996) has become one of the ISPMs under the IPPC (ISPM 3: IPPC 2005). This guidance was revised to cover release of a range of beneficial organisms and is widely followed by national frameworks (Quinlan et al. 2003 and Kairo et al. 2003). Protection of biodiversity and management of intentional release for conservation or reintroduction objectives is also covered by the World Animal Health Organisation (OIE after its original name of Office International des Epizooties: http://www.oie.int/). The OIE is the oldest of the rule making bodies referenced in the WTO SPS, having begun in 1924; it includes 180 member countries.

Food safety is largely governed at an international level by the Codex Alimentarius Commission (CAC: http://www.fao.org/fao-who-codexalimentarius/en/), since its inception in 1963 when it was established by FAO and the World Health Organization (WHO). The CAC (1992), the third rule setting body referenced by the WTO SPS, and its 187 members produce standards, guidelines and codes of conduct on food safety issues. Their procedures for risk assessment are described in general terms in guidelines (e.g. CAC 2007) as well as specifically for microbiological contamination. Codes of practice to avoid other contaminants are more specific to particular commodities, industries or contaminants. The majority of countries have food laws that reference CAC standards, either implicitly or explicitly. The risk outcomes of food safety and quality are the primary objectives of this body.

The risk paradigm of security is not as clearly linked to only one particular international agreement (see Chap. 6). The Convention on the Prohibition of the Development, Production and Stockpiling of Bacteriological (Biological) and Toxin Weapons and on their Destruction (more commonly known as the Biological and Toxins Weapons Convention (BTWC 1976) or simply Biological Weapons Convention (BWC)), entered into force in 1975, presently includes 173 parties and other signatories. The BTWC covers biological weapons, devices that disseminate disease-causing organisms or poisons to kill or harm humans, animals or plants. Historical efforts to produce biological weapons have included, among others, aflatoxin and the rice blast pathogen, *Magnaporthe grisea*. The BTWC also recognises that these agents can be enhanced from their natural state to make them more suitable for use as weapons. The multilateral treaty is aimed at States committing to disarmament and avoiding this category of weapons. It is not written with individual perpetrators in mind. It also does not have close linkage mechanisms with the other conventions relating to biosecurity risks, mentioned above, although some coordination does take place.[3] Therefore, the BTWC is an international framework for security but it does not explicitly address biosecurity.

Overall, however, the biosecurity risks identified in Fig. 18.1 are covered, at least to some extent, by an existing international legal structure and there are shared concerns across the international legal instruments and bodies about the introduction of

[3] https://www.ippc.int/en/liason/organizations/biologicalweaponsconvention/

alien agents that would threaten biosecurity. Furthermore, there are shared principles for analysing risks and planning and communicating their management. The outcomes elaborated in Fig. 18.1 are authorities generally assigned to States, as a public good, albeit less directly in terms of satisfaction of the public. Cross-border cooperation is critical to any effective preparedness and response strategy, for this reason an approach at the European Union (EU)-level is necessary and appropriate.

18.1.2 The European Context

Whatever the cause of an introduction of a novel agent or outbreak, a comprehensive biosecurity system is essential to protect domestic agriculture, natural resources and food supplies. Whereas numerous representatives of the United States Administration (Executive branch) and Congress have publicly expressed concern regarding the threat of biological terrorism, in Europe an attack on crop and food biosecurity is not yet discussed as a real threat in the political agenda. Current EU capabilities to detect and respond to agro-terrorism and bio-criminal acts are modest at best, spread amongst many organisations, normally handled by regional or national bodies, and of limited coordination. At the same time, legal regimes concerned with biosecurity, trade, biodiversity and food safety are connected neither at international nor European levels.

Actions related to plant and food biosecurity have a strong transboundary component in most regions. In Europe, a bioterrorist attack or introduced harmful organism could affect several Member States and spread fairly easily across borders. Factors contributing to the vulnerability of European agriculture include the continuing trends of intensive production techniques, the increasing production of genetically uniform crops, a notable amount of imported propagation material, the vertical integration of the production continuum, an increasing industrial dependence on the export market and the limited presence of resistance to disease agents in key crop and livestock species.

Food safety (discussed further in 18.4) and biodiversity have their own European legal and institutional contexts; the former focusing on food hygiene and the latter covering invasive species, habitat preservation and protected areas. The release of beneficial organisms in Europe is regulated through a patchwork of national approaches with links to more than one EC Directive, as discussed in a recent workshop (EPPO/COST SMARTER 2015). There is not as clear a decision pathway for intentional releases as there is for traditional plant health issues. Security has not been closely coordinated with these sectoral regimes in the past.

Considering plant biosecurity, at a regional level the European and Mediterranean Plant Protection Organization (EPPO) acts, in alignment with FAO, as a recognized regional plant protection organization under the IPPC. EPPO has 50 members, covering almost all countries of the European and Mediterranean region and extending to Russia, the Middle East and North Africa. EPPO advises member governments on the technical, administrative and legislative measures necessary to prevent the introduction and spread of pests and diseases of plants and plant products and provides coordination on risk assessments of alien species of concern. It also works on plant protection products such as pesticides and biological control agents for beneficial release.

For European Member States, all of whom are also members of EPPO, conventional plant health risks are managed by national regulatory authorities, all of which are recognised as competent authorities by the European Commission and the IPPC. The Council Directive 2000/29/EC established the Community Plant Health Regime that aims to protect the EU against the harm caused by the introduction and spread of harmful organisms and thus ensuring food security and plant health protection through sustainable production. Through this framework, all EU Member States are obliged to prohibit the import and internal movement of specified quarantine organisms, to notify the Commission and other Member States of the presence within their territory of these harmful organisms and finally they are obliged to take measures to eradicate or, if this is not possible, contain and prevent their spread. The Commission may also seek scientific advice from the European Food Safety Authority (EFSA), such as from the Scientific Panel on Plant Health.

The Community Plant Health Regime is under revision following an in depth review (EU 2015; FCEC 2010). Some of the recommendations for enhancement of the plant biosecurity system appear in Box 18.1. The results are informative for other international regimes, as well. Some mechanisms have been proposed for better coverage of invasive alien species.

Box 18.1 Some Recommended Enhancements of the European Plant Biosecurity System

The European Academies Science Advisory Council (EASAC), in its report "*Risks to plant health: European Union priorities for tackling emerging plant pests and diseases*" (EASAC 2014), clarified what is needed to achieve EU goals in the analysis and management of plant health risks covering three priority areas:

1. Surveillance systems

   EASAC recommends to improve monitoring of pests with establishment of early warning systems; to enhance linkage between databases; to use new forms of monitoring, to extend surveillance to natural habitats and to consider possible new challenges, for example bioterrorism. Meanwhile the EU Biodiversity Strategy to 2020 (EC 2011) proposes a dedicated EU legislative instrument (Regulation on Invasive Alien Species-IAS) to tackle outstanding challenges relating to Invasive Alien Species pathways (routes of biological invasions and. the mechanisms and vectors that allow the introduction and spread of IAS).

(continued)

Box 18.1 (continued)

2. Research and training

   EASAC also recommended that the research agenda address diagnosis; biology, ecology and epidemiology of plant pests and pathogens and their relationships with hosts and vectors; plant pest resistance; biological and cultural strategies for sustainable pest management; modelling, prediction and extrapolation. In addition, networking among disciplines and sectors should be improved.

3. Innovation

   The translation of knowledge from research to practical applications is recommended by EASAC, in particular to develop durable control approaches to overcome current limitations of pesticides and to breed improved plants, durably resistant to biotic stresses.

   EASAC also pointed out that protection and promotion of plant health cannot be tackled successfully without raising political and public awareness of the importance of plant health and resilience for sustainable agriculture, food security and environmental protection.

A Consortium of European researchers has been exploring the topic of crop biosecurity since 2004, taking into account the risks that the deliberate introduction of plant pathogens poses to European agriculture and forestry. This work has been carried out through several EU and NATO-funded research projects, such as the following:

- FP[4]6 CROPBIOTERROR – "Crop and food biosecurity, and provisions of the means to anticipate and tackle crop bioterrorism" (2004–2007);
- NATO[5] Security through Science – "Tools for Crop Biosecurity" (2005–2006);
- NATO Science for Peace and Security (2008); and
- EuropAid[6] – "Tackling BIOSECurity between Europe and Asia: Innovative detection, containment and control tools of Invasive Alien Species potentially affecting food production and trade" (2007–2010).

The EU Network of Excellence PLANTFOODSEC (2011–2016) renewed and reinforced the established partnership by enlarging it to include new countries, institutions and topics, with the ambition of contributing, through a multidisciplinary

[4]EU Framework Programme for Research and Technological Development (FP), numbered by round of funding.

[5]North Atlantic Treaty Organization (NATO) http://www.nato.int/

[6]EuropeAid (also shown as EuropAid in many websites) is a new Directorate General (DG) responsible for designing EU development policies and delivering aid through programmes and projects across the world. Its formation is described at: http://ec.europa.eu/europeaid/historical-overview-eu-cooperation-and-aid:en

mission-oriented research, to building up the capabilities to address food and crop biosecurity threats.

Making reference to the Commission green paper on bio-preparedness (EC 2007) the PLANTFOODSEC Joint Programme of Activities aimed to enhance European preparedness for deliberate or accidental introduction of the most threatening organisms harmful to plants, thus covering prevention, protection, response and recovery capacities (eradication and containment). The project has been designed to combine and functionally integrate in a durable way a substantial amount of partners' activities in the field, including:

- Actions to identify and update the biology, epidemiology and impacts of high priority pathogens, as well as through the optimization of detection and diagnostic tools;
- Actions to develop effective responder strategies by defining specific protocols on emergent pest and disease management;
- A comprehensive strategy to enhance knowledge of target groups and to inform relevant stakeholders; actions aimed to enhance networking, to overcome the fragmentation of partners' research, and to facilitate and coordinate cooperation within and among the working groups.

These projects have been changing the biosecurity context for Europe. Their experiences demonstrate the need for collaborative or coordinated approaches across national boundaries. Lessons learned could be valuable for solving both current and future global-scale challenges in biosecurity.

18.2 Trade and Other Sources of Unintentional Introductions

The risks of unintentional introductions of alien pest organisms through trade or natural spread are assessed at a national level by the competent authority in each receiving (importing) country. For plant health, this is carried out through a risk assessment process in line with the pest risk analysis standard established by the IPPC, ISPM 11 (IPPC 2013). These risk assessments[7] are often available publicly and document the significant threats that particular organisms pose to agricultural production. The risk assessments are required to identify specific vulnerable crops, defined locations and other conditions (sometimes including time frames) that make preventative action against the organisms essential. These factors do not need to be defined if the risk is acceptable to the receiving country, and no further steps would need to be taken.

[7] Risk assessments are discussed, although the documents might be entire Pest Risk Analyses. The difference is whether the possible management options are included in the document, or appear separately. For this chapter, when referring to a risk assessment it may be included in an overarching analysis document, or be standalone.

Table 18.1 Similar, though not equivalent, terms used in various biosecurity risk paradigms

	Trade/plant health	Beneficial release/ biodiversity	Food safety	Security
Instigator of the event (knowingly or unknowingly)	Exporter Handler/shipper Broker/Importer Smuggler Agroterrorist	Introducer Applicant (when done with permit) Polluter Ecoterrorist	Source Disgruntled (i.e., angry or vengeful employee) Poisoner	**Motivation** (-ed person) Perpetrator Terrorist Criminal
Introduced organism or agent (generally harmful)	Pest Regulated pest Quarantine pest Disease Weed (plants that are pests)	Agent (beneficial or not) Invasive species Weed	Hazard Contaminant Pathogen Food poisoning Toxin Human	**Agent** Pathogen Hazard Toxin Biological weapon
Relationship to location where introduced	Exotic Non-indigenous	Exotic Non-native Alien (e.g. species)	Foreign matter Non-food material	Alien (e.g. substance)
Means or mechanism for introduction	Pathway Diversion from intended use Introduction	Pathway Mechanism Causal pathway Introduction (re-)	Exposure Vector	**Pathway** Means (for introduction)
Environment including conditions into which the agent is introduced	Endangered area Containment area (when spread is being limited through official control measures)	Receiving environment Impacted area Non-target organisms Keystone species Ecosystems Habitat	Food Feed Processing plant	**Receptor** Target

An example of an innovation for conventional risk planning is the UK National Plant Health Risk Register (DEFRA 2016; Mumford et al. 2013b), which covers over 900 agents from around the world, describes their potential origins, pathways, impacts and management. Any plant health risk agent that warrants consideration through conventional pathways (i.e. trade of commodities associated with the pest) is also worthy of attention as a potential agent to be used in an intentional introduction.

Conventional trade-related pest risk assessment involves an **Agent-Pathway-Receptor** system, although using other terminology such as pest (defined in the categorization phase)-pathway-endangered area and susceptible crops, as shown in Table 18.1. When considering pest risk from trade pathways, these are always specific to both the pathway (commodity or other pathway, e.g. dunnage, shipping containers, waste) and the origin, since the pest status of the various countries or regions of origin for trade can vary. The final calculation of the level of risk depends, then, on the receiving environment or Receptor. All of these factors taken together define the pest risk from trade.

Assessments may be focussed on particular pest agents already known to cause impact in areas where they have been introduced through trade. In that case, the focus is on the likelihood of the agent being introduced, as well as options for mitigation, containment or eradication after the fact. Alternatively, assessments may change to reflect growing concern about the protection of the receiving environment. For example, expansion of production of vulnerable crops, increase of volume of the pathway (trade) or newly imposed limitations on control options are all legitimate reasons to reconsider assessments of risk.

According to Latxague et al. (2007), pest risk analysis schemes aligned with the related international standard for plant health in trade should be amended for security risks to account for at least five further criteria: (1) the ease of use of the agent, (2) the epidemic potential of the agent, (3) the importance of the target crop in relation to the motivation (may be economic or social), (4) potential obstacles to swift and effective response, and (5) potential regional or global consequences of a planned attack. Despite aspects which are missing, these trade related risk assessments provide a baseline against which intentional malicious introductions of pest organisms should be assessed. There is substantial experience in assessing and managing such accidental introductions and this expertise should be brought to bear on intentional introductions. In conclusion, existing pest risk assessments may reveal a large set of possible threats for malicious, intentional introductions. The agents of concern in the security paradigm, however, extend further to include some that would not normally be associated with trade pathways.

18.3 A Framework for Assessing Intentional Introductions

The **Motivation-Agent-Pathway-Receptor** paradigm for biosecurity risks extends the conventional Agent-Pathway-Receptor concept of plant health risk to include the possibility of intentional agent introduction (Fig. 18.2, and see Chap. 6). The motives, agents and pathways all have potential global aspects.

The PLANTFOODSEC project has developed an extensive set of risk scenarios for assessment and management based on this risk chain approach.

Vulnerability is a key concept related to impacts (Kareiva and Quinlan 2002; see also Chap. 1). In terms of ecological vulnerability, agricultural systems are generally more vulnerable to alien organisms than are more diverse natural systems. In terms of management vulnerability, prevention, detection and control are likely to be less effective for agents that are unfamiliar and unexpected than for known or endemic threats. Those sourced from abroad are more likely to be unfamiliar and unexpected. And for social and institutional vulnerability, unfamiliar risks are likely to have greater impact when they occur than do normal, local problems.

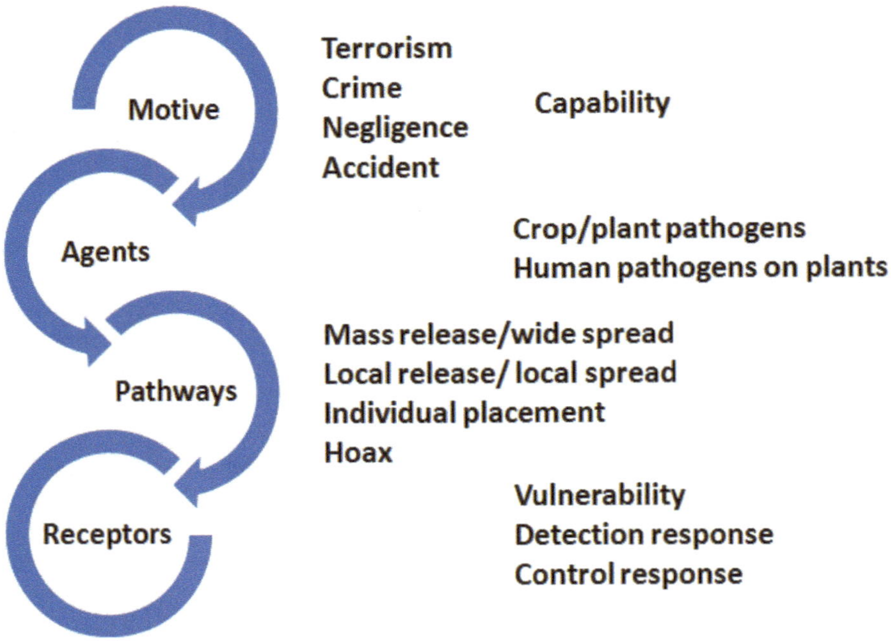

Fig. 18.2 A paradigm for biosecurity risk assessment, involving a risk chain

Within the PLANTFOODSEC project (see Chap. 2) plant disease epidemiology has been applied to crop biosecurity to develop a list of 555 target plants and crop products relevant to Europe, which includes field crops, vineyards, orchards, vegetable crops, nursery and ornamental horticulture, medicinal and aromatic plants, forest products, beverage crops, straw, tree sap and seeds. In addition, 570 pests were identified, updating the list of candidate pathogens established under the EU project CROPBIOTERROR, including harmful organisms likely to threaten crop biosecurity. Criteria for prioritisation were also identified, leading to a short list of 21 crops strategic for Europe and to a short list of pests including 63 pathogens.

Increasing preparedness against biosecurity threats requires authorities to increase the knowledge of the disease initiation and spread, because the earlier a disease is detected, the quicker counter-measures will be implemented. To mimic the early dynamics of an epidemic following a deliberate introduction, the course of epidemics occurring naturally in crops has been investigated: the scientific knowledge framework and appraisal tools developed may be applied to important crop systems across Europe.

A tool was developed to enable rapid assessments of agro-terrorism scenarios. The tool is based on pest risk assessment schemes, but includes agro-terrorism threats. The usefulness of the tool has been demonstrated on almost 100 scenarios covering a wide range of potential motivations, biological agents, pathways and receptor systems in order to provide a comparative measure of risk. By re-evaluating the ratings of appropriate criteria to reflect a managed situation, the tool makes it possible to assess the effects of potential prevention and mitigation measures. The results indicate how the threat posed by different scenarios might be reduced and how responses might be improved. PLANTFOODSEC contributed to the effectiveness and efficiency of surveillance and response programmes by allowing models of management systems to be tested against representative bio-terrorism threats.

18.4 A Framework for Consistent Response

Whatever the cause of an introduction of a novel agent or outbreak, a comprehensive biosecurity system is essential to protect domestic agriculture, natural resources and food supplies. The management of both conventional and malicious biosecurity risks requires responses to be proportionate to risks in order to prevent and limit damage to legitimate activities (Mumford 2013). It is important that both sources of risk (generally unintentional, and intended and probably malicious) are consistently managed on any shared components to ensure that this proportionate response is achieved. Schrader et al. (2012) considered the issue of consistency in assessment and response in relation to conventional plant health risks. Authentic consistency, however, recognises real differences.

While some dimensions of risk are quite different in intentional introductions there remain common elements in the analysis. The ISPM 11 (IPPC 2013) describes a framework for pest risk analysis that provides a basis for addressing concerns, assessment, management and communication of biosecurity risks. The PLANTFOODSEC project has adapted this framework (see Chap. 6) to conduct malicious biosecurity risk assessments in a similar way. So, risk components related to motivation of perpetrators, handling of agents, and public susceptibility and reaction are added to the basic trade biosecurity analysis. While other components of trade related risks, such as volume, seasonality and distribution of trade, are removed because they are less relevant to an intentional introduction.

After the risk is assessed, management actions should address the factors that contribute to risk. The measures to be taken in order to prevent the establishment and spread of harmful crop pathogens or pests have been established by identifying activities and responsibilities following pathogen introduction (see Chap. 5). In particular, PLANTFOODSEC identified international expertise for setting up contingency plans; listed resistant cultivars and alternative crops for a given pathogen; and developed containment and eradication protocols for selected pathogens to increase convergence of responder networks. Agencies and individuals have similar responsibilities to prevent and control an introduction of agents that pose a threat, although additional security agents would also be involved to deal with perpetrators and public reactions. Communication about security threats and responses to incidents needs to be culturally appropriate and allay fear rather than exacerbate it.

As biosecurity risks transcend national and regional boundaries, we must monitor, assess and manage these risks in a coordinated way across the EU, as well as other regions. Strong plant biosecurity programs should integrate elements of penalties and incentives, prevention and detection measures, and response and recovery planning by including the following: early detection and diagnostic systems; epidemiological models for predicting pathogen spread; reasonable but effective strategies and policies for crop biosecurity; distributed physical and administrative infrastructures; a national response coordination plan; and strategies for forensic investigation and attribution in cases of intentional or criminal activity. Most of these recommendations hold true for biodiversity and food safety biosecurity programs as well.

18.5 Enhanced Preparedness for Foodborne Introductions

In today's global economy, a head of lettuce produced in summer on a farm in the southern hemisphere can be harvested, packed, and shipped to a winter-bound country in the northern hemisphere on the other side of the world in a time frame equivalent to a small fraction of its shelf life. Consumers in developed nations now expect to find fresh, high quality fruits, vegetables and grains of all types in their grocery stores year-round, and rely upon sanitary standards at all points along the food distribution chain, as well as at border inspections, to safeguard these critical commodities. Their trust is generally well placed. However, although serious outbreaks of foodborne illnesses due to contamination by either human enteric bacteria or fungus-produced mycotoxins continue to be relatively infrequent in first world nations they are on the increase, in part due to vulnerabilities of the global marketplace. Furthermore, consumers in developing nations cannot be as confident of consistency in either food safety or food security.

Food biosecurity specialists (see Chaps. 3 and 4) have examined and identified the most critical food safety issues/vulnerabilities for the EU partners and associated countries and provided a baseline assessment of forensic capability within the EU to trace the sources of foodborne pathogens and toxins. The sheer volume and diversity of critical food-associated issues was addressed by the development of a prioritization strategy based on the political, economic, social, technological, legal and environmental (PESTLE) factors of a threat scenario. Currently available methods for the detection of human pathogens in foods were reviewed, and areas for improvement identified. Practical tools for applications in the food safety arena were also developed. PLANTFOODSEC partners developed a decision tool, intended to aid case investigators to discriminate, in the early stages of an investigation, between deliberately caused and accidental incidents. A new molecular detection assay capable of fine discrimination among strains of pathogenic, foodborne *E. coli* was developed to support forensic trace back efforts. Finally, PLANTFOODSEC project partners reviewed technologies available for detection and identification of mycotoxins in food items, with a focus on applications and gaps in biosecurity related investigations.

In addition to producing the specific deliverables noted above, the interactions and collaborations among PLANTFOODSEC partners and others has led to the realization of the larger goal of creating a network of excellence that crossed national, continental and hemispheric lines in a manner that allowed for the incorporation of global perspectives and examples, international training and exchange experiences for students, postdocs, and mature scientists, and lasting collaborative partnerships among the partners. However, to minimize foodborne illnesses caused by microbial contamination of foods in the future, larger and more comprehensive international networks will be needed for continuing progress in addressing challenging vulnerabilities and gaps in food safety and biosecurity associated with food production, processing, shipping, trade and marketing, and consumption. Targeted research is needed to generate new knowledge related to the biology and genetic variability of foodborne pathogens, over time and in diverse plant hosts and geographical location. We need to know more, also, about fundamental interactions between human pathogens, plant associated microbes, the host, and the environment. Furthermore, we need to better understand how on-farm production and harvesting standards impact food safety.

Finally, new information and understanding gleaned from both fundamental and applied research must be translated into practical recommendations for implementation by food producers, distributers, and marketers. Training opportunities targeted to each of these specific groups, as well as to consumers, should be developed. Access by food producers and handlers to assistance and incentives for on-site implementation would likely boost rates of acceptance and adoption of new regulations or recommendations. Each nation, in the EU or elsewhere in the global food arena, has unique farming practices and governmental requirements addressing both consumer protection and environmental impacts, but coordination and communication across borders, within and outside of the EU, will be needed for consistent and reliable outcomes.

18.6 Detection and Diagnostics

Few nations have effective customs and border inspection practices that eliminate the risk of inadvertent introductions from external sources, for example trade. This is due, in part, to the limits on resources and sustained support for trained border agents and the massive flow of people and goods. In addition, adequate detection technologies are not available, although some are emerging as discussed below. To rely on border control for detecting malicious introductions from outside of a country or region is not wise.

The PLANTFOODSEC project developed a web-based diagnostic network (see Chap. 11) as one approach to strengthening detection and diagnostic systems. The virtual diagnostic network allows information to be gathered, searched and reported, and also makes possible information flow between experts and field workers. The primary components are a database of diagnostic laboratories and expertise in the EU Member States; a community resource detailing plant pathogen news, updates on diagnostic techniques, and training and workshop information; and a structure to allow the uploading of diagnostic records and their interrogation. The network thus provides a unique tool and allows Member States to access summary information on plant pest and food safety outbreaks in Europe. While its uptake will depend on adoption of recommended practices in all Member States, and a promotion of the concepts beyond EU members, the durability of the tool will be ensured in the framework of new EU-funded initiatives for plant health, such as the Horizon2020[8] Project "EMPHASIS". In addition, techniques for mycotoxin analysis have been reviewed (see Chap. 7).

Recent advances in the technologies that underpin detection and diagnostics of pests and pathogens (see Chaps. 7 and 10) offer both opportunities and challenges for international cooperation. These technologies have dramatically lowered the detection limits for many pathogens and pests while at the same time increasing the accuracy of identification. The use of partial gene sequencing for pest and pathogen identification has become routine in many diagnostic laboratories. Recent developments in transitioning these technologies into point-of-care applications will make it easier to support biosecurity surveillance in the field and at ports of entry. One such portable technology under development is a disposable gene sequencing system the size of a thumb drive that plugs into a laptop computer (Benowitz, https://www.genome.gov/27555651). Clearly, this type of system will revolutionize our approaches to surveillance and in-field detection, although challenges also remain as considered in Box 18.2.

Box 18.2 Challenges and Opportunities from Emerging Diagnostics Technologies

As such technologies develop, users must also choose the best software systems to analyse the data generated by these technologies and determine the dependability of the databases necessary to interpret the data. Both the software systems and the databases have been steadily improving and in time they will have been vetted to a level to support regulatory decisions. Calls for sequencing all organisms (plants, pests, and pathogens) detected at points of entry may be premature (Roberts 2013). Although that may be technologically feasible, it is uncertain as to whether the requisite databases to support data interpretation are ready for the legal scrutiny of regulatory actions. If the resources are available to support such sequencing, the data acquired could be very useful to support the generation of the requisite databases and the validation and verification of those databases.

We are now able to detect organisms at population levels far below what traditional detection methods have allowed. This should reduce the risk of

(continued)

[8] EU Framework Programme for Research and Technological Development (FP), funding from 2014 to 2020 is now in the Horizon 2020 program.

Box 18.2 (continued)
inadvertent introductions associated with trade and travel. But do we have the knowledge base in place to interpret the results that we generate with increased sensitivity and specificity in detection and diagnostic tools? What level of confidence can be applied to a "new" detection? Government agencies will be asked if the newly detected organism really is new to an area or has always been present but below the detection limit of traditional methods. How reliable are the existing geographic distribution maps of pests and pathogens? Historical boundaries may not be relevant due to the changing demographics of plants, pest, and pathogens as a consequence of climate change, trade, and travel; not to mention possible intentional introductions. Geographical boundaries (e.g., oceans, mountains) are not congruent with phylogeographic boundaries used for more ecologically based analyses.

Will these new, more sensitive detection technologies encourage a shift to tolerance-based phytosanitary and sanitary decisions? Will they generate a need for revisiting international standards that are based on presence/absence? This is a worthy discussion to promote. Some of these were raised already by the International Barcode for Life project (iBOL 2011), and answers were not yet available.

18.7 International Cooperation in a Broader Context

Alien pest organisms or novel food contaminants can be among the most significant threats to domestic plant and public health due both to the novelty of the agents, whether introduced by accident, such as through trade, or intentionally and to the lack of preparation for identifying and managing them. The motivation for intentional release may have a basis in international terrorism or bio-crime, adding a global dimension not considered in the current international legal structures for plant biosecurity. The pathways for delivery of harmful agents to a targeted area may involve foreign actors and origins, and handling and transport not identified in traditional pest risk analysis, for example, as a pathway for risks. While some intentional releases of harmful organisms may involve local actors using local agents, these cases are likely to result in lower levels of impact and will be less disruptive than those with foreign elements.

Risk analysis processes for unintentional introductions of pests are well established for international trade and environmental conservation. The same is true for accidental, but anticipated, introduction of pathogens to food supplies. Release of organisms into the environment has always been more complicated in terms of agreement on risk analysis criteria and conclusions, but global protocols and standards do exist to guide this analysis. Although the international convention most related to security, the BTWC, is not designed for similar pathway and receptor analysis, the infrastructure exists for notification and information sharing.

Additional risk analyses related to intentional introductions motived by terrorism or other criminal acts should build on currently recognised systems to ensure that security risks involving similar agents, pathways and receptors can take advantage of existing legal and institutional structures and expertise for other risks to crops and food supply chains. The motivation and capabilities of perpetrators and the vulnerability of ecological and social systems to terrorism and criminal acts are particular points that distinguish malicious biosecurity risks from more conventional risks, and these aspects need further development for robust risk analysis to take place. Therefore, rather than focusing solely on the biological agents or pathways, it is appropriate to develop additional components within these conventional risk analysis processes based on scenarios related to perpetrator motivation, feasibility of access, to complement the biological potential for introductions and receptor details. While the specificity of a target or endangered area indicates what information to use regarding the receptor environment, there is likely to be an international dimension to the motivation, agents and pathways related to these biosecurity risks. One component of that is the use of alien organisms.

Risk assessment processes modified to include security components can be used by risk assessors familiar with agricultural trade, biodiversity or food safety risk analysis. These processes, along with security analysis, will be able to cover a broad range of potential agroterrorism and bio-crime scenarios. The process should be compatible with existing international standards or regional regimes as far as possible, to enhance effectiveness of the assessment and management measures. A system is needed to integrate these new components for malicious biosecurity scenarios.

The considerable amount of research promoted by the European Union – which has also involved non-EU countries such as the United States, Israel and Turkey – has made possible the development of a comprehensive set of tools to this end. Project achievements included the identification and regulatory analysis of biosecurity challenges; experimental and modelling approaches applied in plant disease epidemiology; advanced molecular diagnostics; and, more generally, training, dissemination and networking activities to increase awareness of plant biosecurity and food safety among agronomists and food producers and within the scientific, policy and inspection sectors.

PLANTFOODSEC performed the first step needed to set up a virtual Centre of Competence on Biosecurity, the aim of which is to become the backbone for the EU Plant and Food Biosecurity Scientific Community. PLANTFOODSEC acted as a network of research centres, universities and other stakeholders to enhance preparedness and response capabilities to prevent, respond and recover from both intentional and unintentional biosecurity threats to EU agriculture, farming and the agro-food industry. The PLANTFOODSEC Consortium is currently working to make the virtual European Centre of Competence on Plant and Food Biosecurity a sustainable reality by building the capability to ensure the broad implementation of the project results.

The WTO and the IPPC at the FAO have long recognized the many threats to plant health posed by international trade. WTO encourages the harmonization of phytosanitary standards among trading partners to facilitate trade while safeguarding plant health. However, the WTO's primary focus is to promote safe trade. The IPPC provides science-based recommendations for safe trade, but does not have enforcement authority. The WHO and CAC at the FAO have decades of experience in providing guidance for assessing, detecting and managing threats to food safety. For both the CAC and the IPPC, compliance with standards is essentially voluntary unless incorporated into national or regional legal regimes. Thus mechanisms and entities are in place to foster international cooperation to protect biosecurity, two of the primary treaties being deposited at FAO in Rome, and yet FAO itself has reported that inspection and interception systems regularly fail to intercept non-novel and predictable introductions due to a variety of human and material resource shortcomings (FAO 2007a).

National governments should consider biosecurity in the broadest context to include the potential for intentional introductions (FAO 2007b; Myerson and Reaser 2009; Stack et al. 2010). As such, the responsible authorities within each national government should include security experts with the responsibility for identifying threat agents, assessing risks, and developing mitigation and recovery plans for intentional introductions of plant pests and pathogens. By integrating security into the existing plant protection and food safety networks, creating new platforms and networks would be unnecessary and synergies can be achieved. This has already been recognized between human food safety and animal health networks in regard to zoonotic diseases and certain classes of contaminants of food and feed.

Additional funding will be needed to engage countries with emerging market economies. With a few exceptions, they lack the resources necessary to develop and maintain the physical infrastructure, policy framework, and human capital required to support an effective biosecurity system from the sectoral systems that are already strained from traditional threats. One last challenge results from a global food security strategy that is dependent upon emergency food aid. Often, emergency food aid is distributed across large geographic distances and geopolitical boundaries in very short time periods that preclude adequate inspections prior to movement. A possible consequence of such well-intentioned actions is the introduction of pests and pathogens into countries or regions that are already stressed with respect to food production capacity. Although the most likely biosecurity risk from this scenario would be the inadvertent introduction of a pest or pathogen, this scenario also provides a difficult to manage pathway for an intentional introduction.

Our global biosecurity strategy needs more thinking. Much more work is needed to develop phytosanitary strategies that account for the rapid movement of plant-based foods and feeds, plant products, and containers in response to food emergencies should be developed to provide post-shipment monitoring in areas that receive emergency food aid and post detection rapid response capabilities in affected areas. Further thought is needed on adapting food safety practices to be applied at points in the food system chain where hazards could be introduced but are not traditionally present. This will require a more comprehensive biosecurity framework built and sustained on international cooperation.

It may be difficult for some to imagine intentional introductions of pests, pathogens or other agents with the intent to harm plant systems or cause large food safety incidents as a real or significant threat to the public. It is true that there are very few documented cases of intentional introductions of pests or pathogens to harm plant systems, and that most foodborne disease outbreaks have not been traced to intentional introductions but rather to accidental lapses. It is equally true that determining whether an introduction was intentional or accidental is exceedingly difficult. The lack of demonstrated intentional introductions may be due to the inability to identify introductions as intentional, better forensic systems may resolve this dilemma. Meanwhile the absence of evidence is not evidence of absence (Altman and Bland 1995). It would seem prudent to err on the side of "preparedness with prioritization," recognizing that we cannot protect everything at all times but at the same time enhancing and expanding existing systems to face prioritized novel threats. From a "win-win" perspective, such enhanced systems will provide improved response to the traditional threats even if novel threats never arrive. The most effective way to accomplish these goals is through international cooperation, across sectors and disciplines, with the perspective of global biosecurity showing the way.

Note: All adopted international standards for phytosanitary measures (ISPMs) should be checked against latest versions and are available at www.ippc.int.

References

Altman DG, Bland JM (1995) Absence of evidence is not evidence of absence. Br Med J 311:485

BTWC (1976) Convention on the prohibition of the development, production and stockpiling of bacteriological (Biological) and toxin weapons and on their destruction. Opened for signature 10 April 1972. Vol. 1015, 1-14860 United Nations – Treaties Series 1976

CAC (1992) Codex alimentarius. Joint FAO/WHO food standards programme. FAO, Rome

CAC (2007) Working principles for risk analysis for food safety for application by governments. Document CAC/GL 62-2007. Joint FAO/WHO Food Standards Programme. FAO, Rome. 4 pp

CAC (Codex Alimentarius Commission) (2015) Procedures manual. Joint FAO/WHO Food Standards Programme. FAO, Rome

Defra (2016) UK plant health risk register. Web-based search tool. Department for Environment, Food & Rural Affairs. Last updated 17 Nov 2016. https://secure.fera.defra.gov.uk/phiw/riskRegister/

EASAC (2014). Risks to plant health: European Union priorities for tackling emerging plant pests and diseases. European Academies Science Advisory Council policy report 24. Jägerberg, Germany. 22 pp. ISBN: 978-3-8047-3251-3

EC (2006) Meeting the challenge – the European security research agenda. A report from the European Security Research Advisory Board. Luxembourg: Office for Official Publications of the European Communities. 79 pp. ISBN 92-79-01709-8

EC (2007) Green paper on bio-preparedness. Presented by the Commission. COM (2007) 399 final. Brussels. 17 pp

EC (2011) Our life insurance, our natural capital: an EU biodiversity strategy to 2020. Communication from the Commission to the European Parliament, the Council, the Economic and Social Committee and the Committee of the Regions. COM (2011) 244 final. Brussels. 16 pp

EPPO/COST-SMARTER (2015) Report on workshop on the evaluation and regulation of the use of biological control agents in the EPPO region. European and Mediterranean Plant Protection Organisation with European Cooperation in Science and Technology (COST) initiative for Sustainable management of *Ambrosia artemisiifolia* in Europe (SMARTER). Workshop 3–24 November 2015. Budapest. http://archives.eppo.int/MEETINGS/2015_conferences/biocontrol.htm

EU (European Union) (2015) New EU plant health rules. Web page for all official reports. http://ec.europa.eu/food/plant/plant_health_biosecurity/legislation/new_eu_rules_en. Last updated 20 Dec 2016

FAO (1997) International plant protection convention. 1997 revision. FAO, Rome

FAO (2007a) Biosecurity toolkit. Food and Agriculture Organisation, Rome, 128 pp

FAO (2007b) Development of an analytical tool to assess Biosecurity legislation. FAO Legislative Study 96. Food and Agriculture Organisation, Rome, 261 pp. ISBN 978-92-5-105871-8

FCEC (2010) Evaluation of the community plant health regime. Framework Contract for evaluation and evaluation related services Lot 3: Food Chain. Final Report. Food Chain Evaluation Consortium (FCEC)

iBOL (International Barcode of Life) (2011). The Consortium for the Barcode of Life (CBOL) held a side event on DNA barcoding during the 6th session of the Commission on Phytosanitary Measures (CPM) in Rome, March 14–18, 2011

IPPC (1996) Code of conduct for the import and release of exotic biological control agents. ISPM no. 3. FAO, Rome (later revised to IPPC, 2005)

IPPC (2005) Guidelines for the export, shipment, import and release of biological control agents and other beneficial organisms. ISPM no. 3. FAO, Rome

IPPC (2013) Pest risk analysis for quarantine pests, including analysis of environmental risks and living modified organisms. (Originally adopted in 2001, revised in 2004 and 2013) ISPM no. 11. FAO, Rome

IPPC (2015) Glossary of phytosanitary terms. ISPM no. 5. FAO, Rome

IPPC Secretariat (2005) Identification of risks and management of invasive alien species using the IPPC framework. Proceedings of the workshop on invasive alien species and the International Plant Protection Convention, Braunschweig, Germany, 22–26 September 2003. Food and Agricultural Orgsanisation, Rome, 301 pp

Kairo MTK, Cock MJW, Quinlan MM (2003) An assessment of the use of the code of conduct for the import and release of exotic biological control agents (ISPM No. 3) since its endorsement as an international standard. Biocontrol News and Information 24(1):15N–27N

Kareiva P, Quinlan MM (2002) Review of methodologies for environmental risk assessments. NAPPO PRA Symposium, Puerta Vallarta, Mexico, March, 2002. 22pp. http://www.nappo.org/PRA-Symposium/PDF-Final/Kareiva.pdf

Latxague E, Sache I, Pinon J, Andrivon D, Barbier M, Suffert F (2007) A methodology for assessing the risk posed by the deliberate and harmful use of plant pathogens in Europe. EPPO Bullet 37:427–435

Mumford JD (2002) Economic issues related to quarantine in international trade. Eur Rev Agric Econ 29:329–348

Mumford JD (2013) Biosecurity management practices: determining and delivering a response. In: Dobson A, Barker K, Taylor S (eds) Biosecurity: the socio-politics of invasive species and infectious diseases. Earthscan, London, pp 105–119. ISBN 978-0-415-53477-2

Mumford JD, Suffert F, Leach AW, Holt J, Sache I, LeFay-Souloy C, Barbier M, Hamilton RA (2013a) Quantification and interpretation of risk for security, trade, food and environment. Acta Phytopathologica Sinica 43(Supplement):409

Mumford JD, Holt J, Leach AW, Quinlan MM (2013b) National risk registers for plant health: lists, priorities and performance. Acta Phytopathol Sinica 43(Supplement):81–82

Myerson LA, Reaser JK (2009) Biosecurity: moving toward a comprehensive approach. Bioscience 52:593–600

Quinlan MM, Mumford JD, Waage JK, Thomas M (2003) Proposals for revision of the code of conduct. Biocontrol News and Information 24:1N–14N

Roberts WP (2013) Should quarantine action be based on DNA sequence rather than species? Acta Phytopathologica Sinica 43(suppl):84

Schrader G, MacLeod A, Petter F, Baker RHA, Brunel S, Holt J, Leach AW, Mumford JD (2012) Consistency in pest risk analysis – How can it be achieved and what are the benefits? EPPO Bulletin 42:3–12. doi:10.1111/epp.2547

Stack JP, Suffert F, Gullino ML (2010) Bioterrorism: a threat to plant biosecurity? In: Strange RN, Gullino ML (eds) The role of plant pathology in food safety and food security. Springer, Dordrecht, pp 115–132. ISBN: 978-1-4020-8931-2

Suffert F, Latxague É, Sache I (2009) Plant pathogens as agroterrorist weapons: assessment of the threat for European agriculture and forestry. Food Security 1:221–232

UNEP (1992) Convention on biological diversity. Opened for signature 1992. Earth Summit, Rio de Janeiro

WTO (1994) Agreement on the application of sanitary and phytosanitary measures. In: Agreement establishing the World Trade Organization: Annex 1A: multilateral agreements on trade in goods. Geneva, Switzerland. Available at www.wto.org

FSC
www.fsc.org
MIX
Papier | Fördert gute Waldnutzung
FSC® C083411

Kolibri 360 GmbH
Ferdinand-Jühlke-Straße 7
99095 Erfurt, Deutschland
produktsicherheit@kolibri360.de